Realizing the Dream of Flight

Biographical Essays in Honor of the Centennial of Flight, 1903–2003

Realizing the Dream of Flight

Edited by
VIRGINIA P. DAWSON
and MARK D. BOWLES

National Aeronautics and
Space Administration
NASA History Division
Office of External Relations
Washington, DC

NASA SP-2005-4112

Library of Congress Cataloging-in-Publication Data

Realizing the dream of flight : biographical essays in honor of the centennial of flight, 1903-2003 / Virginia P. Dawson and Mark D. Bowles, editors.
p. cm.—(The NASA history series)
"NASA SP-2005-4112."
1. Aeronautics—Biography. 2. Aeronautics—History. I. Dawson, Virginia P. (Virginia Parker) II. Bowles, Mark D. III. Series.
TL539.R43 2005
629.13'092'273—dc22

2005018938

Table of Contents

Introduction

WHILE GROWING UP IN CEDAR RAPIDS, IOWA, MILTON WRIGHT, THE WRIGHT BROTHERS' FATHER, LIKED TO PURCHASE TOYS FOR HIS SONS THAT HE HOPED WOULD STIMULATE THEIR IMAGINATION. One of the most memorable gifts was a toy helicopter that was designed by the French aeronautical experimenter Alphonse Pénaud. Milton gave his sons this gift in 1878, and, though it was a simple device with a stick bound to a four-blade rotor set in a spindle, it had the intended effect—it caused them to dream. Twenty-five years separated the gift of this toy and their invention of the airplane, yet the Wright brothers were convinced it had exerted an important influence. Tom Crouch argued in *The Bishop's Boys* that toys like these perfectly illustrated the significance of play for technological innovation. He wrote, "rotary-wing toys were to intrigue and inspire generations of children, a few of whom would, as adults, attempt to realize the dream of flight for themselves."[1]

If the first powered flight on 17 December 1903 represented a childhood dream realized, it was only the first step in the rapid evolution of the airplane from their flimsy kite-like contraption of wood and cloth to jet airliners and rockets in space. And, as extraordinary as the achievement of powered flight seemed in 1903, before the end of the century, space travel also would become a dream realized. Soviet astronaut Yuri Gagarin

1 Tom D. Crouch, *The Bishop's Boys: A Life of Wilbur and Orville Wright* (New York: W. W. Norton & Company, 1989), p. 57.

first circumnavigated Earth in April 1961, and, eight years later, American astronauts took the first steps for humankind on the Moon.

It is with great pleasure that we introduce *Realizing the Dream: Biographical Essays in Honor of the Centennial of Flight*. These essays in celebration of the Wright brothers' first flight 100 years ago grew out of presentations by a group of prominent scholars in 2003 at a conference sponsored by the NASA History Division and held at the Great Lakes Science Center in Cleveland, Ohio. The volume focuses on the careers of some of the many men and women who helped to realize the dream of flight both through the atmosphere and beyond. These accounts are original and compelling because they examine the history of flight through the lens of biography. Collectively, these individuals helped to shape American aerospace history. There are obviously many other individuals that could, and arguably should, have been included in this collection, but we believe that the cross section of diverse individuals contained in this volume is important because it is symbolic of the dream of flight as a whole. These people all devoted their lives, and sometimes even sacrificed them, to the demands required for its realization.

The reasons behind the dreams were diverse. The technological potential first demonstrated by the Wright brothers enabled those who followed them to use flight as a means of racial uplift, gender equalization, personal adventure, commercial gain, military superiority, and space exploration. The history of flight is more than a story of technology; it had important cultural consequences as well, and these are some of the themes that the following biographies explore. We have arranged the essays roughly chronologically, though the careers of the people described here often span more than one period of history. None of the people in this volume were inventors like the Wright brothers, but their contributions to flight were nevertheless significant. They were daredevil pilots, entrepreneurs, business men and women, military strategists, and managers of large-scale technology who advanced the art, science, and business of air and space travel, often through sheer force of character. The final paper serves as an epilogue as well as a tribute to the Wright brothers. It describes a reenactment of their important glider experiments at Kitty Hawk, North Carolina, where the Wrights' childhood dream was first realized.

BARNSTORMERS AND ENTREPRENEURS

In 1900, at the point when they were ready to build a full-scale glider, Wilbur Wright wrote to Octave Chanute, an authority on flying machines, for advice on where to try out their latest glider. Wilbur admitted his obsession with flying, stating, "I feel that it will soon cost me an increased amount of money, if not my life."[2] While neither brother died

2 Ibid., p. 181.

in an airplane crash, many other early aviators lost their lives, including Bessie Coleman, the first African American female pilot and the first of her race to earn an international pilot's license. In "Bessie Coleman: Race and Gender Realities Behind Aviation Dreams" (chapter 1), Amy Sue Bix describes the obstacles Coleman faced because of her gender and race, explaining the social context that allowed Coleman to promote the association of flying with social uplift for her race. Although race worked to her advantage in drawing media attention, it also affected her ability to earn as much as her white counterparts. Nevertheless, Coleman was compelled by the "gospel of aviation" to escape the bounds of gravity. She confronted, like other women and minorities, the challenges of pursuing her dream despite the social assumption that it was inappropriate for marginalized groups to fly and the technological dangers of undependable planes.

Bessie Coleman's contemporary, Amelia Earhart, was able to capitalize on the huge popularity of aviation after Charles Lindbergh's transatlantic flight. In "She Flew for Women: Amelia Earhart, Gender, and American Aviation" (chapter 2), Susan Ware argues that Earhart used her fame as an aviatrix to advance a strongly feminist ideology, demonstrating the capabilities of women in a modern world. Earhart took to the skies not only for the thrill of aviation, but also to use women's competence as pilots as a tool to end prejudice against them. Ware believes that Earhart's disappearance in 1937, and the mystery surrounding this event, has obscured her legacy as a strong voice for feminism. Ware points out that women could earn a living from flying only as long as it was considered a form of entertainment. As the aircraft industry took shape in the 1930s, they encountered gender barriers that prevented them from becoming commercial pilots. The irony of Earhart's disappearance is that by 1937 she was already part of the bygone era of stunt flying, and today she is remembered more for the fact that she is missing than for her piloting accomplishments, such as being the first woman to cross the Atlantic by plane. Though women like Coleman and Earhart were strong individuals at the forefront of the emergence of aviation, they were unable to open the skies for women. The commercial airlines industry froze them out of the business side and also excluded them from flying. Men were the pilots; women were the stewardesses.

The next three essays feature larger-than-life male characters who played signal roles in shaping commercial aviation. In "Sharing a Vision: Juan Trippe, Charles Lindbergh, and the Development of International Air Transport" (chapter 3), William Leary discusses the emergence of international commercial air transport using the vantage point of the relationship between Charles Lindbergh and Juan Trippe. In the 1920s, they dreamed of a time when a commercial aviation industry would carry passengers around the world in both comfort and safety. They worked to achieve this dream by first providing commercial air service to Latin America in the early 1930s. Leary shows how Lindbergh's technical expertise and international fame coupled with Trippe's determination and entrepreneurial skills created Pan American Airways—one of the greatest airlines of the 20th century. Leary describes Lindbergh's influence on the decision to replace the Pan

American fleet with turbojet airliners after World War II (WWII)—the high point in the fortunes of the company. Ultimately, though their disagreement over the development of a supersonic transport caused a rift, their shared vision of the future of commercial air transportation exerted a significant influence on the course of the development of 20th-century aviation. Despite the fact that Pan American could not recover from the oil crisis of the 1970s, Trippe and Lindbergh helped to make their early dream of commercial aviation a reality.

In "The Autogiro Flies the Mail! Eddie Rickenbacker, Johnny Miller, Eastern Airlines, and Experimental Airmail Service with Rotorcraft, 1939–1940" (chapter 4), David Lewis approaches the history of aeronautics by looking at one of its little-known detours. Ace pilot Edward V. Rickenbacker had already transformed Eastern Air Lines into a commercial success by the 1930s. Always a risk-taker interested in futuristic technology, he became interested in the potential of a rotor-driven aircraft called the autogiro. Rickenbacker saw the autogiro as a solution to urban congestion, since it could be used for both road and air transportation. With it the skies would be democratized, and the dream of an "aircraft for everyone" was born. To test the idea that the autogiro could speed the flow of the mail, Eastern Airlines obtained a contract with the U.S. Post Office for delivering mail between the Camden Airport in New Jersey and the roof of the 30th Street Post Office in Philadelphia. Lewis's essay is enlivened by the firsthand account of Johnny Miller, the company's former test pilot, now in his nineties. The autogiro, piloted by Miller, made 2,643 trips between Camden and Philadelphia in 1938. Despite proving the concept, Eastern Airlines failed to win another airmail contract, and the project quietly died. Though never a successful endeavor, the autogiro is symbolic of the multitude of unrealized aerospace dreams rendered unsuccessful due to myriad social, political, technical, or environmental factors.

To Roger Bilstein, the career of Donald Douglas was closely linked to the fortunes of the Douglas Aircraft Company. In "Donald Douglas: From Aeronautics to Aerospace" (chapter 5), Bilstein describes how in the 1930s the company developed the DC-3, which was a plane so outstanding that it became "an icon for 20th-century progress." Douglas was unusual because he was among the first of a new generation to receive formal instruction in aeronautical engineering. Bilstein emphasizes Douglas's relationships with other talented individuals of the period. These included not only the excellent engineering team he assembled at Douglas Aircraft, but also business associates like C. R. Smith, the flamboyant and visionary president of American Airlines. Envisioning coast-to-coast airline service, Smith engineered a key loan from the Roosevelt administration that made the depression-era development of the DC-3 possible. Douglas and his company continued to thrive in the 1940s and 1950s, enjoying lucrative military contracts while continuing to innovate. Bilstein ends his essay by describing the flawed corporate decisions and family problems that contributed to the unraveling of the company in the 1960s and 1970s.

MILITARY STRATEGISTS

By the 1940s, the aviation industry had reached its maturity, and flight technology was ready to play a major role as a weapon of war. In WWII, the airplane played a critical part in the Allied victory. Two individuals, whose careers were shaped by their wartime service as pilots, are discussed in the essays by military historians Alan L. Gropman and Tami Biddle. In "Benjamin O. Davis, Jr., American Hero" (chapter 6), Gropman describes Benjamin Davis's leadership of the Tuskegee Airmen, the first all-black fighter squadron in the Army Air Forces. Davis had faced discrimination throughout his career, but he was determined not to let race interfere with the performance of his duty during WWII. Under his command, the Tuskegee Airmen demonstrated their courage and competence in combat. Gropman argues that integration of the Air Force occurred between 1948 and 1951, well before the other military services, because of the powerful example provided by Davis and the Tuskegee Airmen. Four-star General Davis ended his career as the Director of Civil Aviation Security in the Department of Transportation, where he took a keen interest in reforming airport safety procedures under the Ford administration.

In her essay on Ohio-native General Curtis LeMay, Tami Biddle poses the question of whether LeMay was a hero or a threat. In "Curtis E. LeMay and the Ascent of American Strategic Airpower" (chapter 7), Biddle describes how LeMay became associated with the development of American strategic bombing during WWII. After the war, as the architect of the Berlin airlift and head of the Strategic Air Command, LeMay built a technically superior and well-disciplined Air Force during the Cold War. In assessing LeMay's character, Biddle points out that LeMay's detractors have portrayed him as a warmonger—ready and willing to go to war, possibly even without the direct order of the executive branch. Biddle presents a more balanced and intriguing portrait of the hard-assed, cigar-chomping man. LeMay strongly believed that "deterrence through intimidation" was the only effective way to keep the Soviet Union from attacking the United States. Biddle points out that LeMay's belligerent public persona actually may have contributed to keeping the peace.

ARCHITECTS OF SPACEFLIGHT

The next set of essays centers on four individuals who contributed to making spaceflight a reality. In "Willy Ley: Chronicler of the Early Space Age" (chapter 8), Tom D. Crouch, well-known biographer of the Wright brothers, focuses on Willy Ley and the popularization of the idea of human spaceflight. A native of Berlin, like many of his German contemporaries, Ley was infected by the rocket craze of the 1920s. He joined the

German Society for Space Travel and became one of its most visible members through his publications. However, in the mid-1930s, when the German military became interested in rockets, Ley was banned from writing about them. He left Germany and settled in the United States, where he published his first of many best-selling books on rocketry in 1944. These books explored the idea of lunar conquests and Martian explorations and caught the attention of Walt Disney in the 1950s. Ley contributed to three Disney television programs and served as an advisor to NASA. Unfortunately, Willy Ley died in 1969 just days before the first humans walked on the Moon.

In "Who Was Hugh Dryden and Why Should We Care?" (chapter 9), Michael Gorn argues that Dryden made important theoretical contributions to aviation and played a central role in the management of both NASA and its predecessor, the National Advisory Committee for Aeronautics (NACA). Gorn claims that the reason why Dryden's role in aerospace history has not received more historical attention lies in Dryden's self-effacing personality. While this might have been appropriate during the NACA era, the space race required more political savvy than Dryden possessed. Nevertheless, Dryden's contributions were significant. As Director of NACA between 1947 and 1958, he reoriented the Agency toward research on supersonic flight, an effort that culminated in the successful flight of the X-15. At the same time, Dryden encouraged efforts within NACA to become more involved in space-related activities at the three NACA laboratories. Each of the three, he has pointed out, had significant projects underway before the launch of Sputnik in 1957. When NASA was set up in 1958, Dryden became Deputy Administrator, serving under both T. Keith Glennan and James Webb. In this role, he helped to structure the managerial and technical priorities of the Mercury, Gemini, and Apollo programs.

In "Wernher von Braun: A Visionary as Engineer and Manager" (chapter 10), Andrew J. Dunar interprets the career and personality of Wernher von Braun, one of the most complex and controversial individuals associated with the American space program. Recent scholarship has documented extensively von Braun's Nazi background.[3] Dunar critically examines this scholarship in assessing von Braun's character. As coauthor of a publication on the history of the Marshall Space Flight Center, Dunar's knowledge of this particular NASA Center has given him a unique vantage point for his analysis.[4] He focuses on von Braun's management style. Von Braun's autocratic, top-down approach at Peenemünde (known as the arsenal system) became institutionalized at the Army Ballistic Missile Agency in Huntsville, Alabama. This modus operandi continued into the NASA era with the development of the mammoth Saturn rocket at Marshall Space Flight Center. Dunar dispassionately explores the tensions inherent in von Braun's position within NASA. He points out that possibly because of its association with the Germans,

---

3 Michael J. Neufeld, *The Rocket and the Reich: Peenemünde and the Coming of the Ballistic Missile Era* (New York: The Free Press, 1995).

4 Andrew J. Dunar and Stephen P. Waring, *Power to Explore: A History of Marshall Space Flight Center, 1960–1990* (Washington, DC: NASA SP-4313, 1999).

the arsenal system never became the model for NASA contracting. He notes that despite his enormous contributions to the space program, von Braun never overcame the stigma of his Nazi background.

One of the key individuals in NASA's early space program, Robert Gilruth, is another individual whose obscurity is undeserved. In "Godfather to the Astronauts: Robert Gilruth and the Birth of Human Spaceflight" (chapter 11), Roger Launius, former Chief Historian for NASA and now Chair of the Division of Space History at the National Air and Space Museum, argues that Gilruth is an outstanding example of an engineering entrepreneur—a person defined by his ability to manage large-scale technology. Gilruth demonstrated his gift for molding people into a team when he headed Langley Laboratory's Pilotless Aircraft Research Division. Shortly after the launch of Sputnik, Gilruth organized the Space Task Group, later the nucleus for Project Mercury. Then, after President Kennedy announced the national goal of landing human beings on the Moon within the decade, Gilruth took over responsibility for setting up the Manned Spacecraft Center (renamed Johnson Space Center in 1973). There he was in charge of overseeing the space capsule development effort and the training of the Mercury and Apollo astronauts. Launius points out that the Apollo program achieved its political goal of beating the Soviet Union to the Moon. It also represented an enormous technological achievement by a government agency. Finally, "by sheer serendipity Apollo taught humanity about itself, and in the process altered our perception of the world on which we live." He regards Gilruth as the "unsung leader in the race to the Moon." Like many of the other individuals chronicled in this volume, he firmly grasped what was technically possible and set out without fanfare to turn the stuff of dreams into reality.

EPILOGUE

The final essay brings our volume on the history of flight back to the Wright brothers. Wilbur and Orville shared character traits with many of the people described in this volume. They were risk-takers and careful managers, dreamers and doers, entrepreneurs and architects of flight. Before flying their machine, they sought ideal conditions for their experiments, asking the U.S. Weather Bureau in Washington for a list of places on a beach that met their specifications for wind velocities. As a result they chose Kitty Hawk, an isolated location on the Outer Banks of the North Carolina coast.

The Wright brothers aimed for nothing less than complete control of their craft through a system called "wing warping." To their surprise, their first glider experiments at Kitty Hawk produced what Wilbur described as a "peculiar feeling of instability."[5] In analyzing the problem, they became suspicious of the tables of lift coefficients that others

5 Crouch, *The Bishop's Boys*, p. 212.

had previously taken as accepted wisdom. Once wind tunnel experiments had provided the Wright brothers with new, more reliable data, they designed a new wing profile.

Museum curator Edward J. Pershey takes up the story at this point. In "Celebrating the Invention of Flight in a Hands-On Way: Replicating the 1902 Experimental Glider Flights of the Wright Brothers" (chapter 12), Pershey describes the important role this glider played in the invention of the airplane. He points out that even after the Wrights had redesigned the glider, stable flight eluded them. Wrestling with the problem, Orville came up with the idea of coupling the wing-warping control mechanism with a moveable rudder to increase the pilot's active control of the drag on the tail. Pershey argues that the reenactment of the 1902 glider flights, an effort underwritten by the Western Reserve Historical Society in Cleveland, Ohio, and spearheaded by members of its staff, provides important insights into their special form of experience-based knowledge and creativity. The Wrights flew the glider more than one thousand times in October 1902. These flights succeeded so well that by the time they left Kitty Hawk that year they were so confident they had invented the airplane that they sought patent protection for their control system. They fully expected the first powered flight to succeed the next year.

If aviation's "winged gospel" fanned the enthusiasm of the daredevil pilots of the early years of aviation, it was sustained by people who were convinced of the commercial and military value of the airplane. Aviation had matured by the end of WWII, while rocketry was still in its infancy. Unlike aviation, rocketry developed rapidly in the 1950s—stimulated less by dreamers like Willy Ley than by Cold War exigencies. Spaceflight required management of technology on a scale that exceeded even that of the Manhattan Project during WWII. Yet, as these essays demonstrate, even the extraordinary feat of landing human beings on the Moon was both a technical and a human one, requiring unusual skills and vision. The airplane and the space capsule were never just modes of travel; they expressed aspirations. During the barnstorming era, the airplane represented social uplift—a means of literally transporting people regardless of their color closer to heaven. But even in its infancy, aviation had a darker side. Speaking of the airplane in the 1930s, Antoine de Saint-Exupéry wrote, "the machine does not isolate man from the great problems of nature but plunges him more deeply into them."[6] As we face the perplexities of the 21st century, we would do well to ponder the personal qualities and achievements of the people who did not flinch from dedicating their careers and lives to realizing a dream.

6 Antoine de Saint-Exupéry, *Wind, Sand, and Stars* (New York: Harcourt Brace & Company, 1939), p. 43.

Acknowledgments

THE EDITORS WOULD LIKE TO THANK MANY PEOPLE who were involved in helping to conceive, arrange, and manage the conference that formed the basis of these collected essays. Foremost was Kevin Coleman, in charge of the history program and records management at Glenn Research Center. His expert managerial efforts ensured the success of the conference. Roger Launius, Chair of the Division of Space History at the National Air and Space Museum, and Steve Garber, NASA historian, conceived the topics and invited the authors for this conference. Other individuals who gave generously of their time included Susan Hennie of NASA Glenn Research Center and Bob Azzardi, Margarite Aponte, Bob Arrighi, Deborah Demaline, and Suzanne Kelly of Indyne, Inc. The following individuals also were instrumental in making the publication of this book a reality: in the NASA Headquarters History Division, Steve Dick, Chief Historian; Rebecca Anderson, intern; and Glen Asner, historian; in the NASA Headquarters Printing and Design Office, Michelle Cheston, editor; Shelley Kilmer, graphic designer; and Jeffrey McLean, printing specialist.

1

Bessie Coleman

Race and Gender Realities Behind Aviation Dreams

AMY SUE BIX

OVER THE FIRST THREE DECADES FOLLOWING THE WRIGHT BROTHERS' TRIUMPH AT KITTY HAWK, AMERICANS ACROSS RACIAL AND GENDER LINES BECAME FASCINATED by the rich possibilities of flight. Especially after World War I (WWI), ordinary men and women were enraptured by what historian Joseph Corn has called "the gospel of aviation," popular fascination with the marvelous, even magical, implications of flying. Many thrilled to the sense of leaving behind Earthbound limits, exploring suggestions that aviation had the power to cure disease, avert wars, and literally bring human beings closer to heaven.[1]

Underneath this adoration of airplanes, aviation from 1903 through the 1920s poses a more complex, less rosy picture. Early pilots spoke and wrote about the sheer joy of overcoming gravity, joining the birds in gazing down at towns and the land from a supe-

1 Joseph J. Corn, *The Winged Gospel: America's Romance with Aviation, 1900–1950* (New York: Oxford University Press, 1983).

rior height. Yet while the feeling of flying itself might embody freedom, the process of getting into the sky was by no means free or fair. For many women and minorities, simply gaining access to airplanes and flying lessons posed particular challenges. Along with practical problems of finding the necessary financial resources, these marginalized groups faced the barrier of social assumptions that ruled it inappropriate for them to fly.

Ironically, while soaring into the sky might carry a sense of empowerment, the early equipment was undependable; in some cases, wings literally fell off airplanes. The evolution of aviation in this era often intensified life-threatening risks. Aside from the military applications made evident during the war, the practical value of airplanes had not yet matured by the start of the 1920s. In the absence of consistently viable commercial business or thriving passenger traffic, one of the few civilian functions of aviation was entertainment. In the prewar era, the sheer novelty of flying had been amusement enough; in seeing "birdmen" lift off the ground, first-time viewers could satisfy their curiosity and verify reports that flying machines had indeed been invented. Yet soon, flyers began expanding their repertoire, not only to attract spectators, but also as a result of their own competitiveness and desire to push the boundaries of aeronautics. By 1910, the Wright brothers and Glenn Curtiss set up companies to give aerial exhibitions and schools to train new pilots for these traveling teams. Public demonstrations of "fancy flying," as it was called, proved profitable monetarily, but with a high personal toll. Only one of the four men who signed a two-year contract with the Wright Exhibition Company actually lived to complete it.[2]

The postwar years continued the era of flying as a business of entertainment. The phenomenon of aviation as a spectator sport was indirectly given a boost by the recent conflict. With arrival of peacetime, the federal government put hundreds of surplus military aircraft on the market at relatively affordable prices. One of the most common, a Curtiss JN-4 "Jenny" biplane, could be obtained for just $300 to $500. Pilots, many who had acquired flying experience in WWI, purchased surplus planes to begin careers showing off their flying skills to ordinary Americans. Barnstorming provided an avenue for new enthusiasts to enter the field, including Charles Lindbergh, who performed stunts such as parachute jumping and wing walking to earn money to buy his own plane. Barnstormers toured the country, flying out of county fairs, carnivals, local airstrips, or even farmers' fields. The "flying gypsies" incorporated dogfighting tactics of loops, rolls, and dives into their acts, giving "air circuses" a dramatic intensity highlighted in advertisements. The flirtation with danger became part of the attraction for audience members

---

2 Roger E. Bilstein, *Flight in America: From the Wrights to the Astronauts* (Baltimore: Johns Hopkins University Press, 1984), p. 25; and Dominick A. Pisano, "The Greatest Show Not on Earth: The Confrontation Between Utility and Entertainment in Aviation," in *The Airplane in American Culture*, ed. Dominick A. Pisano (Ann Arbor: University of Michigan Press, 2003), p. 49. See also Curtis Prendergast, *The First Aviators* (Alexandria, VA: Time-Life Books, 1980); and Corn, *The Winged Gospel*, p. 24.

who relished seeing others testing the edge or anticipated witnessing a spectacularly horrific accident.[3]

Barnstorming demanded physical skill, mental sharpness, and the utmost daring, but was otherwise a relatively open occupation, shunning formal qualifications and rules. Where women and minorities were barred from entering the U.S. Air Mail Service (formed in 1918) and military flight units, a few found opportunities in barnstorming during the early 1920s. While white male flyers were no longer a novelty in themselves, female and minority pilots or stunt performers commanded attention by their mere existence. Yet while their rarity value could be commercialized, it also raised pointed questions about their abilities. Because they defied traditional gender and racial expectations, female and minority aviators faced skepticism, ridicule, or outright hostility, sometimes even from members of their own communities. Under this intensified pressure, women and minorities felt compelled to prove their skills over and over. Beyond being forced to justify their right to fly, early women and minority aviators faced other serious frustrations. To the extent that marginalization denied them access to the best training, newest planes, and other resources, inherent risks of their flying rose.

The question of how and on what terms women and minorities could find a place in aviation reflected broader debates in American culture during the 1920s on the subjects of race and gender. While suffragists had finally secured voting rights for women in 1920, the women's movement suffered in the Red Scare conservative backlash and split down the middle over whether to pursue an Equal Rights Amendment. The Jazz Age brought individual women new visibility, transforming actress Mary Pickford, athlete Gertrude Ederle, and writer Dorothy Parker, among others, into celebrities. Female aviators faced a similar situation of entering the spotlight for their individual talent, yet also being evaluated and defining themselves in relation to their entire gender. The first generations of female pilots displayed this ambivalence and awareness. Some, including Amelia Earhart, refrained from calling themselves feminists, wary of the radical connotations of that label. Yet in public statements they repeatedly insisted that women could be excellent aviators and worked to enhance support for female pilots.

Similarly, for African Americans, the 1920s represented a time of racial violence, lynch mobs, and a revival of the Ku Klux Klan. Black artists and writers, such as Langston Hughes, found vital expression for their creativity in the Harlem Renaissance but had to depend on white patrons who romanticized African sensuality while ignoring hardships and discrimination facing blacks. For blacks interested in aviation, it would prove impossible to separate their individual aspirations and achievements from their community's broader political, economic, and social conflicts.

---

3 Paul O'Neil, *Barnstormers and Speed Kings* (Alexandria, VA: Time-Life Books, 1981).

THE FIRST GENERATION

In the years after 1903, as aviation entered public awareness, a few women joined men in expressing interest in the art of flying. Although promoters such as the Wrights and Curtiss hoped to encourage the rapid spread of aviation, there were deep reservations about female involvement. Critics worried that women were inherently unsuited to become pilots due to their feminine temperament. They characterized women as too scatterbrained to master complex technology and so emotional that any crisis would send them into catastrophic panic. Curtiss was extremely reluctant to include women in his pilot training, fearing among other things the repercussions if a female aviator should be killed in a crash.[4]

It would be in Europe in the spring of 1910 that the first woman would qualify to fly. After instruction from the great French aviator Charles Voisin, Raymonde de Laroche received her license in March 1910 from the Aero Club of France and went on to compete in races and other contests against male pilots. Soon other women, such as Hélène Boucher, made solo flights in France. By 1913, there were enough female pilots that France could create a special cup to honor woman aviators. Hélène Dutrieu became Belgium's first licensed female pilot, and, in 1913, France awarded this "Girl Hawk" the Legion of Honor.[5]

In the United States, Blanche Scott, who had already gained fame by completing a strenuous cross-country automobile drive, used this record to overcome Curtiss's opposition to female pilots and began studying aviation at his school in the autumn of 1910. Billed as the "Tomboy of the Air," Scott later worked in exhibition flying, specializing in stunts such as a hair-raising "Death Dive." Bessie Raiche studied aviation in France, inspired by de Laroche, and married a Frenchman; she flew her first solo in the fall of 1910.[6]

In 1911, Harriet Quimby became the first woman in the United States to earn her pilot's license. As a journalist, Quimby saw firsthand the media rush to cover the aviation craze. Reportedly, after noticing that no female pilots had appeared at a big 1910 New York air meet and being caught up in racing excitement, Quimby determined to take lessons herself. After joining an exhibition team, Quimby suggested that female pilots also could join men in commercial aviation, running flying schools, carrying passengers, and delivering packages. In April 1912, Quimby won international renown by becoming the first woman to fly across the English Channel, overcoming bitterly cold weather in her open-cockpit monoplane. Almost three months later, back in the United States, Quimby fell to her death before thousands of spectators when equipment problems caused her plane to flip midflight.[7]

---

4 Valerie Moolman, *Women Aloft* (Alexandria, VA: Time-Life Books, 1981). See also Kathleen Brooks-Pazmany, *United States Women in Aviation, 1919–1929* (Washington, DC: Smithsonian Institution Press, 1983).

5 Ibid., p. 17.

6 Ibid., p. 18.

7 Ibid., pp. 22–28.

Other women proceeded to earn pilot's licenses during the remainder of the 1910s, and, like men, some sought to make a living and a name for themselves through flying. Ruth Law and Katherine Stinson joined the ranks of exhibition teams performing dramatic acrobatic stunts. Beyond that, Law and Stinson competed to win acclaim through altitude, endurance, and distance flight, setting new women's records and sometimes breaking men's records.[8]

THE CASE OF BESSIE COLEMAN

The first generation of American female pilots such as Scott, Quimby, Law, and Stinson faced numerous doubters who considered flying inappropriate for women. Nevertheless, each managed to gain entry, finding some supporter willing to offer training and encouragement. The obstacles would be multiplied for Bessie Coleman, with race joining gender in conditioning the reception of her announced ambition of flying.

At first glance, Coleman would not appear a likely candidate to succeed in becoming the world's first black female aviator and the first of her race, male or female, to secure an international pilot's license. Coleman was born in a one-room cabin in Atlanta, Texas, apparently on 26 January 1892. For African Americans of that era, racial tension, public lynchings, community segregation, and assertions of white supremacy shadowed life in the South. Jim Crow laws barred blacks from sharing public facilities such as railroad cars, restrooms, and drinking fountains with whites. Literacy tests, poll taxes, and institutionalized discrimination denied voting rights to many black men. Allegedly frustrated by race-related economic marginalization, Bessie's part-black, part-Native American father, George Coleman, left the family to move to Oklahoma Indian territory when Bessie was nine.[9]

In what became a single-parent household, Bessie assumed the responsibilities of running the home in Waxahachie and tending to her younger sisters while her mother Susan went out to perform domestic services. Such obligations occasionally interfered with Bessie's attendance at the black one-room schoolhouse 4 miles away. Moreover, in cotton country, black youngsters' education was interrupted every year for them to help their families gather extra income through cotton picking. Though Bessie resisted this draining labor by slowing her pace, family members recall that she also protected Coleman interests by making sure the foreman credited them with full weight for each sack. Although financial constraint burdened the Coleman family, Susan strove to improve her

8 Ibid., pp. 29–32.

9 Doris L. Rich, *Queen Bess: Daredevil Aviator* (Washington, DC: Smithsonian Institution Press, 1993), pp. 3–8. The year of Coleman's birth has been the subject of prolonged confusion; newspaper articles and other popular accounts often reported her age erroneously, and Coleman herself contributed to cloud the date.

children's prospects by emphasizing the importance of cultivated manners, self-respect, and racial pride. According to family memory, Susan ensured that, through reading and oral tradition, Bessie and other children would become familiar with black figures such as Harriet Tubman and Booker T. Washington.[10]

After finishing eighth grade, Bessie used money she earned through laundry work to enter Oklahoma's Colored Agricultural and Normal University in 1910. When lack of funds compelled Coleman to depart after a single term, she returned to Waxahachie and continued working as a laundress.

According to accounts, Coleman's ambition drove her to leave Texas in 1915, joining the flood of African Americans making the Great Migration, heading for Chicago to join older brothers Walter and John. After mastering the beautician's trade, Coleman found employment in black community barbershops as a manicurist. When the United States entered WWI, Walter and John served in France with the segregated Eighth Army National Guard unit. Upon returning, John allegedly taunted his sister by comparing black women unfavorably to the strong Frenchwomen he had encountered overseas, particularly citing the example of female pilots.[11]

Engrossed with the challenge of emulating these daring female flyers, Coleman soon encountered difficulty. While a number of white women had gained aviation training since Scott and Raiche in 1910, Coleman's gender and race counted for two strikes against her. All the white pilots and flying schools she approached apparently rejected her requests. There were no African American aviation institutions or even individuals in position to accept Coleman as a flying student. When Coleman related her plight to Robert Abbott, influential editor and publisher of the black newsweekly *Chicago Defender,* he encouraged her to sidestep that barrier by attending flying school in France. With his assistance and encouragement, Coleman acquired a passport, visas, and basic French language, then sailed out of New York in November 1920.[12]

The first flying school Coleman approached in France refused her, since two of its female students had recently died in accidents. Ultimately Coleman gained admission to the École d'Aviation des Frères Caudron in Le Crotoy, France's most renowned training center. Caudron already had accepted female students, including Frenchwoman Adrienne Bolland, who became the first woman to fly over the Andes Mountains.[13] Coleman undertook a seven-month course of ground lessons and trial-and-error practice flights in the wood-framed, fabric-covered Nieuport biplane. Walking 9 miles daily from lodgings to class, Coleman persisted despite seeing a fellow student suffer a fatal crash, a risk well known to every aviator. On 15 June 1921, Coleman demonstrated her takeoff, land-

10 Ibid., p. 11.
11 Ibid., p. 26.
12 Ibid., pp. 30–32.
13 Moolman, *Woman Aloft*, p. 44.

Barred from flight school in the United States because of her race, Bessie Coleman earned a French pilot's license on 15 June 1921. (Photo number 93-7758, National Air and Space Museum, Smithsonian Institution)

ing, and flying skills to earn a license from the Fédération Aéronautique Internationale, the distinguished world aviation organization.[14]

After continuing flying lessons in Paris over the summer, Coleman sailed home in the autumn of 1921. Upon arrival in New York, reporters received her as a curiosity; her accomplishment was interpreted in both the black and white communities as significant primarily due to her race and gender. Aviation magazines duly recorded her acquisition of flying credentials, without analyzing or condemning the discrimination that had forced Coleman to gain her license overseas. For aviation enthusiasts, Coleman's story seemed to verify the inevitable success of human flight, measured in terms of its spread. However, to them, the existence of one African American female pilot did not necessarily signify an equal distribution of flying ability between whites and blacks or provide

14 Rich, *Queen Bess*, p. 34.

reason to anticipate a massive influx of black pilots, male or female. Coverage of Coleman in specialized aviation publications and in mass-market media reflected the press's general fascination with aviation. Newspapers and magazines of this era ran regular columns and special features on aviation, both catering to and feeding readers' fascination with flight. This coverage particularly highlighted aviation "firsts," such as the first air commuter and first scheduled passenger flight; Coleman represented one among this series of "firsts."[15]

For black-oriented newspapers and for the African American public, Coleman's identification signaled something different, not just another aviation "first," but a political and social landmark. Moreover, her arrival coincided with intense debate within the black community over the relationship between race and adoption of modern air technology. Coleman herself would contribute to that discussion, making tangible the concept of black flight.

As Jill Snider has pointed out in her research on African Americans and aviation history, Coleman's return to the United States as an internationally qualified pilot in the fall of 1921 occurred less than four months following the infamous Tulsa race riot of June. While African Americans fled from homes going up in flames, airplanes soared over the chaos. White authorities apparently used planes to conduct aerial surveillance of areas under siege and to ensure that blacks from nearby were not moving toward the trouble. Blacks later reported having observed white aviators dropping gasoline or bombs and shooting at escaping black men, women, and children. Given the climate of Klan-fueled violence against blacks in this era, African Americans began considering with alarm the potential of airplanes as weapons racists might employ to decimate or even exterminate the black race.[16]

Coleman's presentation of herself as an aviator, then, occurred at precisely the time when members of the black community were heatedly denouncing use of the airplane in Tulsa as a tool for murder and debating how African Americans should react. Marcus Garvey, leader of a growing black nationalist movement, incorporated this discussion of the airplane into his campaign to glorify blackness and strengthen the race for a forthcoming struggle. As part of his campaign for blacks to return en masse to their motherland, Garvey declared that control of military technology would dictate the future of Africa. Just as Tulsa's black community had been helpless against the airplane, he declared, it was futile for Africans to use stones and spears against white colonial masters

---

15 Corn, *The Winged Gospel*. For examples of Coleman's press coverage, see "Chicago Colored Girl Learns to Fly Abroad," *Aerial Age Weekly* (17 October 1921): 125; and "Negro Aviatrix to Tour the Country," *Air Service Newsletter* (1 November 1921): 11.

16 Jill D. Snider, "'Great Shadow in the Sky': The Airplane in the Tulsa Race Riot of 1921 and the Development of African American Dreams of Aviation, 1921–1926," in *The Airplane in American Culture*, ed. Dominick A. Pisano (Ann Arbor: University of Michigan Press, 2003), pp. 105–146.

who possessed planes and machine guns. By learning to build, fly, and maintain airplanes themselves, Garvey suggested, American blacks could protect the race and lift Africa triumphantly toward freedom.[17]

Within this rhetorical context, black nationalists hailed any African American involvement with aviation as a step toward racial victory. Those who remained dubious about Garvey's approach could still rejoice in Coleman's achievement. Black newspapers featured the first African American woman flyer as a front-page celebrity, quoting Coleman herself touting her uniqueness as a nonwhite, nonmale pilot. At a time when numerous white critics openly branded black people with charges of laziness, stupidity, criminality, and other vices, Coleman appeared to disprove stereotypical assumptions that blacks did not have the intelligence or bravery necessary to fly. At a time when women were still second-class citizens in terms of legal status, political position, and economic opportunity, Coleman had shown what a "Negro girl" could do. In interviews with the *Chicago Defender*, Coleman projected her success as a beacon for all African Americans, an opinion perfectly suited to the *Defender's* racial-uplift agenda. In the era of the Harlem Renaissance, when black artists and authors such as Langston Hughes and Zora Neale Hurston displayed their talents as an expression of the race, African Americans were ready to embrace Coleman. The cast of New York's hit black musical *Shuffle Along* presented her with a trophy, and the Metropolitan Baptist Church invited her to address its large congregation. In interviews and speeches, Coleman promoted her ambitions, announcing plans to special-order a plane for herself and promising to help other African Americans learn to fly.[18]

In February 1922, Coleman again left for Europe, where she pursued an advanced aviation course in Paris, visited airplane designer Anthony Fokker in Holland, and made numerous flights in Germany. Returning to America in August, Coleman began capitalizing on media attention, dramatizing (and, it seems, occasionally embellishing) her exploits. Even the *New York Times* noted the arrival of this "Negro aviatrix," whom the reporter said had been hailed by "leading French and Dutch aviators as one of the best flyers they had seen."[19]

Fortunately for her, Coleman by nature was not afraid of the spotlight, since, as a curiosity to blacks and whites alike, she had little chance of avoiding attention. More to the point, Coleman consciously cultivated publicity as a tool to advance her aviation

---

17 Snider, "Great Shadow in the Sky," pp. 120–121.

18 "*Shuffle Along* Company Gives Fair Flyer Cup," and "Aviatrix Must Sign Away Life to Learn Trade," *Chicago Defender* (8 October 1921): 2. For more coverage of Coleman by the *Chicago Defender*, see "Chicago Girl is a Full-fledged Aviatrix Now," *Chicago Defender* (1 October 1921): 1. See also "Chicago Colored Girl is Made Aviatrix by French," *Chicago Tribune* (28 September 1921).

19 "Negro Aviatrix Arrives: Bessie Coleman Flew Planes of Many Types in Europe," *New York Times* (14 August 1922): 4.

A smiling Bessie Coleman shows off one of her military-style flying costumes. She hoped her exhibitions would encourage other African Americans to take up flying. (Undated photo number 84-14782, National Air and Space Museum, Smithsonian Institution)

career. She visited newspaper offices to distribute her own press releases and testimonials, wrote on stationery illustrated with pictures of her stunt flying, and screened newsreel footage of her flights in Germany. Coleman fashioned her public identity as a black woman of special beauty and daring, and the press collaborated equally eagerly in creating her glamorous image.

The *Chicago Defender* gave Coleman a particularly enormous buildup in anticipation of her first American flying show in late summer. Coleman made a personal appearance at the 1922 United Negro Improvement Association (UNIA) New York convention to promote her appearance and was greeted as the attractive personification of the Garveyites' hopes for black pilots. Though her debut exhibition was delayed by inclement weather, the rescheduled appearance on 3 September 1922 was understood to be the first public flight of a black female pilot in the United States. The theatricality of that production was undeniable and deliberate; donning a striking military-style uniform, Coleman lifted off in an on-loan Curtiss airplane to the accompaniment of the national anthem. Her flight itself seems to have been able but unspectacular; organizers did not permit her to perform any stunt flying.[20]

To the white public, Coleman's claim on attention lay in the simple fact that a "Negress Pilots Airplane," as the *New York Times* put it. The dramatic ceremony served to underscore Coleman's link to history, the sheer novelty of a black female flyer. But to black observers, details of the program sent deeper messages about race and technological progress—a politicized promise of black incursions into the white-controlled territory of

20 "Bessie to Fly Over Gotham: Queen Bess to Ride Air Next Sunday," *Chicago Defender* (26 August 1922): 1; "Rain Halts the Initial Flight of Miss Bessie," *Chicago Defender* (2 September 1922): 9; "Negress in Flying Show: Bessie Coleman to Give Exhibition for Fifteenth Regiment," *New York Times* (27 August 1922): 2; "Negress Pilots Airplane: Bessie Coleman Makes Three Flights for Fifteenth Infantry," *New York Times* (4 September 1922): 9; and "Bessie Gets Away; Does Her Stuff," *Chicago Defender* (9 September 1922): 3. See also Snider, "Great Shadow in the Sky"; and Rich, *Queen Bess*.

aviation. While Coleman remained the star of the program, UNIA member Herbert Julian performed a parachute jump as an additional feature. African American pilot Edison McVey made a cameo appearance, presenting Coleman with a beautiful bouquet upon completion of her demonstration, a ceremony of welcome for a fellow African American flight pioneer. The theme of black pride in past, present, and future accomplishment was further underscored by the fact that Coleman's appearance honored the all-black Fifteenth New York Infantry, which had seen service in WWI.[21]

The *Defender* expended even more extravagant rhetoric promoting Coleman's first flight in her adopted hometown of Chicago several weeks later. Articles billed her as "the Race's only aviatrix" and a pilot who had "amazed continental Europe." Advertisements promised a chance to "see this daredevil aviatrix in her hair-raising stunts."[22] Coleman performed a display of takeoffs, glides, turns, and figure eights for a crowd of about two thousand, including her mother, sisters, and other relatives. The *Defender* praised Coleman as part of the newspaper's campaign to win respect for blacks, making Coleman into the personification of progress through self-help, education, and persistence. Equally important, editors hoped that Coleman's attractiveness and colorful adventures could draw readers. The black press gave Coleman nicknames of "Queen Bess" and the "Bird Woman." While American observers labeled white female flyers with cute labels such as "angels," "sweethearts of the air," and "powder puff pilots," press references to Coleman consistently stressed her racial identity. White newspapers headlined her uniqueness as a "Negro aviatrix," while black media emphasized her representation of the race.[23]

With such publicity, Coleman was positioned to establish a career as an exhibition flyer performing at airshows and fairs around the country. While white barnstormers made a reputation through ever-more-daring stunts, for Coleman racial identity became a highly visible component of her reputation. Plans for her flying appearances, however, were complicated by racial politics. Some problems would have been virtually inescapable for a black woman trying to compete in a field controlled by white men. On the other hand, Coleman herself contributed to foil some potential prospects. Black entertainers had a separate traveling circuit, and Bessie unfortunately alienated key organizers by confusion over bookings and by breaking a contract to star in a film whose script she considered to be patronizing.

To maintain her access to flying and raise money to open the African American flight school she envisioned, Coleman sought commercial employment. She positioned herself as an entrepreneur, proposing to drop tire-company leaflets from her plane and appear

21 Snider, "Great Shadow in the Sky," pp. 134–135; and Rich, *Queen Bess*, pp. 49–51.

22 Rich, *Queen Bess*, p. 54.

23 "Bessie Coleman, the Race's Only Aviatrix, Will Make Her Initial Local Flight at Checkerboard Airdrome Sunday, Oct. 15," *Chicago Defender* (9 September 1922): 3; "Bird Woman Arrives," *Afro-American* (18 August 1922): 10; and Blaine Poindexter, "Bessie Coleman Makes Initial Aerial Flight," *Chicago Defender* (21 October 1922): 3.

in their promotions. Yet while white female pilots such as Amelia Earhart would succeed in parlaying their attractiveness and ability into endorsement contracts, Coleman's racial marginality limited her sponsorship possibilities. Moreover, Earhart and other white pilots had connections within the flying community—mentors who recommended them for sales jobs or other aviation-related work. Due to her marginalization and training abroad, Coleman had no comparably wide base of support among established aviators. While exploring financial options in California in February 1923, Coleman arranged to present a flying demonstration as the star draw at a Los Angeles fair. Upon taking off from Santa Monica, the engine of Coleman's newly purchased but old surplus Curtiss Jenny stalled. The crash left Coleman with fractured ribs, a broken leg, and useless airplane wreckage. The African American press played up her fortitude in bearing the physical pain and praised her unshaken determination not only to continue flying herself, but to help other blacks become pilots. Coleman reportedly sent a telegram announcing, "Tell them [my fans] all that as soon as I can walk I'm going to fly."[24]

Determination could not work miracles, however, and Coleman experienced continued setbacks both in her personal flying career and in her efforts to secure capital for an African American flight school. Newspaper articles repeatedly announced that Coleman's training center was about to open or indeed was ready to receive students; Coleman apparently declared on several occasions that she had ordered or purchased one, two, or an entire fleet of airplanes for herself and students. Those aircraft never materialized, and Coleman had to keep borrowing planes for her own flights. After lengthy recuperation from her injuries, Coleman resumed appearances in air shows in Illinois and Ohio in late 1923. But beyond those regional shows, Coleman was unable to expand her bookings. Her headstrong independence had antagonized managers and agents, who branded her as overly temperamental. Without capital, Coleman could not purchase her own plane or finance her flying plans; without performance prospects, she could not count on income. After a year and a half of career stagnation, Coleman finally secured a schedule of lectures and public flying demonstrations for mid-1925 in her birth state of Texas.

For Coleman, as for almost any African American of this era who achieved public stature, racial politics inevitably complicated her daily routine. Coleman had to plan her travels knowing there were public spaces and accommodations in Jim Crow regions where blacks were unwelcome. As a celebrity, she gained a small influence over racial conditions at her own appearances. Coleman's performances around Houston and elsewhere in Texas attracted thousands of spectators, both white and black—a situation that generated difficulties in the era of segregation. In Waxahachie, Coleman threatened to boycott the show if arrangers insisted that whites and blacks enter the grounds through separate gates, and organizers capitulated (though seating remained segregated). In

24 Rich, *Queen Bess*, p. 70.

Florida, Coleman similarly warned that she would cancel her scheduled parachute jump if the Orlando Chamber of Commerce refused to let blacks attend.

Beyond her gender and race, Coleman resembled other barnstorming pilots, who were mainly white men, in her drive to maximize the entertainment value of her flying through flair and apparent personal risk. Coleman regularly posed for photographers as part of her appearances, standing in front of her plane in her specially designed costumes and playing to the waving crowds. She aimed to amaze onlookers with attention-getting acrobatic stunts, including parachute jumps, barrel rolls, loops, and steep dives taking her plane extremely close to the ground. While male pilots usually cultivated a macho image to accompany their daring showmanship, Coleman's popularity rested on a more feminine brand of personal charisma. Her uniforms carried an air of military distinction, yet also fashionably accentuated her graceful petiteness and light-skinned femininity, making her control of a powerful machine all the more impressive. Advertisements described her as "the little girl who has the nerve to fly," highlighting her petite status and youthful appearance in contrast to more mature men whom society usually credited with extra courage.[25]

Even when she followed the same standard barnstorming act as white male pilots, Coleman's routine conveyed a unique set of racial messages. In connection with their exhibitions, performers at air circuses regularly sold adventurous spectators a chance to climb aboard planes for short hops. Coleman joined other barnstormers in selling rides for $3 or $5 apiece, yet, for her African American audiences, the offer carried additional racial significance. Coleman was particularly interested in sharing the experience of flying with black passengers—a dimension of her public performances that justified their superficial theatricality in terms of a larger race mission. African Americans who watched a Coleman show or stepped into her plane to venture up themselves had been guided by the black press and black leaders to think about the racial politics of flight.

During the same years that Coleman was striving to advance her career, Herbert Julian also was working to connect his personal interest in flying to the Garvey movement's black-nationalist aviation agenda. The African American stunt team of parachutist Julian and pilot Edison McVey gave exhibitions during the summer of 1923, primarily in black venues. McVey was temporarily forced into retirement by a crash when his aircraft stalled, but later resumed flying and delivered UNIA lectures promoting aviation. Meanwhile, however, Julian had come under criticism for personal scandals, for having misrepresented his background, and for ostentatious behavior. Snider explains, "By late 1923, Julian's panache had become an embarrassment for some, especially as it increasingly caught the eye of white journalists. The *New York Times* and other white papers frequently made Julian a target of their humor, publishing numerous articles parodying him as a Negro buffoon attempting to master a white man's technology."[26]

---

25 Elizabeth Amelia Hadley Freydberg, *Bessie Coleman: The Brownskin Lady Bird* (New York: Garland Publishing, 1994).

Coleman could in no way be characterized as a buffoon; her public character, though dramatized to suit requirements of the entertainment business, was untarnished by scandal or clownishness. At the same time, her remarks about the importance of increasing black involvement in aviation often seemed to suit the Garvey orientation. Coleman frequently stressed the importance of bringing African American men into flying, saying, "We must have aviators if we are to keep up with the times. I shall never be satisfied until we have men of the race who can fly." It would be men, of course, who in Garvey's world would lead the battle of black liberationists against white colonialists.[27]

Yet as a petite female, Coleman herself never matched Garvey's plan for an army of black pilots prepared to wage race war. Her personal image fell more into line with the integrationists' vision of black pilots as a nonthreatening (at least physically) statement to whites of African American capabilities and as an equally significant message to blacks about the possibilities of individual uplift.

Moreover, within just a few years, the Garvey movement headed toward implosion. For all the attention he received, Garvey had never won full endorsement from the African American middle-to-upper-class establishment. The black press generally remained wary of Garvey's radical nationalism, and *Chicago Defender* editors and writers in particular backed an alternative vision of black individual progress within a white American world capable of social reform. Garvey's pretentiousness alienated other observers, and, not entirely to the dismay of his black critics, the federal government pursued and ultimately imprisoned Garvey on counts of mail fraud. Garvey's fall turned the spotlight toward more mainstream leaders' assessment of aviation as a route toward integration rather than race conflict. Integrationists suggested that by learning to fly, African Americans could counter racist stereotypes of blacks as ignorant, cowardly, or backwards, proving the race's claim to join whites in the skies and in a democratically reformed society on equal status. These commentators also stressed that given the promising future for aviation expansion in the United States, blacks should pursue it as a potential source of employment and economic opportunity.[28]

For this set of optimistic observers, Coleman's success in becoming a pilot reflected credit on all African Americans, particularly those who overcame any qualms and paid a few dollars to join her on a ride. This race-uplift aim in her individual gospel of aviation remained a constant theme in Coleman's public career. She spoke publicly about the importance of improving the national perceptions of African Americans and promoted

---

26 Snider, "Great Shadow in the Sky," pp. 128–131. Julian would come to further grief in 1924, when the African American press greeted his plans for making an ambitious transatlantic crossing with skepticism. Julian remained popular among many black citizens, in large part due to his direct racial appeal. Julian decorated his airplane, christened *Ethiopia I*, with the saying, "This plane is the property of the Negro race, donated by them for their future advancement in aviation." Julian's scheme met an ignominious end on Independence Day in 1924, when his long-awaited takeoff ended almost immediately in a clumsy crash.

27 Ibid., p. 135.

28 Ibid.

aviation as a modern means to that end. According to the *Houston Post-Dispatch,* Coleman declared, "I want to interest the Negro in flying and thus help the best way I'm equipped in to uplift the colored race."[29]

Coleman's example as an aviation entrepreneur also encouraged those who believed that blacks could find a place in this new enterprise. Citing the fact that popularity of automobiles had created jobs for blacks as chauffeurs, Coleman (like a number of other African American observers) predicted that the spread of airplanes could bring a natural progression to black employment as hired pilots. In lectures at black churches, halls, and theaters, Coleman spoke about turning "Uncle Tom's cabin into a hanger" with her dream of setting up a black flying school. The day before what would be her final flight, Coleman paid a visit to each black public school in Jacksonville, Florida, impressing the children with her sharply tailored uniform.[30]

Behind her smart personal appearance, Coleman's career was hampered by continued financial tentativeness and lack of decent equipment. On occasion, Coleman apparently had to cancel scheduled appearances when she was unable to rent or borrow a plane from local owners. Coleman had begun making payments on a used Curtiss Jenny; when she was due to give an exhibition in Jacksonville in April 1926, to benefit the Negro Welfare League, she arranged to have white mechanic William Wills fly the Jenny to Florida from Texas. During the trip, Wills had to make two unplanned landings when the worn-out engine malfunctioned.

On 30 April 1926, the day before the air show, Coleman wanted a preliminary run to get an overview of the area and determine the best location for making her parachute jump. Wills handled the controls while Coleman sat in back, leaving her seatbelt unfastened, perhaps so she could crane her neck over the side to survey the field. After taking off without incident and a short period of climbing and circling, the plane abruptly sped up, nose-dived, and flipped over, throwing Coleman to her death. Wills fought vainly to restore control of the plane before crashing. Police and aviation officials soon arrived at the scene. While nervously surveying the catastrophe, John Betsch, Coleman's sponsor from the Negro Welfare League, lit a cigarette, igniting spilled gasoline and reducing the plane to a burnt mess. Through subsequent investigation, authorities determined that a wrench left inside the Jenny had fallen into the engine, jamming it and sending the plane plummeting. The black community of Jacksonville, which had scheduled a post-exhibition "aerial frolic" to honor Coleman, now canceled the dance and held her memorial service instead. Thousands of men, women, and children waited in line to walk past her casket and filled churches first in Florida and then in Chicago.[31]

---

29 "Texas Negro Girl Becomes Able Aviatrix," *Houston Post Dispatch* (7 May 1925): 4.

30 Snider, "Great Shadow in the Sky," p. 135.

31 Rich, *Queen Bess*, p. 111. See also E. B. Jourdain, Jr., "Bessie Coleman, Aviatrix Killed," *Chicago Defender* (8 May 1926): 1; and Evangeline Roberts, "Chicago Pays Parting Tribute to Brave Bessie," *Chicago Defender* (15 May 1926): 2.

COLEMAN REMEMBERED

Other African Americans, male and female, followed Coleman into the skies in the years after her death. Many discovered almost immediately, as had Coleman, that race narrowed access to training. Around 1927, William Powell was turned away from Chicago-area flight schools that refused to consider blacks. After finally gaining entry to Los Angeles's Warren College of Aeronautics, Powell earned his license in 1928. The next year, he set up a Bessie Coleman Aero Club and School, offering African Americans a location for flight training and aviation development. The Coleman Aero Club became a magnet for African Americans interested in flying such as James Banning, whose curiosity had been sparked by taking a $5 airplane ride at a circus. After failing to locate a flight school willing to accept him, Banning took lessons from a WWI veteran and became the first African American to earn a pilot's license from the U.S. Department of Commerce in 1926. Citing Coleman as his inspiration, Banning proceeded to become a barnstormer, mail pilot, and chief flight instructor at the Coleman Aero Club. In the fall of 1932, Banning became the first black pilot to complete a transcontinental journey, flying from Los Angeles to New York with mechanic Thomas Allen. This diversified face of aviation appeared most publicly with the creation of a "Colored Air Circus," whose attractions included a performance by an all-black troupe named "The Five Blackbirds."

Of course, no one knew more intimately than African Americans how many obstacles remained. While Willa Brown succeeded in setting up an African American flight school in Chicago in the 1930s, black pilots had to build their own airfield outside city borders due to racial restrictions barring them from regular Chicago airports. The black press periodically complained about policies excluding blacks from military aviation service and government-backed flight training. When that barrier was finally broken during WWII with the creation of the Tuskegee Airmen, numerous observers credited Coleman as a pathbreaker for those black pilots.[32]

Over the decades, the African American community periodically remembered and celebrated Coleman as a pioneer. By the 1930s, black pilots in Chicago had instituted a tradition of holding an annual flyover to drop flowers on Coleman's grave, and, after a lapse, that tribute was reinstituted in the 1980s and 1990s. Dr. Mae Jemison, the first African American female astronaut, commented on Coleman's inspiration by noting, "[E]very day we see people making small strides in overcoming obstacles of gender,

32 For background, see Walter T. Dixon, Jr., *The Negro in Aviation* (Baltimore: Clarke Press, 1950); and Von Hardesty and Dominick Pisano, *Black Wings: The American Black in Aviation* (Washington, DC: National Air and Space Museum, 1983). On the Tuskegee Airmen, see Stanley Sandler, *Segregated Skies: All-Black Combat Squadrons of World War II* (Washington, DC: Smithsonian Institution, 1992); and Robert J. Jakeman, *The Divided Skies: Establishing Flight Training at Tuskegee, Alabama, 1934–1942* (Tuscaloosa: University of Alabama Press, 1992).

birthright, race, ethnicity, economics, illness, poor technology, education, societal condemnation, and fear . . . . [H]ere is a woman . . . who exemplifies and serves as a model to all humanity: the very definition of strength, dignity, courage, integrity, and beauty." In 1995, the U.S. Postal Service created a stamp honoring Coleman as part of their "Black Heritage" postage series.[33]

Coleman's most visible public legacy today is as a role model meant to inspire, particularly for children. Coleman's life has been the subject of at least five picture books and at least four juvenile biographies for a slightly older market. This new attention to Coleman has emerged over the last decade, following publication of Doris Rich's 1993 adult biography, which called attention to Coleman and clarified many details. It is telling that there are so many juvenile biographies of Coleman, while the Rich biography and Elizabeth Freydberg's analysis of Coleman as an entertainer are the only book-length scholarly accounts. The school-age books on Coleman seem particularly oriented for use as part of the appreciation of Black History Month and Women's History Month; they add another woman of color to the list of "great Americans" about whom students can learn.

As a group, these juvenile biographies interpret Coleman's history as a moral lesson in the power of overcoming obstacles and pursuing a personal vision. While not quite depicting her as a rags-to-riches story, these books present Coleman as a real-life black female version of Horatio Alger's *Strive and Succeed*, someone making her way against great odds to achieve success—measured not in financial worth, but in individual fulfillment.

These biographies usually open with Coleman's rise from poverty, portraying it as the result of inherent determination. Four picture books give vivid visual depth to Coleman's early hardships, offering illustrations of a young black girl picking cotton, scrubbing or ironing clothes, hanging washing, or carrying a full laundry hamper. Sally Walker's *Bessie Coleman: Daring to Fly*; Lynn Joseph's *Fly, Bessie, Fly*; and Nikki Grimes's *Talkin' About Bessie* all use the story that Coleman challenged field foremen who tried to shortchange her family (or even slid her foot onto the scale to add weight to their sacks), praising her willingness to use initiative to fight an unfair situation. Both the Walker and Joseph books contain illustrations showing a defiant Bessie arguing with a stern or smug-looking man over a scale and account book. Walker writes, "If the white man who paid them tried to cheat, Bessie wasn't afraid to set him straight. She was *that* daring."[34]

These books also highlight illustrations of Coleman studying in school or reading at home, stressing her dedication to education as a measure of her personal values and

33 Mae Jemison, "Afterword," in Rich, *Queen Bess*, pp. 143, 145.

34 Sally M. Walker, *Bessie Coleman: Daring to Fly*, illustrated by Janice Lee Porter (Minneapolis: Carolrhoda Books, 2003), pp. 6–7; Lynn Joseph, *Fly, Bessie, Fly*, illustrated by Yvonne Buchanan (New York: Simon & Schuster, 1998); and Nikki Grimes, *Talkin' About Bessie: The Story of Aviator Elizabeth Coleman*, illustrated by E. B. Lewis (New York: Orchard Books, 2002). As another example from children's literature, see Philip S. Hart, *Up in the Air: The Story of Bessie Coleman* (Minneapolis: Carolrhoda Books, 1996).

strength of mind. Walker explains, "In 1902 . . . [m]ost schools for black children had little money. Bessie's school had few books . . . . Still, Bessie worked hard. She planned to do more with her life than pick cotton. The more she learned, the better her chances would be."[35]

These authors usually convey, with varying explicitness, some sense of the racial discrimination pervading early 20th-century society. Joseph's book contains an illustration of Coleman at a circus getting turned away from a "whites only" ticket booth. Joseph tells readers that even as she returned to the black people's line, Coleman held her head high and comforted a crying sister by reminding her of their mother's assurance that all skin colors were the same before God. Some accounts suggest how Coleman's racial pride defied southern expectations of black deference. The picture book *Talkin' About Bessie,* which builds historical evidence into the form of fictionalized narratives from Coleman's family and acquaintances, contains a supposed comment from a white woman who hired Coleman to do laundry. "She'd come to the back door, like they were supposed to in those days. But when I opened it, there this Colored girl would be standin', lookin' me straight in the eye, like we were just any two people meetin' on a street in town. You know, like we were *equals.* It was odd, I don't mind tellin' you."[36]

In rebellion against this discrimination, out of her mother's encouragement, and from her own interest, these books suggest, Coleman developed an early racial awareness. Walker describes a quest for African American role models, writing that Coleman "loved the story of Harriet Tubman. Now there was a daring black woman! She had led hundreds of slaves to freedom . . . . As she read, Bessie made up her mind. Someday *she* would do something important, too." Grimes portrays Coleman's sister Elois saying, "*Uncle Tom's Cabin* was among Bessie's favorite books . . . [but] she had no respect whatever for the slave girl, Topsy, who seemed incapable of self-improvement, or for Tom, who had too little race pride for Bessie's taste . . . . Bessie's habit of probing others' words came . . . of studying those written by Booker T. Washington and Paul Lawrence Dunbar. Sometimes, I wondered how far their words would take her." This self-instruction in African American promise combined with personal ambition, these accounts indicate, to make Coleman committed to bettering her personal prospects and accomplishing something racially meaningful.[37]

35 Walker, *Bessie Coleman,* p. 8.

36 Joseph, *Fly, Bessie, Fly*; and Grimes, *Talkin' About Bessie.*

37 Walker, *Bessie Coleman,* p. 10–11; and Grimes, *Talkin' About Bessie.*

In their eagerness to convey the importance of determination, several children's books stretch beyond solid historical evidence to maintain that Coleman was set on aviation from the beginning. The cover of *Fly, Bessie, Fly* shows a black girl running through a field with her arms spread, imitating the plane soaring above. Joseph shows Coleman staring skyward, singing, "Far away, far away, up past the clouds. High away, fly away, and never come down," engrossed in daydreams about being able to fly like a bird. Reeve Lindbergh's *Nobody Owns the Sky* uses a similar picture of a black girl gazing longingly at geese and bluebirds overhead, with the following rhyme:

Bessie wished she could rise up and fly, high and low,
Over Texas, a long time ago . . . .
Bessie worked hard at school, and she dreamed about flight.
People said she was crazy; it wouldn't be right.
"You're a girl, not a man, and you're not even white!"
But did she stop dreaming? Not quite!

Another illustration shows a young Coleman having colored a blue cloud-filled sky and with a model airplane resting on her desk (inaccurate, given that Coleman was already approaching teenage years when the Wrights first developed their plane). Lillian Fisher's *Brave Bessie: Flying Free* goes further, suggesting that throughout elementary school and college, Coleman found the sound of airplane engines "music to her ears" and particularly requested books on aviation. "For as long as she could remember, she knew in her heart that someday she would become a pilot. She thought about flying constantly, knowing the only way to reach that goal was through more education." Fisher further asserts that Coleman closely analyzed aircraft technology during WWI. "It amazed Bessie that in four short years the plane changed from a flimsy aircraft, one that looked like a crate with linen wings, into a sturdy-looking plane."[38]

Both Joseph and Fisher imply that for Coleman, from the start, flying represented freedom—an African American dream of escaping cotton picking and discrimination. Fisher creates the following poem meant to represent "Bessie's Song":

38 Joseph, *Fly, Bessie, Fly*; Reeve Lindbergh, *Nobody Owns the Sky: The Story of "Brave Bessie" Coleman*, illustrated by Pamela Paparone (Cambridge, MA: Candlewick Press, 1996); and Lillian M. Fisher, *Brave Bessie: Flying Free* (Dallas: Hendrick Long Publishing Company, 1995), pp. 21, 26–27, 31.

I climb, I soar,
Higher, high
Above the cotton
Fields I fly.
Unleashed,
Unbound,
My spirit free
Beyond the clouds
On wings I flee
Alone with wind,
Sweet air, and sun
I sing the song
Of battles won.[39]

By and large, these authors succeed in conveying the race and gender-located difficulties Coleman faced in gaining access to aviation. Lindbergh writes:

There was a young woman who wanted to fly.
But the people said, "Kiss *that* wish goodbye!
The sky's too big, and the sky's too high,
And you never will fly, so you'd better not try."
But this woman laughed, and she just said, "Why?
Nobody owns the sky!"
"White men can fly. Why can't I?" she would say.
But the flying schools turned her away.
Bessie manicured nails while the barber cut hair.
And she dreamed about flying, but didn't know where.
Then one day someone said, "Fly in France! They won't care
That you're black and a woman." So Bessie went there.
She was young, tough, and smart; she had courage to spare,
And she took like a hawk to the air.

Both the Walker and Joseph biographies use illustrations of a dejected or angry-looking Coleman at an airfield, speaking with a white man who blocks her entry.[40]

These accounts build Coleman's persistence in securing flight instruction in France into lessons about the value of commitment. Several illustrations take liberties in attach-

39 Fisher, *Brave Bessie*.

40 Lindbergh, *Nobody Owns the Sky*; Walker, *Bessie Coleman*; and Joseph, *Fly, Bessie, Fly*.

ing an over-glamorized French romanticism to Coleman's training, showing (inaccurately) the Eiffel Tower looming over her school airport. These books assure children that hard work brings rewards; Joseph writes that upon returning home, "Instantly, she is a star. Everyone wants to take the picture of the only black woman aviator in the world." A couple books note that Coleman applied her new celebrity to encourage fellow African Americans to pursue aviation. Lindbergh writes:

> On the ground Bessie lectured to crowds big and small.
> People gathered in church or inside the town hall.
> "Come and fly, boys and girls! Black or white, short or tall,
> Come and fly, everybody! Come, answer my call.
> The air has no barrier, boundary, or wall.
> The blue sky has room for us all."[41]

Understandably, some books for younger readers are circumspect about Coleman's death. Walker and Joseph both relegate it to an afterword, giving the details of Coleman's crash while tying it to a reassurance that "Bessie's spirit lives on today" in the way she inspired others to become pilots. Louise Borden and Mary Kay Kroeger convey the same message in *Fly High*, which contains an illustration of a flag-draped casket being carried past a crowd of mourners. In a startling contrast, Lindbergh actually emphasizes Coleman's crash with an illustration of smoke rising over the horizon as both black and white spectators gasp in dismay. Lindbergh writes:

> Bessie's life was not long, but she flew far and wide . . .
> But in Jacksonville, Florida, everyone cried,
> Because Bessie's plane failed, and she fell, and she died.
> "Farewell to Brave Bessie!" they sighed . . . .
> Other young men and women soon wanted to fly
> And the people said, "Why don't you give it a try?
> The sky's still big, and the sky's still high,
> But you're bound to get there, by and by.
> Just remember her words 'til the day you die—
> *Nobody owns the sky!*

Universally, these books draw on Coleman's life to instruct young readers to believe in themselves. Dolores Johnson writes in *She Dared to Fly*, "Bessie learned to fly when others

41 Joseph, *Fly, Bessie, Fly*; and Lindbergh, *Nobody Owns the Sky*.

said it was impossible for a woman, never mind a black woman . . . . She set a goal and decided to work very hard until she achieved it. She chose a dream she dared to live, and she made it happen." Walker phrases the moral of Coleman's story as "aim high," while Fisher declares, "The sky is the limit!" Borden and Kroeger tell readers to "keep trying! Fly high!" while a biography by Connie Plantz advises children, "Don't be afraid to take risks. Fly!"[42]

Of course, even among adults, not everyone was prepared to accept that injunction to take Coleman's life as inspiration. In 1994, after *Air and Space* magazine ran an article by Doris Rich on Coleman, one irate reader denounced the article as "a waste of space." He continued, "It is clear to me that Coleman made no contribution to the field and in fact was an unskilled and/or very careless pilot. The only rationale for the story was that she was black and an activist to boot. Spare us! Keep the politically correct agenda for the fourth estate; they cover it exhaustively enough."[43]

This belittling of Coleman's skills belies the clear evidence that Coleman was not even piloting the plane during the fatal flight. Such readiness to attribute fault to a female pilot provides a recent example of a phenomenon appearing repeatedly through the decades—critics' rush to condemn female aviators as incompetent. In the early years of aviation, crashes were quite common occurrences; some stemmed from human error, others from equipment failure, and others from cases where flyers deliberately attempted to stretch the limits. Some of the era's best pilots were involved in crashes, often on multiple occasions; to take just one instance, Wiley Post, who had flown around the world twice, was killed in a 1935 accident. Yet crashes involving female pilots were deconstructed to signify personal weakness in ways that men's accidents were not and became an excuse to repeat that women had no place in aviation. Corn comments, "[E]ven death did not free them from the barbs of prejudice. The lady flier stereotype often surfaced in discussions of fatal accidents, such as the one that killed Muriel Crosson in 1929." Journalists and officials suggested that Crosson lost control of her plane after fainting or "neglected" to open her parachute upon bailing out. In truth, evidence suggests that Crosson indeed attempted to use her parachute and that she had commented before flying that her airplane's engine was operating poorly. "She took off anyway . . . . Had she not taken off . . . she would have opened herself to the criticism of being timid and overly cautious, thereby encouraging yet another invocation of the lady flier stereotype."[44] Other women were indeed criticized as being too cowardly to be true pilots when safety concerns led them to cancel flights.

---

42 Dolores Johnson, *She Dared To Fly: Bessie Coleman* (New York: Benchmark Books, 1997), p. 6; Walker, *Bessie Coleman*, p. 47; Louise Borden and Mary Kay Kroeger, *Fly High! The Story of Bessie Coleman*, illustrated by Teresa Flavin (New York: Margaret K. McElderry Books, 2001); Fisher, *Brave Bessie*, p. 71; and Connie Plantz, *Bessie Coleman: First Black Woman Pilot* (Berkeley Heights, NJ: Enslow Publishing, 2001), p. 111.

43 Letter to editor, J. Don Marion, *Air and Space* (December 1994/January 1995): 8.

44 Corn, *Winged Gospel*, pp. 79–80.

CONCLUSION

Coleman clearly belongs to a specific point in the history of aviation, the barnstorming era. As Corn, Dominick Pisano, and Roger Bilstein have made clear, barnstormers fulfilled a purpose for aviation promoters in literally bringing flying to the masses, yet ultimately the aviation establishment became uncomfortable with barnstorming culture. By the early 1920s, the National Aeronautic Association and the Aeronautics Branch of the U.S. Department of Commerce worried that the very daring which gave barnstorming its entertainment value threatened to undermine aviation's serious future. Some among the crowds came to exhibitions hoping to witness an exciting crash, but businessmen wanted people to associate aviation with safety and reliability, rather than danger. Authorities began to impose restrictions on air circuses, pushing demonstrations back from spectators and limiting specific stunts such as wing walking.[45] At the time of Coleman's death, the peak of barnstorming fever had started receding.

Coleman amazed onlookers with attention-getting acrobatic stunts, including parachute jumps, barrel rolls, loops, and steep dives taking her plane extremely close to the ground. She died in the crash of her secondhand Curtiss Jenny in 1926 before one of her air shows. Ironically, that day she was not at the controls. (Undated photo number 93-16054, National Air and Space Museum, Smithsonian Institution)

45 Pisano, "The Greatest Show Not on Earth," pp. 39–74.

As a barnstormer, Coleman had a certain freedom to design her own show, yet at the same time faced continual pressure to top the spectacle content of other entertainers. Furthermore, that nominal freedom was constrained by her persistent problems in gaining access to aircraft. By definition, there were tensions in what Coleman was trying to accomplish. As an African American pilot attempting to convince other African Americans to ride into the sky with her, she had to convince them that flying was reasonably safe; yet the old planes she managed to obtain were often inherently unsafe, and stunts such as parachuting and dives that she used to attract those spectators also courted risk. As a black pilot encouraging other blacks to enter aviation, she had to make flying appear straightforward, something others could master given the opportunity. But as a black pilot trying to convince whites that her accomplishment disproved stereotypes of racial inadequacy, she had to emphasize the difficulty behind her flying, the skill and intelligence it demonstrated. Moreover, if Coleman made aviation appear too safe and simple, her celebrity would be tarnished as less impressive.

Politically and socially, Coleman clearly placed herself and was categorized by others in racial terms. While Coleman deliberately cultivated an unmistakably feminine appearance as her public image, her identity as a female pilot was more problematic. White female pilots proved useful to the aviation establishment precisely because women were assumed to be less courageous, less strong, and less rational than men. Seeing pilots of "the weaker sex" was supposed to convince reluctant men, potential pilots and passengers, that flying was so safe even a woman could handle it. "[W]omen pilots domesticated the sky, purging it of associations with death and terror," Corn writes, and "paradoxically, prejudice begat opportunity." A number found jobs demonstrating and selling private airplanes, with female pilots tacitly guaranteeing reliability. The curiosity factor also increased the attention factor for female sales personnel, as did perceptions of feminine pleasantness and attractiveness.[46] Yet, due to her race, even if Coleman's career had extended over a longer period, it is unlikely she would have secured the same types of aviation business employment that white flyers such as Ruth Nichols, Louise Thaden, Blanche Noyes, and Amelia Earhart did.[47]

Despite stereotypes restricting them to this "woman's sphere" of aviation, white female pilots associated flying with feelings of independence and enjoyed a certain sense of power in appropriating male-linked technology. In addition, in 1929, female pilots

---

46 Corn, *The Winged Gospel*, pp. 78–79, 88. As Corn emphasizes, negative stereotypes of female pilots ironically opened some doors, but meant that others remained closed. Women pilots would find few opportunities with commercial passenger airlines. In another key case, during the 1920s, Helen Richey was effectively pushed out of work as an airmail pilot after male union members protested that women were too weak to handle planes in rough weather, and the government issued guidelines limiting women to fair-weather flying.

47 It is worth noting that for Earhart in particular, employment opportunity preceded celebrity; thanks to connections in the aviation community and perceptions of her appropriately proper femininity, she began sales work long before setting her most famous records.

formed an organization named the Ninety-Nines (the number of charter members) that campaigned to open more aviation opportunities to women and defeat discriminatory measures such as a proposal to bar women from flying while menstruating. By the end of the 1920s, a number of women secured employment in aerial photography, as flight instructors, and on commercial airline staff (though not generally as pilots). Coleman, of course, passed away before the creation of the Ninety-Nines, but, during her lifetime, she never really entered the aviation community of her white sisters. She too sought personal and group liberation through flying, but, for her, that quest was equally if not more racial than gender related.

For insight, it is worthwhile comparing Coleman's situation with that of Amelia Earhart, her famous white counterpart. Earhart earned her American pilot's license from the National Aeronautic Association on 15 December 1921, roughly half a year after Coleman completed initial training. Earhart received her FAI license on 15 May 1923, two years after Coleman. Having been licensed, Earhart struggled to afford the expense of flying by juggling numerous part-time jobs, including secretarial work, settlement house work, photography, and even gravel hauling for a construction firm. Yet unlike Coleman, Earhart ultimately enjoyed two major advantages. Her mother, though in a precarious financial state, still helped Amelia purchase her first plane, and Earhart was able to secure work as an airplane demonstrator and sales representative from a friend in the business.[48]

Earhart's biggest break came thanks to her appearance as "representative" of the liberated yet still feminine "new woman" in America, an image tacitly coded as white, for which Coleman never could have qualified.[49] Organizers of a new transatlantic venture sought "the right sort of girl" to join them; Earhart, as an educated, socially gracious, attractive yet respectable single woman, seemed perfect. Her pilot's license was useful, given that it showed her comfort with being airborne, yet was not directly necessary; although the successful 1928 "Friendship" flight made Earhart the first woman to cross the Atlantic by airplane, she was given no role in the piloting. Considering that cross-Atlantic flight remained supremely risky, Earhart did deserve credit for her courage and became an international celebrity overnight. Publisher George Palmer "G. P." Putnam expertly packaged Earhart as the exemplary "all-American girl," steering the press and observers to remark upon how similar her strong facial lines, light coloring, and tall slenderness were to the appearance of world aviation hero Charles Lindbergh.[50]

Earhart's schedule became packed with almost nonstop public appearances, charity events, press conferences, interviews, and radio broadcasts. Though soon exhausted by

---

48 Doris L. Rich, *Amelia Earhart: An Autobiography* (Washington, DC: Smithsonian Institution Press, 1989).

49 For background, see Nancy Woloch, *Women and the American Experience* (New York: McGraw Hill, 2000).

50 Rich, *Amelia Earhart*, pp. 46–47.

dealing with a relentless press and throngs of curiosity seekers, she knew publicity was essential to raising the money she wanted to underwrite advanced flying lessons, buy and maintain new planes, and support her family. She wrote aviation columns for *Cosmopolitan*, gained endorsement offers for automobiles and (controversially) cigarettes, became a consultant for coeds at Purdue University, and assumed an airline executive job promoting travel to female passengers. While never relaxed in the spotlight, Earhart began traveling the remunerative national lecture circuit giving talks for audiences of five thousand or more. These speeches, sometimes two per day, commanding fees up to $300 apiece, became her biggest source of income. Putnam kept inventing new ventures for Earhart, including designing and modeling her own lines of women's clothing and luggage. Earhart had a master of public relations promoting her (and the books she wrote for his firm), where Coleman struggled with self-promotion. Earhart and Putnam would later marry, placing her within the ranks of female pilots such as Florence "Pancho" Barnes, Lady Mary Heath, Ruth Nichols, and Jacqueline Cochran who were either born into privileged circumstances or married into wealth. Coleman, of course, did neither.[51]

For Coleman, financial difficulties limited flying opportunities and blocked her aim of opening a black flight school. Earhart enjoyed substantial income, yet her personal and professional expenses simultaneously skyrocketed. Earhart became trapped in a vicious cycle; she poured capital into acquiring the faster, more powerful new planes that she needed to meet new aviation goals, which became the subject for new articles, books, and lectures, bringing her money she needed to repeat the process. Each publicity campaign drained Earhart's energy and took valuable time away from flying practice, yet she still compiled an imposing list of accomplishments, including new FAI women's world speed records in 1930, the first women's solo transatlantic flight in 1932, and the first solo flight from Hawaii to California in 1935. Her disappearance on 2 July 1937, as her airplane was lost trying to land on tiny Howland Island in the Pacific Ocean, came in her attempt to set a spectacular new record, a flight circumnavigating the globe near the equator.[52]

Despite all the differences in their circumstances, one similarity linked the paths of Coleman and Earhart. Each found flying fun in itself, but also applied the attention it brought them to promoting a heart-felt cause. In her public appearances, Earhart frequently spoke in favor of women's rights, pacifism, and the expansion of aviation. Coleman, of course, championed the case for African American equality.

---

51 Ibid.
52 Ibid.

Like white fans of aviation, America's black community commonly envisioned flying as a centerpiece of the future, defined by the technological progress already apparent through advances in radio, automobiles, movies, and electrical machinery.[53] Yet while white citizens took for granted their possession of the power of airplanes, African Americans were largely looking at that power from outside. Blacks were not naïve, passive observers of aviation's development; for them, airplanes were not innocent objects of wonder, but rather a potent symbol of contested economic, social, and political control. Coleman concurred that aviation encapsulated the promise of the future and therefore wanted to see blacks share its potential for employment and rewarding accomplishment. This aim was frustrated by her awareness that for all the passion aviation technology attracted, its human institutions were not equally welcoming to all. As the first African American female to venture into this field, Coleman had to devise strategies for simply gaining access to airplanes and training. At the same time, she actively pursued even deeper challenges of raising money and communicating her ideas about aviation. In children's literature, her success is linked to character as an inspirational lesson in the power of dreaming and determination. For historians, her biography can serve as the pathway toward richer interpretations in the social and political history of aviation, showing how Americans' enthusiasm for flight was complicated by racial and gender realities.

53 See Dominick A. Pisano, "New Directions in the History of Aviation," in *The Airplane in American Culture*, pp. 1–15; and Roger Bilstein, "The Airplane and the American Experience," in *The Airplane in American Culture*, pp. 16–35.

2

She Flew For Women

Amelia Earhart, Gender, and American Aviation

SUSAN WARE

ON 20 MAY 1927, IN ONE OF THE DEFINING MOMENTS IN THE HISTORY OF AVIATION, CHARLES LINDBERGH, THEN ONLY 25 YEARS OLD, SOLOED THE ATLANTIC and became the most famous person in America overnight. Just a little more than a year later, another Atlantic crossing also captured the popular imagination, although not quite on the scale of the Lindbergh cult, when 30-year-old Amelia Earhart became the first woman to cross the Atlantic by plane. Even though she was merely a passenger and did none of the flying (she did have a pilot's license), the flight catapulted Earhart into the kind of public celebrity that had greeted Charles Lindbergh on his arrival in Paris the year before.

From that point on, Lindbergh and Earhart were represented in the media as complementary idols of individual achievement, and she was bestowed with the nickname

1 This article is adapted from Susan Ware, *Still Missing: Amelia Earhart and the Search for Modern Feminism* (New York: W. W. Norton, 1993).

"Lady Lindy." (Note: He was never called "Lord Earhart" or the "male Amelia Earhart.") These repeated comparisons were not just a creation of a hyperactive media; the physical similarities between the two aviators were striking. Both shared a tall, slender build, a high forehead, and similar coloring, a code for their joint Protestant Anglo-Saxon stock. Furthermore, Amelia's short, tousled hair, toward which the press developed almost a fetish, made her look boyish, an adjective that always was used to describe Lindbergh. Both shunned cigarettes, alcohol, and profanity; both exuded an aura of personal restraint and good taste.

The two aviators also shared certain personality traits. This may not be particularly surprising since there must be a process of self-selection that turns some people into long-distance risk-takers and keeps the rest of the population firmly on the ground, but the media played up the similarities to the hilt. They both were courageous and unafraid of danger. They both were solitary individuals who welcomed the opportunity to go it alone. Each was self-made, not relying on inherited wealth or position, and both professed disinterest in capitalizing commercially on their flights, a stand that brought them widespread commendation. And yet Amelia Earhart, like all other fliers, male or female, never had quite the blank check that Lindbergh enjoyed.

In the end, the most succinct way to compare the two most famous aviators of their times is to note the following: both Amelia Earhart and Charles Lindbergh flew because they loved the freedom of the skies, and both flew to promote public acceptance of commercial aviation. Both captured popular attention because of their feats of individual courage and bravery in the air. But Amelia Earhart had an additional reason for flying; she flew for women. Unlike Charles Lindbergh, who saw his fame as a hindrance,[2] Amelia Earhart saw hers as an opportunity to do something constructive for her sex. From the moment of her 1928 flight until her disappearance on a round-the-world trip in 1937, Amelia Earhart worked to portray her individual achievements as an example of women's capabilities in the modern world and as steps forward for all women. As Eleanor Roosevelt said of the pioneering aviator, "She helped the cause of women by giving them a feeling there was nothing they could not do."[3]

Amelia Earhart shared the expansive and optimistic vision articulated by swimmer Gertrude Ederle when she set off to conquer the English Channel in 1926, stating, "All the women of the world will celebrate, too."[4] With her widely publicized individual accomplishments and clearly articulated feminist ideology, Amelia Earhart demonstrated that women could be autonomous human beings, could live life on their own terms, and could overcome conventional barriers. This message, while not always specifically labeled feminism, provided a highly individualist route for exceptional women to excel.

---

2 See A. Scott Berg, *Lindbergh* (New York: G. P. Putnam's, 1998).

3 Eleanor Roosevelt quoted in Shirley Dobson Gilroy, *Amelia: Pilot in Pearls* (McLean, VA: Link Press, 1985), p. 53.

4 Ederle quoted in "How a Girl Beat Leander at Hero Game," *Literary Digest* 90 (21 August 1926): 52.

Amelia Earhart truly believed that if women proved themselves competent in aviation, and by extension in all aspects of modern life, prejudices would fade and barriers fall. It did not prove to be so simple, as the course of women's history over the 20th century demonstrated. But the strand of individualistic feminism represented by popular heroines like Amelia Earhart at least tentatively kept women's advancement on the national agenda at a time when mass-based feminist movements were unlikely to coalesce. An 18-year-old factory girl from Tennessee once said of Eleanor Roosevelt, "Say, she's swell. Why, I'm not ashamed of being a girl any more."[5] Amelia Earhart had that same effect.

Besides her significance to the history of modern feminism, Amelia Earhart also holds an important place in aviation history. In many ways she straddled two eras in aviation: that of the barnstorming, record-breaking, front-page-news pilot whose exploits received as much coverage as the World Series or a championship boxing match, and the next stage that was represented by the emergence of a commercial aviation industry in the United States. Many of the opportunities that came her way, and some of the constraints, were grounded in the broader developments of aviation in the 1930s.

One of the most significant was the way in which gender, what Earhart once referred to as "the accident of sex,"[6] affected the course of modern aviation. Aviation was a new profession, and women were getting in near the beginning, yet the deck was stacked against them. It was clear that the industry developed along sex-segregated lines that marginalized women. They were welcomed as stewardesses, but banned as pilots. They could demonstrate light sports craft, but were denied access to heavier commercial and military aircraft. Except for an occasional woman in the front office to deal with the "woman's angle," they were frozen out of the business side. Individual women found opportunities in this new profession, with the 1920s and 1930s representing something of a golden age for women pilots. But at the end of the decade, women pilots had been excluded from the next stage of development (commercial aviation), and their marginalization was cemented by WWII. The postwar world of aviation was very much a man's world, although women continued to participate in most of its facets.[7]

In the 1930s, however, Amelia Earhart and the other women pilots did not know what the outcome would be, and they were optimistic about the future. If women proved themselves equal to men in the air, then opportunities would expand and prejudice would recede. That was the message that Amelia Earhart promoted with words and deeds in her short but extremely significant career.

---

5 Quoted in Ruby A. Black, "Is Mrs. Roosevelt a Feminist?," *Equal Rights* (27 July 1935): 164.

6 Amelia Earhart, *20 Hours, 40 Minutes: Our Flight in the Friendship* (New York: G. P. Putnam's Sons, 1928), p. 201.

7 For an excellent overview, see Deborah G. Douglas, *American Women and Flight Since 1940* (Lexington: The University Press of Kentucky, 2004).

A MODERN WOMAN IN THE MAKING

Since there was no clear aviation "career path" for women in the early 20th century (or for men, for that matter), attempts to locate yearnings for flight in Amelia Earhart's girlhood were retrospective at best, but her entire life did have a certain restless quality to it, as if she were searching for her role as a woman in modern America. She recalled "growing up here and there" (a chapter title in her second book, *The Fun of It*) and always felt a bit like a rolling stone. For the rest of her life she could never stay long in one place or be content doing one thing; she was always on the move. As she once boasted, "I've had 28 different jobs in my life, and I hope I'll have 228 more."[8] At heart she was an experimenter who liked to try everything at least once, and her early years gave her a curiosity about life, an ability to adapt, and a desire to stretch and explore. "I got out and did *something*," she remembered, and then drew the following lesson: "In fact, I think it is just about the most important thing any girl can do—try herself out, do something."[9]

She was born Amelia Mary Earhart on 24 July 1897 in Atchison, Kansas.[10] Her early years were shaped by her parents, Edwin and Amy Otis Earhart, and her younger (by two years) sister, Muriel. Her father worked as a railroad claims agent, but money was often tight. Additionally, the family moved around a lot, in part because her father's alcoholism made it difficult for him to hold a steady job. Amelia spent extended periods of time with her grandparents in Atchison, who often tried to damp down her tomboy tendencies. "Unfortunately I lived at a time when girls were still girls. Though reading was considered proper, many of my outdoor activities were not." She later remembered with "special glee" putting on a bloomer-type costume and going out "to shock all the nice little girls. It seems a trivial thing now, but it was tremendously daring in those strictly conventional days." Amelia Earhart never could have lived a conventional female life. More than anything else, her childhood embraced a sense of experimentation and physical freedom, which served as a forerunner of her later interest in aviation.[11]

Whenever a girl grows up to be famous, it is always tempting to ascribe future choices to the circumstances of childhood or family. Retrospectively, Amelia explained her attraction to aviation by three threads: the many trips she made with her father on the railroad, her love of sports and games usually reserved for boys, and her lifelong propensity to experi-

---

8 Amelia Earhart, "Flying the Atlantic," *American Magazine* 114 (August 1932): 72.

9 Helen Ferris, ed., *Five Girls Who Dared* (New York: Macmillan, 1938), p. 5.

10 Standard biographical sources include Doris L. Rich, *Amelia Earhart: A Biography* (Washington, DC: Smithsonian Institution Press, 1989); Mary S. Lovell, *The Sound of Wings: The Life of Amelia Earhart* (New York: St. Martin's Press, 1989); and Susan Butler, *East to the Dawn: The Life of Amelia Earhart* (Reading, MA: Addison-Wesley, 1997). See also G. P. Putnam, *Soaring Wings: A Biography of Amelia Earhart* (New York: Harcourt, Brace, and Company, 1939).

11 Amelia Earhart, *The Fun of It: Random Records of My Own Flying and of Women in Aviation* (New York: Harcourt, Brace, and Company, 1932), pp. 8, 11.

Earhart (right) with early flying instructor Neta Snook. (Photo number 71-1055, National Air and Space Museum, Smithsonian Institution)

mentation. Her choices definitively set her apart from her mother, who lived the life of a typical Victorian wife solely devoted to the needs of her family. And she took a different path from her sister, who eventually married, had children, and worked as a schoolteacher in a Boston suburb. As Muriel said many years later, "Amelia had her planes, and I had my children. That was the difference."[12]

Though the Earhart family was often strapped for cash, they found the resources to send Amelia to the college preparatory Ogontz School outside Philadelphia. But while Muriel later attended Smith College, Amelia showed no inclination to go on to college full time, although she did take special courses at Columbia University and at the Harvard extension school. (She later bragged about climbing to the top of the dome at Columbia, which was further evidence of her adventuresome spirit and physicality.)

There were two defining moments for American women in the 1910s: the woman suffrage campaign and WWI. Earhart was not an active suffragist (although it is hard to imagine her not supporting votes for women), but WWI affected her personally, philosophically, and professionally. In 1917, while visiting her sister in Toronto, Amelia volunteered as a VAD (voluntary aid detachment) nurse's aide at the Spadina Military Convalescent Hospital, where she was captivated by the stories told by military pilots who were recuperating from their war injuries. It was several years, however, before she was able to follow through on this new interest. After the war and now settled in California with her parents, she determined to learn to fly, taking lessons and then purchasing her own airplane, a Kinner Canary. She deliberately sought out a female flight instructor, Neta Snook, because she felt she would be less self-conscious taking lessons from a woman. She made her first solo in June 1921 and was soon participating in the camaraderie of the airstrip and hangar with other pilots and mechanics, mainly men. In this

12 Muriel Earhart Morrissey quoted in the *Boston Globe* (1989), clipping in the possession of Susan Ware.

period, she also cut her hair (a symbol of female liberation in the 1920s) and sported the bobbed haircut of a modern woman.

In these early years, there really wasn't any way that she could make a career out of aviation, so it was more a hobby while she supported herself with odd jobs. As she put it once, "no pay, no fly and no work, no play."[13] When her parents' marriage ended, she and her mother drove cross-country (a fairly unusual undertaking for two women at that time) to Boston, where Muriel was settled. Amelia's boyfriend, Sam Chapman, followed her East, but she kept putting him off when he suggested marriage because she feared being tied down. She registered with a placement bureau in Boston run by the Women's Educational and Industrial Union, where the interviewer deemed the prospective candidate as "an extremely interesting girl—very unusual," and added in amazement, "holds a pilot's license!"[14] By 1927, she had landed a job as a resident at Denison House, a settlement house in Boston's South End. In her spare time she continued to fly, once dropping fliers for a Denison House event from her airplane.

It was while she was at Denison House that she was approached about being a passenger on the transatlantic crossing, a flight subsidized by a wealthy Philadelphia woman, Amy Phipps Guest, whose family had vetoed her plan to make the trip herself. Earhart was interviewed by Hilton Railey and G. P. Putnam, her future husband, and could not avoid seeing the comic ironies of the discussion. "I realized, of course, that I was being weighed. It should have been slightly embarrassing, for if I were found wanting on too many counts I should be deprived of a trip. On the other hand, if I were just too fascinating the gallant gentlemen might be loathe to drown me."[15] She approached the opportunity as an exciting (albeit dangerous) vacation and assumed that she would be able to slip back into her life as a social worker when she returned. Ironically, she became such an instant celebrity that she could never return to the field.

Just a year after Lindbergh's flight, the preparations for the first transatlantic flight by a woman were front-page news, especially since several teams were racing for the distinction of being first. After a series of weather-related delays, the *Friendship* team of pilot Wilmer (Bill) Stultz, mechanic Slim Gordon, and log-taker Amelia Earhart took off from Trepassey Bay, Newfoundland, on 17 June 1928. Twenty hours and 40 minutes later, they landed in Burry Port, Wales. Their pontoon craft bobbed in the water for an hour before attracting the notice of local residents. This was practically the last time that Amelia Earhart went anywhere unnoticed. Soon pictures of her in what became her trademark

---

13 Earhart, *The Fun of It*, p. 26.

14 The application, dated 18 August 1926, is found in the Women's Educational and Industrial Union Papers, Schlesinger Library, Radcliffe Institute for Advanced Study, Harvard University.

15 Earhart, *20 Hours, 40 Minutes*, pp. 99–100.

outfit—leather flying jacket and cap, short unruly hair, jodhpurs and high boots, plus the much-copied silk blouse and necktie—circulated throughout Europe and America. Frances Perkins, soon to be named Franklin Delano Roosevelt's industrial commissioner in the state of New York and later his secretary of labor in the New Deal, observed at a New York reception in Earhart's honor that all women "got a vicarious thrill out of Miss Earhart's flight."[16]

Amelia Earhart was at something of a crossroads in the summer of 1928. Even though the press often referred to her as a girl, she was actually almost 31 years old. The success of the 1928 flight brought her many opportunities—a book contract, lecturing and promotional work, endorsements, and an affiliation as aviation editor for a New York magazine—which removed for the moment the necessity to hold a steady job. It was still just a glimmering, but there was an increasing possibility that she might be able to make a living out of what had become her passion for flying. Mainly, it seemed, what she wanted to do was to fly around.

Even though Amelia Earhart was at that time the most famous female pilot in the country, she was not necessarily the most experienced or competent. She used her freedom and new financial stability to buy her own plane and embark on what turned out to be the first transcontinental trip by a woman. Quietly building her skills as a pilot and establishing her bona fides among the growing sorority of women fliers, she acquired her transport license in 1930, only the fourth one issued to a woman. She joined with other women pilots to found the Ninety-Nines (the name representing the number of charter women pilots who answered the call), serving as its first president from 1929 to 1933. And in a move that had a dramatic impact on her career and public visibility, she married G. P. Putnam in February 1931. Theirs was as much a business partnership as a love match, but it suited them both fine. From that point on, he devoted his considerable skills in public relations to making sure that she remained the best-known female pilot in the country, if not the world.[17]

A little more than a year after her marriage, Amelia Earhart asked her husband if he would mind if she flew the Atlantic again. She had never forgotten the experience of being merely a passenger on the 1928 flight, or as she referred to it, "a sack of potatoes,"[18] and wanted to prove to the world, and more importantly to herself, that she deserved some of the credit and adulation that had been showered on her. On 20 May 1932, not coincidentally the fifth anniversary of Lindbergh's flight, she took off from Harbour

16 Frances Perkins quoted in the *New York Times* (8 July 1928): 16.

17 For information on G. P. Putnam, see *Wide Margins: A Publisher's Autobiography* (New York: Harcourt, Brace, and Company, 1942). Also of interest is Sally Putnam Chapman, *Whistle Like a Bird: The Untold Story of Dorothy Putnam, George Putnam, and Amelia Earhart* (New York: Warner Books, 1997).

18 Amelia Earhart remarks reprinted in *National Geographic* 62 (September 1932): 363.

Grace, Newfoundland, in her single-engine bright red Lockheed Vega. Flying through dismal weather and experiencing various mechanical mishaps—such as a failed altimeter, which meant she could not gauge her altitude—she landed in Londonderry, Northern Ireland, 14 hours later. Characteristically, Earhart played down the significance of her flight, repeatedly stressing, "I flew the Atlantic because I wanted to." To her mind this was not a selfish reason. "To want in one's heart to do a thing, for its own sake, to enjoy doing it, to concentrate all one's energies upon it—that is not only the surest guarantee of its success; it is also being true to oneself."[19]

Of all of Amelia Earhart's aviation accomplishments, her 1932 Atlantic solo was probably her most noteworthy and most widely praised feat. She was the first person since Lindbergh to fly the Atlantic solo; her crossing was done in record time; and she was the first person to have crossed the Atlantic twice by air. Among other honors, President Herbert Hoover presented her with the Gold Medal of the National Geographic Society, putting her in the same company as polar explorers Robert E. Peary and Roald Amundsen, Admiral Richard Byrd, and Colonel Charles Lindbergh. At the White House ceremony honoring her for her flight, she proudly but with characteristic modesty drew links between aviation and feminism. "I shall be happy if my small exploit has drawn attention to the fact that women are flying, too."[20] Never just out for her own personal and professional advancement, Amelia Earhart consciously identified her individual accomplishments as victories for women as a whole. The thousands of telegrams, tributes, and letters that poured in after that solo flight testified that women in the United States, and throughout the world, did indeed take Amelia Earhart's individual triumph as a triumph for womanhood.

WOMEN IN MODERN AVIATION

Americans in the interwar years were fascinated by flying and aviation. The term "air-minded" was coined to describe this new orientation. This fascination, which predated Charles Lindbergh's 1927 solo, translated into what one historian has called the "winged gospel," a secular creed about the promise of the future which was promulgated with the intensity of evangelical religion.[21] In this worldview, aviation would reorder human society and promote world peace by breaking down isolation and distrust. When everyone took to the air, society would be transformed along the lines of democracy, freedom, and equality—

19 Earhart, "Flying the Atlantic," pp. 15, 16.

20 Amelia Earhart quoted in the *New York Times* (22 June 1932): 3.

21 The best introduction remains Joseph J. Corn, *The Winged Gospel: America's Romance with Aviation, 1900–1950* (New York: Oxford University Press, 1983).

and perhaps even women's liberation. The well-known British aviator Lady Heath recited this ditty in 1928: "Woman's place is in the home, but failing that, the airodome."[22]

Amelia Earhart believed devoutly in air-mindedness, and she used her public platform to advance the creed. Like so many Americans who subscribed to the winged gospel, she envisioned a future in which every citizen flew as a matter of everyday life. In addition to aviation as a form of transportation, she saw flying as a sport, just like tennis or horseback riding. She felt tradition-bound women must work especially hard to become air-minded, if only to keep up with their children. "The year 1929 is ushering in the flying generation," she reminded readers in her *Cosmopolitan* column. "And the stratagem of it all is that elders must not let themselves be left behind."[23] Earhart also envisioned large potentialities for women's advancement through aviation. In a period without an active feminist movement, the woman pilot was an excellent symbol of women's emancipation in the postsuffrage era.

This wonderfully optimistic, and in retrospect remarkably naïve, view of human progress lasted for only a generation or two. Technological breakthroughs in the manufacturing of aircraft and the expansion of commercial aviation supplanted the individual heroes and heroines of earlier decades. The rise of militarism worldwide in the 1930s focused attention on the airplane not as a model for beneficent social change, but as a potential weapon of destruction, a view confirmed by the course of WWII. With the ebbing of the dream of "mass personal flight," the unrealistic belief that everyone would be as familiar with flying an airplane as with driving a car, the winged gospel lost its appeal.[24]

What is most remarkable, in fact, is the speed with which the aviation industry outgrew its youth and adolescence and, in the metaphor favored by most aviation experts at the time, settled into a young maturity. It took railroads half a century to complete a cycle of pioneering, merger, regulation, and stabilization; the airlines did it in just over a decade between 1925 and 1938.

Amelia Earhart was very much involved in the industry side of commercial aviation. Not surprisingly, several fledgling airlines sought the endorsement and expertise of the first woman to fly the Atlantic and a public figure closely associated with the excitement of aviation. But her roles were more ceremonial than substantive (she joked about being a "chronic vice president").[25] Her activities usually were limited to goodwill promotion tours. Since airlines were struggling financially in these years, the compensation she received was usually along the lines of a token dollar a year plus the opportunity to fly free on the line, which was actually a substantial benefit during the depression.

22 Quoted in *Equal Rights* (8 December 1928): 349.

23 Amelia Earhart, "Shall You Let Your Daughter Fly?," *Cosmopolitan* (March 1929): 88–89.

24 See the epilogue to Corn, *The Winged Gospel*.

25 Quoted in undated clipping (c. 1934), "Flyer Designs Sports Clothes," found in Amy Otis Earhart Papers, Schlesinger Library.

Earhart and Eleanor Roosevelt (Photo number 87-2489, National Air and Space Museum, Smithsonian Institution)

Between 1929 and 1934, Earhart was associated with the following three airlines: TAT (Transcontinental Air Transport, later absorbed into TWA), the so-called "Lindbergh Line"; the Ludington Line, a forerunner of modern shuttle service between Washington and New York; and commercial air service between Boston and Bangor as a subsidiary of the Boston and Maine Railroad. None of these airlines survived in its original incarnation, which was indicative of the turnover and widespread failures common to the airline industry in the uncertain business climate of the 1930s. As aviation struggled to convince potential passengers that flying was safe, convenient, and not prohibitively expensive, in many ways it had to disabuse the public of the very things—romance, excitement, daredevil feats—that made this form of travel fascinating in the 1910s and 1920s. This proved no simple task. In the early 1930s, very few Americans were flying at all—perhaps 500,000 out of a total population of around 125 million. Fear, as much as fare, kept Americans (and American airlines) on the ground.

Many of Amelia Earhart's public activities between 1928 and 1937 were directed at overcoming popular fears about commercial aviation, especially women's supposed resistance to flying. (The industry referred to this problem as "Father won't fly if Mother says he can't.") Earhart's job with the traffic department of TAT was specifically created so that she could provide "the woman's angle." "Vocal salesmanship" was how she described her extensive speaking tours to promote aviation to women's clubs, college women, professional organizations, men's groups, and anyone else who would listen. She often began her lectures by asking for a show of hands of how many in the audience had flown. (She once did this in an appearance at a prison, and a surprising number of the inmates' hands went up.) To any doubters, she issued a simple challenge—try flying.[26]

The best example of the air-mindedness Amelia Earhart had in mind was that of Eleanor Roosevelt. The first lady made a point of traveling by air on official or personal business whenever she could, always making sure to follow up by casually plugging air travel in her next press conference or daily newspaper column discussing the event.

26 Earhart, *The Fun of It*, p. 106.

Eleanor Roosevelt even consulted with her good friend Amelia Earhart about getting a pilot's license, but FDR nixed the idea.[27]

As she traveled around the country, Earhart often was asked about career opportunities for women in the field of aviation. She developed a multilayered set of answers that were generally optimistic but with a decided undercurrent cognizant of the difficulties women faced in the aviation field, as elsewhere. Earhart's entire attitude toward advancement in the professions, aviation and beyond, was summed up in this statement: "I believe if a woman showed the ability that is required of a man, she would be employed."[28] From statements like that, it is clear that the aviator had a much broader agenda in mind than just getting a few thousand women jobs in the midst of the Great Depression. This emphasis on opening opportunities for women, showing that women were capable and asking that they be given a chance, ran like a refrain through almost all of her public statements on aviation. She was not talking about a battle of the sexes competition; she just wanted a chance to show what women could do.

One key to women's future in aviation was whether they would be allowed to fly the heavier aircraft developed for commercial aviation in the 1930s. Some pilots such as Ruth Nichols believed that women should stick to flying mainly for sport, but Earhart and most other women aviators tried to pressure the airlines into hiring women as commercial pilots. A step in that direction occurred in 1934 when Helen Richey beat out seven men for a position as copilot on the Central Airways mail route from Washington, DC to Pittsburgh. But the pilots' association convinced the Aeronautics Bureau of the Commerce Department to limit women to fair-weather flying on regularly scheduled routes, which made it impossible for Richey to do her job and caused her to resign in frustration in 1935. Stonewalling by male pilots, backed up by the aviation industry, made it clear that women would not be accepted if they dared to try to break in. None did for almost 40 years.[29]

Widespread skepticism greeted the notion that women were capable of flying, despite all the evidence to the contrary. Earhart noted that if a man and a woman emerged from a plane, the public always assumed that the man had really done the flying. When male aviators crashed, it was usually seen as an equipment problem or bad luck; when women crashed, it was their fault, confirmation that they were inept in the air in the first place. If gender stereotypes had not proved so resilient, one might have expected that technology would have neutralized or made moot the differences between the sexes. Flying a modern airplane required keen eyesight, quick reactions, and manual dexterity, not brute strength. To those who questioned whether flying heavy transport planes required more muscular strength than a woman could provide, even as a copilot, aviation editor Carl B.

27 For the friendship between Earhart and Roosevelt, see Blanche Wiesen Cook, *Eleanor Roosevelt*, vol. 1 (1992) and vol. 2 (1999) (New York: Viking).

28 Amelia Earhart to Miss Mintern (28 February 1929), Amelia Earhart General File, National Air and Space Museum Archives, Washington, DC.

29 Douglas, *American Women and Flight*, p. 176.

Allen had a good reply: "[N]o plane that can't be controlled by either sex as far as physical strength goes is a safe plane for passenger operation."[30]

Women aviators increasingly found themselves in a trap not of their own making. In addition to their exclusion from commercial aviation and the military, the two major career paths for pilots, women faced continuing difficulty obtaining access to the best equipment and newest technology, which manufacturers were loathe to put at the disposal of mere women who had no future in the field. As a result, women did not get enough training time on new techniques such as night flying, instrument flying, or the use of radio. Instead they concentrated on lighter, less technologically advanced sport craft. Here they excelled with their solos, endurance flights, and competitions for women's records, but they were mainly competing against other women. And the chance to compete in something like the Bendix race was open only to the very elite of fliers, male or female. Women could be glamorous record setters, they could fly for sport or personal satisfaction, but they were not integrated into the emerging aviation industry. Or to put it another way, success by individual women posed no threat to the emerging gendered patterns that structured the development of aviation.

Yet women aviators did not feel inferior or downtrodden. For the most part, women fliers, including Amelia Earhart, dealt with the ongoing discrimination and double standard by a combination of two tactics: ignoring it and just going about their own business, or trying to use their own examples of individual success as a way of breaking down prejudices and stereotypes. They simply wanted to be taken seriously as aviators, to be allowed to excel and compete on an equal basis with men. "Some day, I daresay, women can be flyers and yet not be regarded as curiosities!" hoped Earhart.[31] In her darker moments, she was well aware that accumulated years of prejudice and discrimination made it difficult for women to get ahead in aviation, no matter how hard they tried, but she refused to be cowed. Women, she noted optimistically, "have opened so many doors marked 'impossible' that I don't know where they'll stop."[32]

AMELIA EARHART AS POPULAR HEROINE

Because popular heroines usually burst on the scene with a dramatic feat of individual courage, they often appear curiously disconnected from their ensuing fame and adulation. Each was simply the right person in the right place at the right time. Perhaps this is what happened to Charles Lindbergh, but, for all the rest, male and female, celebrity was not handed to them on a silver platter, and many found it difficult to capitalize on and sustain their initial success.

30 Allen quoted in *Newsweek* 6 (16 November 1935): 22.

31 Earhart, *The Fun of It*, p. 95.

32 Earhart, "The Feminine Touch," *Aero News and Mechanics* 2 (April–May 1930): 35.

Amelia Earhart's career confirms that there was no easy route to being a popular heroine. She had to make it happen, working on it practically every day of her life after her June 1928 flight thrust her into the limelight. G. P. Putnam, her husband, manager, and biggest promoter, captured this well when he drew attention to "the sheer, thumping hard work of conscientious heroing."[33] For Earhart it meant 14-hour days of lecturing and receptions, answering hundreds of letters a week from fans, cranking out instant books, dealing with newsreel and camera photographers the very moment a grueling flight finished so they could make their deadlines, and always being on display wherever she went. By dint of hard work, skill, and luck, she was able to make a viable living out of promoting herself as an aviation celebrity. "I'm really very fortunate," she admitted, "because flying is both my business and my pleasure. I've got a job I love."[34]

There are many parallels between Amelia Earhart's life as a public figure and those of Hollywood movie stars, such as cultivating a public image, attending carefully staged promotional events, endorsing selected products, being widely photographed and written about, and working very hard at what one does. Like Katharine Hepburn or Greta Garbo, Amelia Earhart could not walk down the street unnoticed. (When she wanted to be invisible, she pulled a cap over her hair.) Like Hollywood stars, details about her personal life and opinions were picked up by the eager media, including practically every time she got a speeding ticket, which was often. (She loved to drive fast.)

Another similarity was receiving a huge amount of mail. Just as fans wrote to their favorite actors in Hollywood, female and male admirers, young and old, wrote to the aviator about their flying (and other) aspirations. After her 1928 flight, four secretaries were necessary to deal with the telegrams, letters of congratulations, commercial offers, proposals of marriage, and crank mail that poured in. For the rest of her career, Earhart kept up an enormous correspondence, especially after record-breaking flights such as the 1932 solo. Inundated by so much mail, she tried to keep her sense of humor. She called her personal papers her "peppers" and had files marked "bunk" (for all the songs, poems, and accolades that people sent in) and "cousins" (for those who tried to establish an often fictive family connection).[35]

Of all the duties surrounding being a public figure, the aviator seemed to have derived the least satisfaction from writing, despite the fact that she had written poetry all her life. Unfortunately, she was unable to transfer her affinity for verse into the autobiographical articles and general nonfiction required of a public figure trying to keep her name in the news. Moreover, she found the deadline conditions under which she had to churn out her prose especially disruptive to good writing. As a result, she was rarely satisfied with anything she produced.

---

33 G. P. Putnam, *Soaring Wings: A Biography of Amelia Earhart* (New York: Harcourt, Brace, and Company, 1939), p. 190.

34 Amelia Earhart quoted in Universal Service clipping (9 May 1935), found in Clarence Strong Williams Papers, Schlesinger Library.

35 Putnam, *Soaring Wings*, p. 210.

She was much more at ease on the lecture circuit. In the days before television, the lecture circuit was the principal means by which popular figures reached the public. Here was a chance for citizens in Des Moines, Iowa or Tacoma, Washington to see what polar explorer Richard Byrd, humorist Will Rogers, or First Lady Eleanor Roosevelt was like in person. Rather than the vicarious experience of radios or newsreels, lectures provided an opportunity to connect in person with the most influential political, social, and cultural leaders of the times, at least for one night.

Of course for a public figure to draw a crowd, he or she had to have something to say; speakers were only as good as their most recent exploit or accomplishment. Whenever Amelia Earhart's standard lecture on "Adventures in Flying" began to go stale, she would do something like make another record-breaking flight or write a new book. Her talks were geared toward a general audience and avoided controversial subjects such as pacifism or politics, but she always included references to women in aviation. Earhart earned between $250 and $300 for each lecture, although she was willing to take less for a good cause. Seven or eight lectures a week totaled close to $2,400, a tidy sum in depression-era America. Lecturing very quickly became her major source of income, but it was an exhausting way of life. No wonder Amelia loved the solitude of flying—the one time she was free from the demands of her life as a public figure.[36]

By 1935, Amelia Earhart had taken her place as a major aviation celebrity and as one of America's best-known and admired women. She had been in the public eye for seven years. Few of the popular heroes, male or female, who flashed briefly into public consciousness in the 1920s and 1930s and then just as quickly disappeared, could make such a claim to longevity. But few of them devoted as much energy and hard work to keeping themselves in the public eye as she did. She presented herself, and was presented in the press, in almost iconographic representations of both *woman* and *aviator*. Indeed, her image has stood the test of time remarkably well. If she walked into a room today, she would look perfectly at home, dressed in her comfortable but flattering slacks and silk blouse with an easy-to-care-for hairstyle and trim, athletic physique.

This message was, and is, remarkably powerful. In photographs and newsreels Amelia Earhart is in motion. She flies planes, greets crowds, gives speeches, and meets famous people. Even when photographed standing still, she is often surrounded by symbols of action and power usually associated with men—sleek cars, large airplanes, and crowds of admirers. By her appearance, manner of presentation, and propensity to stare forthrightly into the camera instead of shyly averting her eyes in the more traditional female gaze, Amelia Earhart opened a window, albeit a small one, toward more autonomy and individual freedom in women's lives.

In sum, Amelia Earhart offered the public a highly individualistic yet compelling new way to be a woman. Her image challenged received notions of femininity and women's

36 Lovell, *The Sound of Wings*, p. 205; and Rich, *Amelia Earhart*, p. 159.

identification with home and family without abandoning womanliness entirely. Here was a woman in public, acting, doing things, surrounded by props and symbols that left no doubt she had left the kitchen and entered the public realm. More to the point, the very confidence and poise she displayed in her public roles suggested that women had a right to such freedoms, that it was perfectly natural for women to be doing nontraditional things. She claimed the public space not just as an individual, but as a woman. And she claimed it not just for herself, but for womankind as a whole.[37]

THE LAST FLIGHT

Two events in 1935 capture aviation in transition. In January 1935, Amelia Earhart was the first person, male or female, to solo between Hawaii and the mainland, landing in Oakland on a field mobbed by well-wishers. The feat received front-page coverage throughout the country, a reminder of Earhart's unchallenged status as aviation's best-known female figure as well as the ongoing appeal of record-breaking flights and daring exploits well into the 1930s.

Just 10 months later, on 22 November 1935, more than twenty thousand people turned out in San Francisco to watch Pan American Airways' *China Clipper* take off on the first direct transpacific flight. This flight, offering regularly scheduled commercial air service, was the wave of the future. In a fanciful image, the two aircraft—Amelia's small single-engine Lockheed Vega and Pan Am's huge four-engine flying boat complete with sleeping berths, lounges, and dining areas—could almost have passed in the skies over the vast Pacific Ocean, the one heading east, the way of the past, and the other heading west, the way of the future.

Earhart signing stamp covers before her round-the-world flight. (Photo number 71-1056, National Air and Space Museum, Smithsonian Institution)

Two years later, Amelia Earhart had a bigger plane, her own Lockheed Electra that she dubbed "the flying laboratory," and she was ready to set out on an adventure of her own—a round-the-world flight. As she told the press, "I have a feeling that there is just

37 For further discussion of the links between popular culture, feminism, and women's history, see Susan Ware, *Letter to the World: Seven Women Who Shaped the American Century* (New York: W. W. Norton, 1998).

Earhart standing in front of the Lockheed Electra in which she disappeared in July 1937. (GRIN database number GPN-2002-000211)

about one more good flight left in my system."[38] Yet it was much harder work to pull off this 1937 flight than the 1928 and 1932 Atlantic crossings. The novelty of ocean flying had worn off, the commercial backers and endorsements had all but disappeared, and the syndication deals were drying up. All the major routes had been spanned, and many were now serviced by regularly scheduled air transport. Even flying around the world was not really newsworthy anymore. Between 1924 and 1933, six expeditions had circled the globe, including two by Wiley Post in his Lockheed Vega. The only thing left was to do old routes faster or with another twist or gimmick. Earhart's novelty was flying around the globe close to the equator, a distance of 27,000 miles, which was 10,000 more than Wiley Post's solo trip in 1933. And, of course, being a woman made a difference. What was her main motivation? "I am going for the fun," she said, adding, "Can you think of a better reason?"[39]

Although her publicist husband could certainly have orchestrated an enthusiastic reception when she returned from her world flight (lecture dates at $500 a shot were already under contract before she left), it was hard to imagine Amelia Earhart receiving

38 *New York Herald Tribune* syndicate press release (3 July 1937) found in the Amelia Earhart Papers, Purdue Special Collections, Purdue University.

39 Amelia Earhart quoted in the *New York Times* (30 May 1937): 16.

the kind of heroine's welcome in 1937 that she had gotten after her 1928 or 1932 flights. She was now 39 years old, hardly the girl flier anymore, and she had been steadily in the public eye for almost a decade. The publicity value of a successful month-long leisurely circumnavigation of Earth at the equator remained unclear. Despite G. P. Putnam's promotional talents, this one might have been very hard to pull off.

An *unsuccessful* round-the-world flight, however, was a different matter, especially when the plane and its occupants were never found. Without doubt, the mystery of Amelia Earhart's disappearance brought the aviator far more publicity than any stunt the intrepid promoter Putnam could have dreamed up. Ironically, the unresolved circumstances surrounding her death have kept her in the news to a far greater extent than if she had returned safely. And the search still continues.[40]

Yet this posthumous cult, which is quite distinct from the celebrity she enjoyed during her lifetime, threatens to obscure her legacy. While she was alive, she was celebrated for what she accomplished and for what her example meant to women and aviation. Once she was presumed missing, Amelia Earhart, the role model for women, was increasingly replaced by Amelia Earhart, the lost aviator, shifting attention away from her strongly articulated feminism to speculation about the circumstances of her fateful last flight. Yet Amelia Earhart the role model and inspiration for women has never really died. In fact, her appeal will probably outlast all attempts to mine the vast expanses of the Pacific Ocean for clues to her final fate.

Amelia Earhart was not afraid of death; she said so many times. She would have faced death with the same unflinching courage and honesty with which she lived her life. The words she had bravely written to her husband before another dangerous flight now became a prophetic epitaph: "Please know I am quite aware of the hazards. I want to do it because I want to do it. Women must try to do things as men have tried. When they fail, their failure must be but a challenge to others."[41] Her friend and great admirer Eleanor Roosevelt was certain that Amelia Earhart's last words were "I have no regrets."[42]

40 Recent books on her disappearance and the search for her plane include Elgen M. Long, *Amelia Earhart: The Mystery Solved* (New York: Simon and Schuster, 1999); Thomas F. King, ed., *Amelia Earhart's Shoes: Is the Mystery Solved?* (Walnut Creek, CA: AltaMira Press, 2001); and Randall Brink, *Lost Star: The Search for Amelia Earhart* (New York: W. W. Norton, 1994).

41 Quoted as a postscript to Amelia Earhart, *Last Flight* (New York: Harcourt, Brace, and Company, 1937).

42 Eleanor Roosevelt quoted in the *New York Times* (20 July 1937): 24.

3

Sharing a Vision

Juan Trippe, Charles Lindbergh, and the Development of International Air Transport

WILLIAM M. LEARY

THE "GOLDEN AGE" OF AVIATION DURING THE LATE 1920s SAW JUAN TRIPPE AND CHARLES LINDBERGH FORGE A RELATIONSHIP TO PROMOTE U.S. INTERNATIONAL COMMERCIAL AIR TRANSPORT—and the fortunes of Pan American Airways—that would last more than 40 years. When the partnership began, Pan American was a tiny airline that operated three wooden trimotored Fokkers on a 90-mile airmail route between Key West and Havana. Four decades later, it had become a giant company flying routes around the world, with 143 transports in service, $1.5 billion worth of airplanes on order, and more than 40,000 employees. The energy and vision of these two men would be key components in the remarkable growth of one of the world's great airlines of the 20th century.

FIRST ENCOUNTERS

In many ways, it was an unlikely pairing. Both men tended to be extremely private individuals. Although Trippe could turn on the personal charm when necessary, he seemed more comfortable in solitude. According to John Leslie, a Pan American executive who worked hard to preserve the history of the airline, Trippe possessed a powerful analytical intelligence that combined with an intuitive sense to produce a rare farsightedness. He was extremely tenacious and persistent. He could be prudent, cautious, patient, evasive, and would often equivocate when he did not want to be pressured into a commitment. All in all, Leslie concluded, Trippe was an extraordinarily complex man.[1]

Charles Lindbergh and Pan American Airways founder Juan Trippe stand near one of the company's planes on 9 August 1929. (Photo number 78-5520, National Air and Space Museum, Smithsonian Institution)

1 John C. Leslie to Wesley P. Newton (20 March 1973), the records of Pan American World Airways, Box 30, Otto G. Richter Library, University of Miami, Miami, FL. The major collection of Pan American material is at the University of Miami. There is a smaller collection of Trippe material at the National Air and Space Museum, Smithsonian Institution, Washington, DC.

If Trippe was complex, Lindbergh would best be described as enigmatic. Everyone knew about Lindbergh—wing walker, pilot, hero, scientist, author, conservationist, and mystic—but few, if any, knew him well. "I encountered many brilliant people in the airline industry," wrote longtime Pan American manager S. B. Kauffman, "but the one who stands out as the most intriguing is Charles Lindbergh." The two men worked closely together for many years, yet Lindbergh always remained a puzzle to Kauffman. Nonetheless, Kauffman came to value Lindbergh's relationship with Trippe. "I often used him to get ideas to Trippe," Kauffman noted, "who always listened to anything Lindbergh had to say."[2]

Trippe and Lindbergh met for the first time in 1926. Trippe was the 27-year-old head of Colonial Air Transport, a struggling pioneer airline on the airmail route from New York to Boston. Fascinated with aviation since boyhood, Trippe had left Yale University at the end of his freshman year to join the U.S. Navy. He learned to fly in 1918 and was ready for overseas service when WWI ended. He returned to Yale to finish his education, worked briefly as a Wall Street bond salesman, then entered the aviation business in the spring of 1923. During the summers of 1923 and 1924, he operated Long Island Airways. Although Trippe failed to turn a profit, he gained invaluable experience. Following passage of the Kelly Air Mail Act of 1925, Trippe joined with a group of New England investors and secured the airmail contract for the route from New York to Boston. Colonial Air Transport, with Trippe as General Manager, began flying mail and express service on 1 July 1926.[3]

Lindbergh—three years younger than Trippe—also was involved in flying the mail. The son of a prominent Minnesota congressman, Lindbergh had spent two largely indifferent years at the University of Wisconsin before taking a few flying lessons, acquiring a Curtiss Jenny, and embarking on a career as a barnstorming pilot. Seeking more professional training, he joined the U.S. Air Service in 1924. As no active-duty positions were available after he finished flying school, Lindbergh secured a job with the Robertson Aircraft Corporation. When Robertson won the Chicago to St. Louis airmail contract in October 1925, Lindbergh surveyed the route as the company's chief pilot. Using DeHavilland DH-4s, Robertson inaugurated the service on 15 April 1926.[4] Trippe and Lind-

2 S. B. Kauffman, *Pan Am Pioneer: A Manager's Memoir, from Seaplane Clippers to Jumbo Jets* (Lubbock: Texas Tech University Press, 1995), p. 122. The Papers of Charles A. Lindbergh are housed at the Sterling Memorial Library, Yale University, New Haven, CT. Access to this collection is controlled by the Lindbergh family. There is additional material at the Missouri Historical Society, St. Louis, MO.

3 For Trippe's career, see Robert Daley, *An American Saga: Juan Trippe and His Pan Am Empire* (New York: Random House, 1980); and the more critical account by Marylin Bender and Selig Altschul, *The Chosen Instrument: Pan Am, Juan Trippe, The Rise and Fall of an American Entrepreneur* (New York: Simon and Schuster, 1982).

4 A. Scott Berg, *Lindbergh* (New York: G. P. Putnam's Sons, 1998), is the standard biography. Berg had full access to the Lindbergh Papers at Yale University. For a briefer account of the aviator's life, see Walter L. Hixson, *Charles A. Lindbergh* (New York: Addison Wesley Longman, 2002).

bergh came together at New Jersey's Teterboro Airport. Trippe recalled that they discussed the work of Hugo Leuteritz, a young engineer with the Radio Corporation of America who was experimenting with directional finding radio equipment. Lindbergh had no recollection of the meeting. They next came together on the early morning of 20 May 1927. Trippe was among a large group of individuals at Roosevelt Field, Long Island, as Lindbergh prepared to lift off the soggy grass and head for Paris.[5]

Trippe's fortunes had not fared well since their first encounter. By March 1927, Colonial was losing $8,000 a month and had only $100,000 in capital remaining. Trippe had struggled to secure additional funds but without success. As Robert Ward Johnson, a target of Trippe's fundraising, had replied to a solicitation: "I do not think a large investment in an air transport system is wise at this time." Johnson went on to explain the airplanes were not sufficiently safe, lacked adequate speed, and had an unsatisfactory load-carrying capacity. Trippe had hoped to bid on the New York to Chicago route, but this scheme failed to attract the interest of the New England group in Colonial. By the time Lindbergh departed for Paris, Trippe was trying to interest several of his Yale and Wall Street acquaintances in investing in a new company, this time to bid on the Key West to Havana airmail contract.[6]

Lindbergh's successful transatlantic flight came at an opportune time for Trippe. "Overnight," Trippe later pointed out, "the United States became air conscious. Our nation realized that civil aviation had arrived." On 2 June 1927, less than two weeks after Lindbergh's epochal flight, Trippe and 12 investors formed the Aviation Corporation of America, capitalized at $300,000, to carry out plans to move into the Caribbean through the Key West to Havana gateway.[7]

Trippe had the opportunity to see Lindbergh following the young airman's return from Paris and ticker-tape welcome in New York City. Along with a number of other favored individuals, Trippe was invited by Raymond Orteig, Jr., to a breakfast for Lindbergh held at the Hotel Brevoort on 17 June. It was unlikely that Trippe had time to extend more than congratulations to Lindbergh, who had no recollection of the meeting.[8]

Flooded with some $5 million in commercial offers during the first months of his new fame, Lindbergh deferred a decision on his future. Instead, he embarked upon a three-month tour of the United States, sponsored by the Guggenheim Fund for the Promotion of Aeronautics. Intended to stimulate popular interest in aviation, the tour certainly

---

5 Trippe, "Charles A. Lindbergh and World Travel," the Fourteenth Wings Club "Sight" lecture (20 May 1977), copy in the Pan American Records, Box 48.

6 Letter from Robert Ward Johnson to Trippe (7 September 1926), Pan American Records, Box 30. See also John C. Leslie, "Pan Am History Project—1975," Pan American Records, Box 507.

7 Trippe, "International Air Transportation," address before the General Session of the Chamber of Commerce of the United States (19 April 1937), Pan American Records, Box 460.

8 Letter from Trippe to Raymond Orteig, Jr., (15 June 1927), Pan American Records, Box 30.

achieved its objective. Lindbergh stopped in 82 cities. He met with enthusiastic crowds everywhere he traveled. As Lindbergh biographer Scott Berg observed, "People behaved as though Lindbergh had walked on water, not flown over it."[9] Upon returning to New York, Lindbergh received an invitation from Dwight Morrow, newly appointed Ambassador to Mexico, to visit Mexico City. "The Ambassador's invitation," Lindbergh recalled, "gave me an opportunity of accomplishing several objectives on a single flight. In addition to the gesture of friendship toward Mexico he desired, I could demonstrate still more clearly the capabilities of modern aircraft." Lindbergh, also with thoughts of promoting the expansion of commercial aviation to South America, accepted the invitation from his future father-in-law.[10]

To dramatize the event, Lindbergh decided to fly nonstop from Washington, DC to Mexico City, a distance of some 2,100 miles. He departed from Bolling Field on the morning of 13 December 1927 and reached Mexico City 27 hours and 15 minutes later. This flight, historian Wesley P. Newton pointed out, "ranks only slightly behind the May epic in its technical accomplishments and its reception."[11] Following his enthusiastic welcome in Mexico City, Lindbergh decided to continue southward, believing that a route to South America was an essential first step in the development of international commercial aviation. He flew through Central America, stopping in Guatemala, Honduras, El Salvador, Nicaragua, Costa Rica, and Panama; he then continued to Columbia and Venezuela in South America. His return flight took him through the Caribbean to Cuba, via the Dominican Republic, Haiti, and Puerto Rico. "In each country," Newton has written, "he received a hero's welcome: statesmen, lavish in their affection, decorated him and sang his name. Crowds of citizens sometimes threatened to smother him in sincere adoration."[12]

The main purpose of Lindbergh's trip, at least as far as he was concerned, was not to promote international goodwill but to explore the possibilities for commercial air service to Latin America. Although he had been advised not to attempt the flight to South America in a landplane because of a lack of airfields, he had been confident that he could carry sufficient fuel to return to his point of departure. Facilities, he found, were indeed primitive. Not only were adequate landing areas few and far between, but there were only four weather-reporting stations—at Mexico City, the Canal Zone, San Juan, and Havana—in the entire region. Obviously, much work on a route infrastructure would have to be done before service to Latin America would be possible.[13]

---

9 Berg, *Lindbergh*, p. 170.

10 Charles A. Lindbergh, *Autobiography of Values* (New York: Harcourt, Brace, and Jovanovich, 1977), p. 83.

11 Wesley P. Newton, *The Perilous Sky: U.S. Aviation Diplomacy and Latin America, 1919–1931* (Coral Gables: University of Miami Press, 1978), p. 131.

12 Ibid., p. 132.

13 Lindbergh, *Autobiography*, p. 88.

Lindbergh reached Cuba in early February 1928. While spending four days in Havana, he recalled, "I met a young man named Juan Trippe." Trippe no doubt informed Lindbergh about the many developments that had taken place since the formation of the Aviation Corporation of America in 1927. Together with two other groups, Trippe's company had competed for the airmail contract between Key West and Havana. As Trippe had earlier secured exclusive landing rights in Havana from President Gerardo Machado, he believed that the Aviation Corporation deserved the award. The Post Office, however, insisted that the three groups cooperate. As a result, Pan American Airways emerged as the operating company for the contract, with Trippe as President.[14]

PAN AMERICAN AIRWAYS TAKES OFF

Pan American had gotten off to a shaky start. Although it had ordered Fokker trimotors for the route, none had arrived in time to meet the Post Office's deadline of 19 October 1927 to begin mail service. Fortunately, Trippe was able to charter a Fairchild FC-2, being ferried to Haiti, to inaugurate the route. The following week, Pan American began mail and freight service with an eight-seat Fokker F-VIIa/3m.[15]

Lindbergh expressed interest in Trippe's plans to expand into the Caribbean and Central America. "More than anybody Lindbergh had met," biographer Berg noted, "Trippe had the passion and power to make that happen." The young flyer inspected Pan American's base in Havana and test-flew one of the company's Fokkers. He came away impressed with the efficiency of the company's operation. By the end of his visit, it was clear that the two men shared a vision of the future of international commercial aviation, and they agreed to meet again later in the year.[16]

Lindbergh's flight received widespread praise. Assistant Secretary of Commerce William P. MacCracken, Jr., pointed out that the southern tour had emphasized "the feasibility and importance of extending our commercial airline activities to the West Indies and Central and South America . . . ." The United States government, MacCracken told the *New York Times*, was taking steps to secure airmail service to the south. Secretary of State Frank B. Kellogg also hailed the flight for spreading goodwill and for demonstrating "the feasibility of aviation as a means of communication between the American states."[17]

---

14 Ibid., p. 107.

15 For this inaugural flight, see Gene Banning, *Airlines of Pan American Since 1927* (McLean, VA: Paladwr Press, 2001), pp. 9–11.

16 Berg, *Lindbergh*, p. 191. See also Newton, *Perilous Sky*, p. 149.

17 Newton, *Perilous Sky*, p. 150.

In the summer of 1928, Lindbergh made an appointment to see Trippe in New York City. Pan American had just been awarded two Foreign Air Mail contracts: FAM 5, extending from Key West to the Canal Zone via Cuba, Mexico, British Honduras, Nicaragua, and Costa Rica; and FAM 6, from Havana to San Juan via Haiti and the Dominican Republic. Lindbergh recalled that his first conference with Trippe and other officers of Pan American focused on the operation of the new routes. He noted:

> New airplanes had to be developed, radio stations set up, ground and water facilities installed. Where should we use flying boats? Where should we use amphibians, even with their penalty of added structure weight? Could landplanes carry passengers on long overwater flights—Havana to Panama, for instance—and should they be built of wood or metal? Some officers argued that wood planes would be buoyant and cited the fact that a Pan American Fokker had once ditched in the Florida Straits and stayed afloat until passengers could be rescued. As for navigational aids, would it be practical to anchor buoy lights at intervals across the Caribbean Sea, or would radio beacons at lighted terminals be enough? We had to consider the passengers' needs in good flying conditions and bad; how would we accommodate them at stops where no good hotels or restaurants existed? What emergency-rescue facilities should be set up?[18]

Lindbergh obviously had a lot to contribute to Pan American's planned expansion, both in terms of technical expertise and publicity. On 17 January 1929, he signed a contract to serve as the airline's technical consultant at an annual salary of $10,000 with the right to purchase 1/10th of Pan American's shares at half their current value. This agreement marked the beginning of a relationship with Trippe's airline that would last over 40 years.[19]

Lindbergh soon began to earn his money. On 3 February 1929, he was at the controls of a Sikorsky S-38 amphibian that left Miami at 6 a.m. to inaugurate FAM 5 to the Canal Zone. Trippe accompanied Lindbergh to Havana. Lindbergh continued on to Le Fe in western Cuba, refueled, then took off for Belize, British Columbia, via Cozumel, Mexico. The next day he flew on to Managua, Nicaragua. Lindbergh reached Cristobal's France Field in the Canal Zone on 6 February, where he was escorted to a landing by a squadron of Navy pursuit planes. Because authorization had not yet been received to carry the mail northbound from the Canal Zone, Lindbergh spent the next three days meeting with a variety of military and government officials.[20]

---

18 Lindbergh, *Autobiography*, p. 108.

19 Berg, *Lindbergh*, p. 191. Lindbergh earlier had signed a similar agreement with Transcontinental Air Transport.

20 Banning, *Pan American*, pp. 21–25.

While awaiting the necessary permission to fly a large load of mail that had accumulated in the Canal Zone, Lindbergh cabled a series of recommendations to Trippe. He believed that multi-engine amphibians should continue to be used between Miami and Honduras, but that trimotored landplanes should fly the Honduras-Panama segment of the route. A three-times-a-week schedule, he estimated, would require three amphibians and four landplanes, with one of each in reserve. It was imperative, he urged Trippe, to rush construction of a field in the vicinity of Port Morales in Nicaragua. Other construction necessary before the start of regular service included fields at San Pedro (Dominican Republic) and Puntarenas (Costa Rica); hangars at France Field, Managua, and San Pedro; and radio equipment, including directional finding stations, at La Fe, Port Morales, and Puntarenas. Obviously, a great deal of work remained to be done on the infrastructure of the route to the Canal Zone before regular operations could begin.[21]

Finally obtaining approval for the northbound mail, Lindbergh was waiting in the predawn darkness of 11 February for Postmaster Gerald Bliss to hand over the mail. When the postal official arrived, however, prop wash from the S-38 caught a mailbag and scattered its contents over the field. Lindbergh and Bliss had to retrieve the letters before the flight could depart at 6 a.m. Two days later, he arrived in Miami, the same day on which Ambassador Morrow announced the engagement of his daughter Anne to Lindbergh.[22]

Trippe took Lindbergh's advice on equipment for the route. In early April, Pan American opened regular service from the Canal Zone to Tela, Honduras, via Managua, using two new Ford trimotored landplanes. During the spring and summer, Pan American directional finders and ground radio stations were opened along the route. Trippe already had his sights set on an expansion of service to South America and had obtained authorization to extend the mail route from San Juan to Paramaibo, Surinam, an essential first step to Brazil. On 20 September 1929, Lindbergh and Edwin Musick set out from Miami in a Fokker F-10A to inaugurate this route. Onboard were Anne Morrow Lindbergh, Trippe and his wife Betty, and Glenn Curtiss.

Betty Trippe, who kept a diary during the flight, was impressed with the Fokker. "The airplane," she wrote, "the largest built at that time, flew at 100 mph and could carry 10 passengers. It was quite comfortable, but you could not move around as the ceiling was so low, the aisle was very narrow, and the little wicker seats were very close together. The noise from the three engines was so loud that conversation was almost impossible. No one seemed to mind, however, as everyone was so excited to be on this flight." The airplane made a brief stop in Havana, where it was met "by a great unwashed, shouting,

21 Letter from Lindbergh to Trippe (9 February 1929), Pan American Records, Box 48.

22 Newton, *Perilous Sky*, pp. 215–216. On Anne Morrow Lindbergh, see Susan Hertog, *Anne Morrow Lindbergh: Her Life* (New York: Nan A. Talese, 1999).

wild-eyed crowd." After government officials presented bouquets to Betty Trippe and Anne Lindbergh, the flight departed for Camaguay. En route, steward Raphael Vega, wearing a white duck uniform, served the passengers fried chicken, salad, and coffee, inaugurating Pan American's in-flight meal service.[23]

The Fokker had to take on fuel at Camaguay. The whole town turned out for the event. There were speeches, bouquets, and "pushing, cheering crowds, who literally mauled the Lindberghs." They arrived at Santiago de Cuba to "the same wild, tumultuous crowds." They stood for the playing of the national anthems of Cuba and the United States, listened to welcoming speeches, and received the usual bouquets. The Lindberghs were then "rushed off in a flag-draped open car of the governor," while the Trippes and an escort of mounted police followed behind. The streets swarmed with people shouting "Viva Lindbergh!" A reception followed in the governor's palace. Anne Lindbergh did not care much for the stale cake that was served, and she discretely shoved her piece in her glove. When Lindbergh was called out on the balcony to wave to hundreds of cheering people below, he insisted that Trippe—"as President of Pan American"—join him. After all, this was just an inaugural flight by the airline!

The next day, Lindbergh flew to Port-au-Prince for the same kind of reception at the airfield and a meeting with President Borro at his palace. The flight reached Santo Domingo in the afternoon, refueled, then continued to San Juan. As the route to Dutch Guiana lacked airfields, Lindbergh switched to a Sikorsky S-38 for the balance of the trip. Because of limited space in the amphibian, the Trippes followed behind in a second S-38 piloted by John Tilton.

The two airplanes continued their triumphal tour along the island chain to the south, stopping briefly at St. Thomas, St. Kitts, Antigua, and St. Lucia en route to Port-of-Spain, Trinidad. "The welcome there for Lindbergh," Betty Trippe observed, "was more dramatic and thrilling than any of the other places we had stopped. The crowd literally went wild with joy shouting, 'Lindbhor.'"

On 23 September, they departed for Georgetown, British Guinea, refueled, then flew to Paramaribo, Dutch Guiana. Two days later, they embarked upon the return flight to Miami via Curacao, the Canal Zone, Managua, San Salvador, Guatemala City, and Havana. In all, they covered 9,000 miles in three weeks, gaining valuable operational data and creating tremendous publicity and goodwill for Pan American. As historian Wesley Newton has observed, it was difficult for Pan American to secure landing rights in countries like Mexico, Guatemala, Venezuela, Columbia, Chile, and Ecuador. "These were

23 Betty Stettinius Trippe, *Pan Am's First Lady: The Diary of Betty Stettinius Trippe* (McLean, VA: Paladwr Press, 1996), pp. 26–47.

Sikorsky S-40 Clipper, Pan American Airways. (Photo number 78-7097, National Air and Space Museum, Smithsonian Institution)

places," he emphasized, "where the United States was seen in an especially bad light, where nationalistic forces favored boosting native airlines." Trippe employed a variety of methods to achieve his objectives, often using former U.S. diplomats and powerful Latin American insiders to argue the case for Pan American. But the favorable publicity generated by Lindbergh, Newton concluded, was "singularly important" in achieving a favorable outcome for Trippe's route extensions to the south.[24]

It was clear to Lindbergh that there was "desperate need" for longer-range aircraft for the new airmail routes. In the fall of 1929, Lindbergh, Trippe, operations manager Andre Priester, and other Pan American officials met frequently with aircraft designer Igor Sikorsky to discuss requirements for a new aircraft. "I remember one conference in particular," Lindbergh recalled. "Sikorsky unrolled several drawings on the table in front of us.

24 Newton, "Pan American Airways," in *The Encyclopedia of American Business History and Biography: The Airline Industry*, ed. William M. Leary (New York: Facts on File, 1992), pp. 343–349; and Newton, *Perilous Sky*, pp. 125, 131–136.

They portrayed his latest design, a four-engine sesquiplane amphibian, to be designated the S-40. It looked like a scaled-up S-38, with similar struts, wires, and huge pontoons—air resistance everywhere." Lindbergh objected to the awkwardness of the design, telling Sikorsky that "it would be like flying a forest through the air." Sikorsky agreed, but he argued that the S-40 was the wisest next step in the evolution of aircraft design.[25]

Although Lindbergh remained convinced that the design was "of the past rather than the future," the pressing need for an improved airplane forced him to accept Sikorsky's argument. On 9 December 1929, he recommended to Trippe that Pan American purchase the S-40 for its Caribbean routes. Trippe agreed and signed an order for two of the large amphibians with an option for a third.[26]

Lindbergh's technical advice on the purchase of the S-40 was his first contribution to Pan American's acquisition of new equipment, but it would not be his last. As Trippe later pointed out, Lindbergh "was in on virtually every decision of a technical nature that Pan American made." During the 1930s, he participated at every stage in the later development of the S-42, Martin M 130, and Boeing 314. Although not a graduate aeronautical engineer, Trippe noted, "he understood flying totally, and he was a man of extraordinary vision."[27]

On 19 November 1931, Lindbergh and Basil Rowe departed Miami in the first S-40—christened *American Clipper* by Mrs. Herbert Hoover—for Barranquilla (Columbia) and Cristobal via Kingston. While en route to Kingston, Lindbergh turned the controls over to Rowe and came back into the cabin to chat with designer Sikorsky, who was onboard for the inaugural flight of his new airplane. Lindbergh and Sikorsky spoke about the next step in aircraft design, a conversation that continued during an overnight stop in Kingston. At one point, Lindbergh made several sketches of a more streamlined seaplane that would be capable of flying at least 2,500 miles. The result of these and later discussions would be the advanced S-42, a machine that looked less like a "flying forest" than the S-40.[28]

The next day, the *American Clipper* flew nonstop from Kingston to Barranquilla, the longest nonstop overwater commercial passenger-carrying flight ever attempted. It took place without incident; the passengers were treated to a hot meal en route. The return flight took place on 26 November. Anxious to reach Miami, Lindbergh departed Barranquilla at dawn with 28 passengers and 15,000 pieces of mail. He landed at Miami after dark, marking the first time that anyone had made a one-day flight between South and North America.[29]

---

25 Lindbergh, *Autobiography*, p. 112; and letter from Lindbergh to Wolfgang Langewiesche (19 November 1967), Pan American Records, Box 14.

26 Letter from Lindbergh to Trippe (9 December 1929), Pan American Records, Box 48.

27 Trippe, "Lindbergh and World Travel."

28 Daley, *American Saga*, pp. 102–103.

29 Banning, *Pan American*, p. 60.

ARCTIC ADVENTURES

As Pan American expanded through the Caribbean and Central America to Brazil, Trippe never lost sight of his goals to cross the Atlantic and Pacific Oceans. Lindbergh and Trippe often discussed the means by which these objectives might be accomplished. Given existing technology, Lindbergh believed that the best possibility for reaching both Europe and Asia lay through the Arctic. "When you looked at a globe of the world," he recalled, "the Arctic routes were tantalizing, ideal for air routes." Only the Bering Strait (50 miles wide) separated Alaska from Russia. Also, the largest water gap on a northern route to Europe was the 700 miles from Labrador to Greenland. Although aerial operations were being conducted in the Far North, Lindbergh realized he " . . . did not know enough about the Arctic to form intelligent opinions." In the summer of 1931, he set out to explore the viability of an Arctic air route to Asia.[30]

Lindbergh selected a Lockheed Sirius for his Arctic adventure. Powered by a recently developed 575-horsepower Wright Cyclone engine, it had a range of 2,000 miles, thanks to 150-gallon fuel tanks in each of the airplane's pontoons. Pan American supplied a new, lightweight radio transceiver that had been developed by Hugo Leuteritz, the former RCA engineer who was now head of Pan American's communications department. Anne Morrow Lindbergh was set to accompany her husband and serve as radio operator for the trip. Leuteritz trained her on the equipment, explaining the use of the six transmitting coils and how the length of the trailing wire antenna determined frequency. Anne soon became proficient in transmitting and receiving Morse code messages.[31]

The Lindberghs departed Long Island on 27 July 1931. They flew across Canada and the Northwest Territories to Barrow, Alaska. Heading to the southwest, they stopped at Nome, then crossed the Bering Strait to Karaginski Island, a tiny fur station on the coast of Soviet Kamchatka. They continued down the Soviet coast to the Japanese islands, touching down in Hokkaido in late August. The Lindberghs received a warm welcome when they reached Tokyo, where they spent two weeks. They next flew south, crossed the Yellow Sea, and landed outside Nanjing. After a number of adventures in China, they arranged for their airplane to be crated and returned to California, while they crossed the Pacific by steamer.[32]

Upon his return to New York, Lindbergh reported to Trippe that an Arctic route to the Orient was feasible but would be difficult. "Flying boats could not be operated from ice-

---

30 Lindbergh, *Autobiography*, pp. 108–109; and letter from Lindbergh to Langewiesche (19 November 1967), Pan American Records, Box. 14.

31 "Leuteritz on Lindbergh: Survey Flights," n.d., Pan American Records, Box 30. See also Anne Morrow Lindbergh, *North to the Orient* (New York: Harcourt Brace, 1935).

32 Daley, *American Saga*, pp. 111–116.

covered water," he noted. "Airports for landplanes would be expensive to construct and maintain in sub-zero temperatures. Strange electronic phenomena created new problems for radio communications." Trippe, nonetheless, remained interested in the Arctic route. In the fall of 1932, he purchased Alaskan Airways and Pacific International of Alaska and began negotiations with the Soviets for landing rights on their territory. Ultimately, the political and technological difficulties forced Trippe to look elsewhere for a transpacific route.[33]

Lindbergh next turned his attention to the possibility of reaching Europe via a northern route. Trippe had long had his eye on the lucrative transatlantic market and had initiated discussions with European officials for landing rights as early as 1928. Additional talks had taken place in November 1930, at which time Trippe had signed an agreement with Britain's Imperial Airways and France's Aeropostale to cooperate in establishing a transatlantic airmail service. In the summer of 1931, a Pan American subsidiary, Boston & Maine Airways, had operated between Boston and Halifax, Nova Scotia, affording some limited flying experience on the first leg of a northern transatlantic route. The following year, Pan American sponsored two expeditions to Greenland to study the topography and weather conditions in the Far North.[34]

Lindbergh set off to explore an Arctic route to Europe on 9 July 1933. He employed the same Lockheed Sirius he had used on the flight to Asia. This time, however, the seaplane was equipped with a more powerful 710-horsepower Wright Cyclone F engine, a two-position Hamilton Standard propeller, and a newly developed Sperry directional gyroscope. Again, Anne was along to operate the radio. Leuteritz had a special interest in radio conditions along the route—how radio frequencies behaved at various distances at various times of day and night. Trippe chartered a Danish steamer, *Jelling*, to deposit fuel and supplies along the route, while a Pan American mechanic provided servicing for the Sirius.[35]

The Lindberghs reached Cartwright, Labrador, in mid-July, where they rendezvoused with the *Jelling*. After waiting a week for the weather to clear, they departed for Greenland. They spent three weeks flying back and forth across the Greenland icecap, then flew on to Iceland. During the course of a week, in Reykjavik, Lindbergh took time to write a preliminary report to Trippe on the northern route. Flying difficulties along the Arctic route, he advised, had been greatly exaggerated. Airplanes using this route had to be reliable, have plenty range, and have high speed. They had to be capable of flying nonstop from the western side of the Greenland icecap to Iceland. Radio stations also were needed along the route for weather information and directional finding. Unless a large amount

33 Lindbergh, *Autobiography*, p. 110.
34 Daley, *American Saga*, pp. 105–106.
35 Ibid., pp. 126–127.

of work was done to build landing fields, airplanes would have to be able to set down on water in summer and on snow in winter. Before a final decision to fly this route was taken, he cautioned, it would be necessary to assemble and study meteorological data covering a period of several years.[36]

The Lindberghs continued on to Ireland, then to Scandinavia. In mid-September 1933, while in Stockholm, Lindbergh again wrote to Trippe about the northern route. This time, Lindbergh was more pessimistic about the viability of the route, at least as a year-round mail service. "I believe," he wrote, "that a transatlantic air route by way of Greenland and Iceland can be operated satisfactorily during at least part of the summer months" with existing equipment and be competitive with steamship schedules. However, it was questionable that operations could be conducted during the winter with sufficient regularity to expedite the mails to northern Europe.[37]

After visiting various countries in Europe, the Lindberghs flew south to cross the Atlantic from the Azores. Failing to find a sheltered harbor, they continued south to Bathhurst in British Gambia via the Canary Islands. On 6 December 1933, they covered the 1,875 miles to Natal, Brazil in 16 hours. Thirteen days later, the Lindberghs arrived back in Long Island, having traveled 30,000 miles and visited 21 countries. "Lindbergh's detailed records of conditions at all places visited," Captain Gene Banning—Pan American pilot and historian—has written, "enabled him to make an important contribution to Pan American's development of Atlantic routes. He was also to advise on political situations and provide information on possible associations with European airlines."[38]

ENTERING THE WAR

In the short term, at least, the political problems along the route proved to be the major frustration for Trippe's plans to span the Atlantic. In 1934, his tripartite agreement with the British and French airlines collapsed, and he turned his attention to the transpacific route. Igor Sikorsky had developed the longer-ranged S-42 that had been discussed with Lindbergh earlier. Tests of the new seaplane by Lindbergh revealed that it was capable of reaching Hawaii from San Francisco, but only under ideal conditions. Lindbergh also had reservations about the S-42's handling techniques. "I do not believe it is a dangerous plane," he advised operations manager Andre Priester in July 1934, but it "necessitates pilots who are also engineers." Trippe, nonetheless, decided to press ahead.[39]

---

36 Letter from Lindbergh to Trippe (18 August 1933), Pan American Records, Box 507.
37 Letter from Lindbergh to Trippe (15 September 1933), Pan American Records, Box 507.
38 Banning, *Pan American*, p. 466.
39 Daley, *American Saga*, p. 225.

While Trippe focused on the transpacific route in 1935, Lindbergh faced the trauma associated with the sensational trial of Bruno Hauptmann, the man accused—and later convicted—of kidnapping and murdering Lindbergh's first son. Driven to Europe by the publicity generated by the trial, Lindbergh resigned his position as technical adviser to Pan American. His close association with Trippe nonetheless continued. In 1935, Pan American released a travel film describing the company's routes under the title *Flying the Lindbergh Trail*. Over three million people viewed the film during the year. Lindbergh also kept Trippe and other Pan American officers apprised of European aviation developments, and he kept an eye on the progress of the airline. In October 1936, for example, he wrote about Pan American's interest in new equipment. "I am glad that you are developing a landplane [the Boeing 307] in addition to the new flying boats [Martin M 130 and Boeing 314]," he commented. "I believe that it is probable that the landplane will replace the flying boat on all important routes in the future." In order to protect the future of Pan American, he counseled, it was "extremely important" that the landplane be capable of flying the Atlantic routes with a reasonable payload and large fuel reserve. Otherwise, Pan American would be vulnerable to companies operating landplanes of considerable higher performance than the Boeing flying boats. "It is important to keep in mind," he emphasized, "that planes can now be built better, by a fairly large margin, than any we have yet ordered."[40]

Lindbergh's public image suffered in the late 1930s as he became too closely associated with the German Luftwaffe, especially after receiving (unexpectedly) a decoration from Hermann Goering in October 1938. The problem grew worse in 1939 when Lindbergh returned to the United States and became a spokesman for the anti-interventionist cause. The "Lone Eagle," radio commentator Walter Winchell quipped, had become the "Lone Ostrich." Lindbergh's name disappeared from Pan American's annual report for 1940, and TWA, which also had a close association with the flyer, dropped the motto "The Lindbergh Line" from its advertising campaign.[41]

Trippe also had his problems. On 14 March 1939, 12 days before the long-awaited inaugural transatlantic service got underway, Pan American's board of directors replaced him as the airline's Chief Executive with Cornelius Vanderbilt Whitney. Lindbergh was philosophical about the change at the top. He had lunch with Trippe at the Cloud Club in New York City on 23 May 1939. Whitney, Lindbergh noted, had moved into Trippe's office. While Trippe continued to be listed as President, he had much less influence in the

---

40 Banning, *Pan American*, p. 78; letter from Lindbergh to Trippe (28 October 1936), Pan American Records, Box 507; and Berg, *Lindbergh*, pp. 353–354.

41 Berg, *Lindbergh*, pp. 377, 401. For Lindbergh's alleged pro-Nazi stance and his opposition to U.S. intervention in WWII, see Wayne S. Cole, *Charles A. Lindbergh and the Battle Against American Intervention in World War II* (New York: Harcourt Brace Jovanovich, 1974).

company. "In many ways I am sorry to see this," Lindbergh wrote in his journal, "for I like Juan and have always felt he had great ability." After lunch, Lindbergh spent the rest of the afternoon in Pan American's offices "meeting old acquaintances and talking over company affairs."[42]

Trippe's exile from power did not last long; by the end of the year, he was back in control of Pan American. Although the company kept its distance—at least in public—from the now controversial Lindbergh, the private relationship remained cordial. On 15 January 1940, Lindbergh and Trippe again met for lunch at the Cloud Club, where they spoke about the war in Europe, new equipment for the airline, and Trippe's plans to reorganize the company. Lindbergh then visited Chief Engineer Andre Priester and Vice President Franklin Gledhill to discuss Pan American's future. Lindbergh wrote:

> It seems they have eventually come around to the ideas I have been advocating for so many years in regard to using landplanes for the North Atlantic route! Priester was violently opposed to this policy not so long ago. I have for many years advocated the start of transatlantic routes with flying boats, but changing to landplanes after the pioneering years were passed. I have written many letters and reports to Pan American in regard to this.[43]

Lindbergh also was interested to learn that his controversial reports on the Azores had turned out to be correct. "I took the stand," he noted, "that the Azores were not suitable for scheduled year-round operation of flying boats, because they had no harbors of sufficient size." A Pan American expedition had reported that it would always be possible to find adequate operating conditions on the lee side of the islands. "I disagreed with this report," Lindbergh wrote. "Recent operating experience has demonstrated that I was correct in my stand."[44]

When the United States entered WWII in December 1941, Lindbergh attempted to join the Air Corps, but his anti-interventionist stance had so alienated President Franklin Roosevelt that a military connection proved impossible. Lindbergh visited the War Department on 13 January 1942 for discussions with Secretary Henry L. Stimson, Assistant Secretary Robert A. Lovett, and Air Corps chief General H. H. Arnold. "I told Arnold and Lovett," Lindbergh wrote in his journal, "that in view of the feeling which existed, it seemed to me it would be a mistake for me to return to the Air Corps at the present time and that I thought it would be better for me to try to make my contribution to the war

---

42 Charles A. Lindbergh, *The Wartime Journals of Charles A. Lindbergh* (New York: Harcourt Brace Jovanovich, 1970), p. 204.

43 Ibid., p. 306.

44 Ibid. On the problems with the Azores, see Banning, *Pan American*, pp. 487, 490.

through the aviation industry." Lindbergh asked if there would be any objections if he took part in the work of some commercial company, such as Pan American. Lovett told him that the War Department had no objections and would support such a connection.[45]

The following week, Lindbergh called upon Trippe at the Chrysler Building to follow up on his discussions with the War Department. Since "it seemed best for me to make my contribution to the war through the aviation industry," he told Trippe, "my first choice would be Pan American." Before taking any action on this, however, Lindbergh believed that it would be a good idea for Trippe to check with President Roosevelt to see if he had any objections to such a connection. Trippe, Lindbergh reported, "said there were many things I could do for Pan American and that he thought he would have a chance to talk to the President later this week."[46]

On 3 February, Lindbergh received bad news from Trippe. "He said," Lindbergh wrote in his journal, "he had talked to the War Department and that 'they' were entirely willing that I take an active part in Pan American projects. <u>But</u>, he said, when he talked to the White House, 'they' were very angry with him for even bringing up the subject and told him 'they' did not want me to be connected with Pan American in any capacity." The feeling at the White House, Trippe reported, was "extremely bitter" toward Lindbergh. "Juan seemed very much chagrined about the entire situation," Lindbergh noted. "He agreed with me that it would be inadvisable for me to make any active connection with the company at present . . . ."[47]

Lindbergh explored the possibility of working for United Aircraft or Curtiss Wright. Although company officials were receptive, the hostility of the White House proved impossible to overcome. Only Henry Ford, who detested Roosevelt, was willing to brave presidential wrath. Lindbergh became a consultant to Ford, who was manufacturing B-24s at a massive new facility in Willow Run, Michigan. Later in 1942, Lindbergh was able to join United Aircraft to assist in the development of the Vought Corsair F4U.[48]

In the spring of 1944, Lindbergh traveled to the South Pacific as a United technical representative to deal with operational problems as the Corsair was employed in combat. He also flew combat missions in twin-engine Lockheed P-38s to compare its performance with the single-engine Corsair. "On my first P-38 combat mission, from Hollandia, New Guinea," Lindbergh later recalled, "I used the engine settings I normally would for long-

45 Lindbergh, *Journals*, pp. 583–584.

46 Ibid., p. 587.

47 Ibid., p. 590. Professor Newton suggests that Trippe may have been paying Lindbergh back for his neutrality during Trippe's quarrel with Pan American's board of directors in 1939. Professor Hixson, however, believes that Trippe was accurately conveying Roosevelt's anti-Lindbergh position. See Newton, "Juan T. Trippe," in William M. Leary, ed., *Airline Industry*, pp. 464–476; and Hixson, *Lindbergh*, pp. 127–130.

48 Berg, *Lindbergh*, pp. 447–448.

range cruise—not knowing what the squadron's usual procedures were." As it happened, the pilots had been taught to keep their speed up and revolutions high so that they could maneuver quickly in the event of a sudden enemy attack. While these settings "might be highly advantageous in combat areas," Lindbergh noted, they were "disadvantageous while cruising from base to combat areas." At the end of Lindbergh's first mission, his crew chief reported to the unit's commander "that I had an inexplicable amount of fuel remaining in the tanks. I think it was probably as much of a surprise to me as it was to him."[49]

Lindbergh's inadvertent discovery would have an important impact on General MacArthur's plans in the Southwest Pacific. When reports reached MacArthur's headquarters in Australia that it was possible to increase the combat radius of the P-38s to 700 miles, Lindbergh was called to Brisbane to speak personally with the general. MacArthur wanted to know if the reports were true. Lindbergh assured him that they were. "MacArthur said it would be a gift from heaven if that could be done," Lindbergh noted, "and asked me if I were in a position to go back up to New Guinea to instruct the squadrons in the methods of fuel economy which would make such a radius possible." Lindbergh agreed. MacArthur then took him to a map of the South Pacific and spoke about the limitations that were imposed by the present fighter radii and the strategic benefits that could be derived from the longer range.[50]

Lindbergh made good on his promise to MacArthur. At first, the pilots were reluctant to adopt the new technique, fearing that a higher manifold pressure and lower rpm would damage their engines. Lindbergh gave lectures on cruise control, compiled engine-setting tables to post on instrument panels, and flew extra combat missions to demonstrate the technique. Gradually, the P-38 pilots came to accept the new engine settings, which added over 100 miles to their range. Lindbergh even managed to shoot down a Japanese Zero on one of his demonstration combat missions.[51]

ARRIVAL OF THE JET AGE

At the end of the war in Europe in May 1945, Lindbergh joined the Naval Technical Mission to Europe as a representative of United Aircraft. He spent a month inspecting German wartime developments in aeronautics and rocketry. Especially interested in German work on jet aircraft, he spoke to Willy Messerschmitt and to engineers at the German technical institutes about the past and future of this new form of flight. Messer-

49 Letter from Lindbergh to Langewiesche (19 November 1967), Pan American Records, Box 14; and Lindbergh, *Journals*, p. 872.

50 Lindbergh, *Journals*, p. 873.

51 Lindbergh, *Autobiography*, pp. 361–362

Boeing 747, Pan American Airways. (Photo number 2000-9716, National Air and Space Museum, Smithsonian Institution)

schmitt was optimistic about the prospects for jet power, and he told Lindbergh that it would be possible to build a transoceanic passenger jet within four years.[52]

When he returned to the United States in mid-June, Lindbergh discussed the possibility of a jet transport with Trippe and Priester. "Trippe," Lindbergh recalled, "was immediately open to it." He assigned Philip B. Taylor as a consulting engineer to investigate further both German and British work on jets. It was soon clear that the British were intent on developing a jet transport to capture the postwar aviation market. In 1949, the prototype for the Comet made its first flight. Lindbergh, who had gone to England to inspect the new airplane, concluded that it was too small and had too short a range to be commercially viable for transoceanic routes. Trippe agreed. Both men wanted a larger aircraft that could fly nonstop across the Atlantic with a decent passenger load. Trippe, nonetheless, took an option on three Comets—just in case.[53]

Douglas, Boeing, and Lockheed all had plans for jet transports to compete with the Comet, but Trippe and Lindbergh were convinced that they were too small to be economical. Trippe, biographer Robert Daley has emphasized, had made up his mind "that he would buy nothing less than a true transatlantic jet, which neither the Boeing nor the Douglas design was, nor would he buy a plane less profitable to operate than the DC-7C."[54]

Trippe, Lindbergh, and other Pan American officials made frequent trips to the aircraft and engine manufacturers between 1952 and 1955, constantly applying pressure

52 Lindbergh, *Journals*, pp. 952–960.
53 Daley, *American Saga*, pp. 397, 401.
54 Ibid., p. 404.

on them to get the kind of airplane that Pan American envisioned. In 1955, Trippe placed an order with Douglas and Boeing for $269 million for a fleet of DC-8s and Boeing 707s. Pan American led the way into the new era of international air transport, thanks largely to the efforts of Trippe and Lindbergh.

Trippe and Lindbergh also worked closely together to bring about the next stage of jet transport—the jumbo jet. In the spring of 1965, Trippe approached Boeing about the possibility of stretching the 707 to carry 400 passengers over 3,000 miles. When this proved technologically impossible, the discussions came to focus on an entirely new airplane. The result was a sales contract, signed in April 1966, for 25 aircraft costing $550 million—the largest single airplane order ever made by any airline. It took nearly three years for the Boeing 747 to get into the air. As the giant airplane took shape, there were constant discussions between Pan American and Boeing about every facet of its development. Lindbergh participated in many of these meetings, often taking a leading role in promoting Trippe's vision for the new airplane. "I don't think the plane would ever have been built," Trippe acknowledged, "if Slim [Lindbergh] hadn't been there adding the weight of his integrity, his insights, and his prestige to so many of the crucial negotiations." Certified on 30 December 1969, the first Boeing 747 went into service with Pan American on 21 February 1970.[55]

Trippe and Lindbergh, who had collaborated on equipment decisions for Pan American from the Sikorsky S-40 to the Boeing 747, finally drew apart over the prospects for a supersonic transport. Both men went to Europe in 1963 to inspect plans by the British Aircraft Corporation and Sud Aviation for an SST to be called the Concorde. Lindbergh came away with a skeptical attitude, both with respect to the economics of the airplane and its impact on the environment. Trippe, however, was enthusiastic about the prospects for the high-speed transport and took options on eight Concordes. A short time later, after President John F. Kennedy had endorsed a U.S.-built supersonic airliner, Trippe announced that he would buy 15 of the 1,800-mile-an-hour machines.[56]

Lindbergh remained convinced that Trippe was wrong about the SST and openly opposed him. The Concorde, he wrote in an editorial in the *New York Times* in 1972, was certainly fast, but it also was "costly, a polluter, and its range is relatively short." Growing more skeptical of technological advances in his later years, Lindbergh asked, "Is the quality of life or the advance of technology to guide us?" As for himself, he answered, "aviation has value only to the extent that it contributes to the quality of human life it gives." In the end, of course, it was economics that doomed the SST.[57]

---

55 Trippe, "Lindbergh and World Travel." On Lindbergh's role in the development of the Boeing 747, see Clive Irving, *Wide-Body: The Triumph of the 747* (New York: William Morrow and Company), pp. 159–160, 180, 271, 355–356; and Laurence S. Kuter, *The Great Gamble: The Boeing 747* (Tuscaloosa, AL: University of Alabama Press, 1973), p. 60.

56 Bender and Altschul, *Pan Am*, p. 501.

57 Charles Lindbergh, *New York Times* (27 July 1972).

CONCLUSION

At Trippe's invitation, Lindbergh had joined Pan American's board of directors in 1965. He would remain on the board until shortly before his death in 1974.[58] Trippe retired in 1968. Before his death in 1981, he watched the decline of his beloved airline as economic factors called into question the viability of the decision to acquire the Boeing 747. History would show that he had made the correct decision, but the timing could not have been worse. A world oil crisis adversely affected airline travel in the 1970s, and Pan American never recovered.[59]

While Pan American World Airways, as it became after WWII, has passed into history, its record of pioneering accomplishments remains as a tribute to the efforts of Juan Trippe and Charles Lindbergh. In the late 1920s, these two men dreamed about a time when giant airplanes would carry people around the world in comfort and safety—then made the dream a reality.

---

58 Hertog, *Anne Morrow Lindbergh*, p. 472, alleges that when Lindbergh chose to fly from Columbia Presbyterian Hospital to Hawaii, so that he could die on Maui, "Juan Trippe at Pan Am flatly refused to help him . . . ." No doubt, this reflected the views of Mrs. Lindbergh. Correspondence in the Pan American Records, however, make clear that Samuel Pryor, retired Pan American Vice President and Lindbergh's neighbor on Maui, was simply unable to reach anyone at the airline when he received the request from the family. He thereupon called the President of United Air Lines and made the necessary arrangement. See "Pryor to Rob Mack" (18 January 1975), Pan American Records, Box 48.

59 On the demise of Pan American, see Robert Gandt, *Skygods: The Fall of Pan Am* (New York: William Morrow and Company, 1995). Gandt, who flew for Pan American for 26 years, concludes that the company's failure was more the result of "the complex life and times" than the culpability of individuals.

4

The Autogiro Flies the Mail!

Eddie Rickenbacker, Johnny Miller, Eastern Air Lines, and Experimental Airmail Service with Rotorcraft, 1939–1940

W. DAVID LEWIS

IN 1939, WHEN I WAS EIGHT YEARS OLD, SOMEONE GAVE ME A BOOKLET WITH SPACES FOR STICKERS SHOWING REMARKABLE NEW INVENTIONS. I COLLECTED THE STICKERS, which came in the mail, and pasted them above captions telling how these marvels of ingenuity would change the world. One sticker showed a roadable autogiro that could swoop down out of the sky, fold its rotor blades against its fuselage, and travel just like a car.

An autogiro, I learned, had a conventional aircraft engine and propeller that provided thrust, but derived lift from a pivoted rotor whose blades were airfoils whirling in the relative airflow with no power required. Contrary to a popular misconception, the blades did not depend on the forward motion of an autogiro to spin; they would do so even if it slipped backward. Autorotation continued as long as the rotor was bearing weight, even if the engine stopped running in midair. In such a case, when airspeed reached about 30 miles per hour, an autogiro began to descend. At zero airspeed, the rotor, still whirling, acted as a parachute while the autogiro settled gently and safely to Earth.

Roadability had been achieved by 1939 when the Pitcairn AC-35 autogiro received a patent, winning a competition sponsored by the Commerce Department to promote convertible air/highway vehicles. It had steerable front wheels, a stable undercarriage, and a compartment for a pilot and passenger that afforded excellent visibility through windows made of safety glass. An air-cooled radial engine located behind and below the cabin (the patent specification stated that a liquid-cooled engine worked equally well) was linked by separate controllable driveshafts to the propeller in front and a single wheel in the rear. When the autogiro was on the ground, the pilot could deactivate the rotor and drive wherever he wanted to go. The AC-35 never reached commercial production, but a prototype created an unusual spectacle moving in traffic on a busy street in Washington, DC.[1]

Roadable hybrids, however, were not the only types of autogiros drawing attention in those days. In 1939, Eddie Rickenbacker, America's "Ace of Aces" in WWI and Chief Executive of Eastern Air Lines, used autogiros in the world's first regularly scheduled commercial rotorcraft service that carried mail between the Central Airport at Camden, New Jersey, and the rooftop of Philadelphia's 30th Street Post Office. Because Rickenbacker was one of my boyhood idols, his faith in the autogiro strengthened my belief that it had a bright future.

Eddie Rickenbacker in the late 1930s at about the time Kellett and Miller approached him to approve having Eastern Air Lines carry mail by autogiro. (Courtesy of Special Collections and Archives, Auburn University)

A glance at Rickenbacker's life, which was dominated by automobiles and airplanes, and full of pioneering activities and ideas, shows why he was immediately excited about the novel concept of using a hybrid vehicle like the autogiro to carry airmail. Rickenbacker's name was well known in every American household. Born in 1890 to Swiss immigrant parents in Columbus, Ohio, he was a living symbol of traits and values that were deeply admired throughout the country—self-reliance, courage, persistence, and an indomitable will to win. He rose from poverty to success like the hero of a Horatio Alger novel. His first love was the automobile, which he encountered in his early teens when a salesman came to Columbus with a Ford runabout predating the famous Model T. After giving a sales pitch to a crowd of curiosity seekers that gathered around the car, the agent offered to give rides to

1 Frank Kingston Smith, *Legacy of Wings: The Story of Harold F. Pitcairn* (New York: Jason Aronson, Inc., 1981), pp. 240–247, 361; U.S. Patent 2,174,946, 3 (October 1939), with specifications; and caption and photograph in Warren R. Young, *The Helicopters* (Alexandria, VA: Time-Life Books, 1982), pp. 66–67.

potential buyers, but found no takers until young Eddie brashly volunteered. The reward for his daring was a spin around a downtown block at more than 10 miles an hour, faster than he had ever gone before.

He was already making a living on his own, having quit school in seventh grade when his father died. At the time of his automotive escapade he was supporting his mother by doing odd jobs that stultified his impetuous nature. The ride in the Ford runabout gave him the idea of looking for work in the infant automobile industry, in which Columbus was already well established. After finding a job at a garage that repaired bicycles and motor vehicles, he signed on with the Oscar Lear Company, a modest establishment that manufactured cars on a limited basis. His insatiable zest for knowledge led him to take mail-order courses in automotive engineering from International Correspondence Schools in Scranton, Pennsylvania. The diligence with which he pored over his lessons on his lunch breaks brought him to the attention of Lee Frayer, a partner in the Oscar Lear Company who had an engineering degree from Ohio State University. In addition to building automobiles, Frayer also designed racecars. Attracted by Rickenbacker's initiative, Frayer took him to the 1906 Vanderbilt Cup contest, a prestigious road race held on Long Island, as his riding mechanic. Serving in such a capacity was an extremely dangerous job requiring Rickenbacker to sit beside Frayer as he careened around the twisting Jericho Turnpike, warning him with hand signals about impending dangers, monitoring tire wear, and pumping gas and oil if the splash and gravity feed systems broke down. Because riding mechanics could not cling to the steering wheel, as drivers could, they were much more likely to be thrown out of a racecar and killed if it overturned.

Reveling in such hazards, Rickenbacker ultimately became a star performer on the American racing tour, winning seven major championships in a career that lasted from 1910 through 1916. He won his first big event, a 300-mile sweepstakes in Sioux City, Iowa, on Independence Day in 1914 before 40,000 fans packing the stands at a dirt racetrack that recently had been a cornfield. In 1915, he became manager of the Prest-O-Lite Racing Team, owned by Carl Fisher and Fred Allison—wealthy sportsmen who made acetylene headlamps and were the principal owners of the Indianapolis Speedway. Rickenbacker's gutsy style, and a flashing grin that clearly indicated the joy he took in an activity even more potentially lethal than bullfighting, made him an idol at the Indy 500 and other championship events sponsored by the American Automobile Association (AAA). After the United States declared war on Germany in April 1917, President Wilson asked a cousin who was active in the AAA to have the organization designate a famous racer to send to France as a driver on General Pershing's staff. The AAA chose Rickenbacker, who enlisted in the Army as a sergeant and sailed for Europe on 28 May 1917 aboard Pershing's troopship, the *Baltic*.

While racing in California in 1916, Rickenbacker had met a young aircraft designer and manufacturer, Glenn Martin, who stirred his imagination by taking him up in one of his planes. At about the same time, Rickenbacker also repaired the engine of a stranded airman, Townsend Dodd, a major in the aviation branch of the United States Signal Corps. Already fascinated by the lure of the sky, Rickenbacker did not intend to remain a driver any longer than necessary. Instead, he had set his sights on becoming a combat pilot. He got the opportunity he wanted when he became Billy Mitchell's driver and was taken on tours of the battlefront in a Packard staff car. Mitchell was so grateful for Rickenbacker's skill in repairing the automobile whenever it broke down that he granted his wish. After Rickenbacker received his *brevet* (pilot's license) at a flying school at Tours and was commissioned as a first lieutenant in the United States Army Air Service, he became chief engineering officer of a large new American training base at Issoudun. From there he went to a gunnery school at Cazaux, where stray bullets could splash harmlessly into the Bay of Biscay. He did well enough in his lessons to qualify as a combat pilot and was assigned to an advanced training base with the 94th Pursuit Squadron at Villeneuve-le-Vertus, a village in the vineyard region south of Chalons.

The rest was history. By the end of the war, Rickenbacker had become America's Ace of Aces by shooting down 26 German aircraft within less than eight months after his first sortie across enemy lines on 28 March 1918 in a Nieuport 28, a highly maneuverable plane with a nasty habit of shedding wing fabric in a power dive. His achievements were all the more remarkable because he spent much of his time in hospitals. Flying at high altitudes and plummeting on unsuspecting opponents hurt his eardrums, resulting in an abscess that had to be lanced. Naturally, he never let on that an accident he had suffered during his automotive career had severely damaged the cornea of his right eye, impairing his depth perception. As in everything else, he overcame his handicap with constant practice until he learned to judge distance. By early June, he had won the five confirmed victories required to certify him as an Ace.

In addition to displaying rare courage and determination, Rickenbacker also used his long experience with pit crews to good advantage in coaching mechanics to repair the complex gearing of Hispano-Suiza "Hisso" V-8 engines in newly assigned Spad XIII fighter planes that he greatly preferred to the Nieuport 28s because of their ruggedness and speed. Because of the high maintenance needs of the Hissos, Rickenbacker's superior officer Colonel Harold E. Hartney had trouble keeping enough Spads in the air to fulfill his orders and desperately needed a man with Rickenbacker's know-how to take charge of the situation. When Hartney decided to sack an ineffective squadron leader, Kenneth Marr, on the eve of the Meuse-Argonne Offensive, he decided that Rickenbacker had the combination of qualities needed to replace him. Hartney prevailed upon the higher brass to put Rickenbacker in command of the 94th Pursuit Squadron despite his lack of formal education. Leading by example—he never asked a man to undertake a mission, however hazardous, that he was not willing to perform—Rickenbacker molded the 94th into the best unit in the Air Service, measured by its total victories, the number of decorations its members won, the sorties it flew, and other criteria of military excellence. Because of its outstanding record, the squadron was part of the Army of Occupation sent into the American sector of Germany after the war. Soon thereafter, wearing rows of medals and campaign ribbons on his chest, Rickenbacker came home in 1919 to receive a triumphant reception as a national hero. His innate ability and infinite supply of grit had carried him far from his humble origins.

Rickenbacker's heart was now in the sky, and he wanted to become an aircraft manufacturer; but opportunities were lacking in a market glutted with surplus wartime planes, chiefly Curtiss JN-4 Jennies, that could be bought for knock-down prices and used in barnstorming. Risking his life in wing walking and stunt flying did not attract Rickenbacker after undergoing the rigors of aerial combat. Reluctantly, he decided to return to what he knew best, the automobile industry. Barney Everitt, a Detroit millionaire who had earlier manufactured a car called the EMF (wags had called it the "Every Morning Fix-it"), capitalized on the young hero's fame by persuading him to become vice president of a new company that Everitt established to produce the Rickenbacker, a high-quality motor vehicle advertised as "a car worthy of its name." Unfortunately, the enterprise could not compete in the turbulent postwar economy and went bankrupt in 1927.

A banker helped Rickenbacker acquire the Indianapolis Speedway, which kept him busy every year in the month before Memorial Day. He spent the rest of his time promoting a new luxury car, the La Salle, for General Motors (GM), to which he had become connected by marrying Adelaide Frost Durant, formerly the daughter-in-law of GM's founder, William C. "Billy" Durant. Nothing, however, could make Rickenbacker forget his love of the sky, and he finally persuaded GM's corporate leaders to diversify by acquiring the Pioneer Instrument Company (which became known as Bendix Aviation), the Allison Engineering Company (which became a leading maker of airplane engines), and the Fokker Aircraft Corporation of America, in which Rickenbacker became vice president for sales. Again, however, bad luck continued to dog his footsteps. The Great Depression and the crash of a Fokker airplane that killed Notre Dame's football coach, Knute Rockne, undermined his effectiveness.

Resigning his position with Fokker, Rickenbacker became vice president for public relations with American Air Lines but made a mistake by backing its biggest stockholder, W. Averell Harriman, in a losing proxy battle with a crafty director, E. L. Cord. Even though Cord asked him to stay, Rickenbacker resigned because he saw that financial disasters suffered by financial wizard Clement Keys in the depression had created an opportunity to persuade GM to take over North American Aviation Corporation (NAAC), a giant holding company that Keys had created in more prosperous times. NAAC's properties included Eastern Air Lines, TWA, and large shareholdings in the Douglas Aircraft Company. Rickenbacker's success in helping arrange a deal in which GM took charge of NAAC was one of his greatest contributions to American commercial aviation, which was in the doldrums and benefited greatly from acquiring GM's patina. When Franklin D. Roosevelt (FDR) canceled airmail contracts amidst scandals that erupted in 1934, Rickenbacker helped GM thumb its corporate nose at the President, whose New Deal had already displeased powerful magnates including Pierre S. Du Pont, John J. Raskob, and Alfred P. Sloan. Together with TWA's general manager Jack Frye, Rickenbacker set a transcontinental speed record in a gleaming new airliner, the Douglas DC-1, on the last day of regular mail and passenger delivery before the Army began an ill-fated venture flying routes previously operated by private enterprises. After FDR was forced to back down in a storm of protest about military pilots being killed in a series of crashes, Rickenbacker became general manager of Eastern Air Lines and started transforming it from an unimpressive enterprise with two potentially lucrative vacation routes (from New York and Chicago to Miami) into the most profitable carrier in America, a position it held by the end of the decade. Among the keys to his success was abandoning Curtiss Condors, Pitcairn Mailwings, and other weary birds, and replacing them with more modern planes, including Douglas DC2s and DC3s. In the process, he created what he proudly called the "Great Silver Fleet."[2]

Rickenbacker had unquestioning faith in technological progress. Even during his triumphant homecoming tour in 1919, when hopes for commercial aviation were dim, he had envisioned a future in which airmail and passenger routes would fill the American sky. Speaking at Banff in the Canadian Rockies during a respite from the endless parades and receptions held in his honor, he foresaw a future need to integrate air and ground transportation to avoid what later became known as "gridlock." His suggestion for dealing with that phenomenon whenever it materialized was characteristically radical. New York City, he urged, could deal with gridlock by eliminating Central Park and covering the site on which it stood with a gigantic high-rise building capped by a flat roof on which aircraft could land. The floors below would provide space for hangars, passenger lounges, maintenance facilities, storage rooms, and offices for airline executives. At ground level would be loading zones for various forms of surface transportation including buses and taxis. High-speed rail lines and subways would run under the streets surrounding the complex. Elevators would carry travelers from one mode of transport to another, obviating delay in meeting their busy schedules.[3]

It was easy to see why a mind capable of conceiving such wonders responded favorably in 1938 when a rotorcraft manufacturer, W. Wallace Kellett, and his test pilot, John M. "Johnny" Miller, urged Rickenbacker to have Eastern Air Lines adopt autogiros to shuttle mail between the Camden airport and the Philadelphia post office, thereby dras-

2 Edward V. Rickenbacker, *Rickenbacker* (Englewood Cliffs, NJ: Prentice-Hall, 1967); and W. David Lewis, *Eddie Rickenbacker* (Baltimore: Johns Hopkins University Press, 2005). See also Lewis, "Edward V. Rickenbacker," in *The Aviation Industry*, ed. William M. Leary (New York: Oxford University Press, 1992), pp. 398–415.

3 "Predictions by Captain Eddie Rickenbacker, Banff, Canada" (19 June 1919), unpublished manuscript in Special Collections Department, Ohio State University, Columbus, OH.

tically reducing the time required for surface transportation to do the job. Even though autogiros were relatively small, they offered ample space for the number of mailbags that would have to be carried by such a service. Autogiros could fly safely over crowded urban areas because of their ability to land gently in small plots of ground in case of emergencies. Rickenbacker also was intrigued by the thought of adding a widely heralded aircraft, only recently invented, to the Great Silver Fleet, epitomizing Eastern's zeal for adopting new ideas and the depth of his commitment to technological progress.

The autogiro was less than two decades old when Kellett and Miller approached Rickenbacker about using it. A predecessor of the helicopter, it was conceived in 1919 by a Spanish inventor, Juan de la Cierva, who, as previously indicated, combined a front-mounted engine and tractor propeller with an unpowered rotor that spun freely in the ambient air to provide lift. The basic difference between an autogiro and a helicopter, which Igor Sikorsky and other inventors were already contemplating in 1938, was that a helicopter's rotors, including the tail rotor that Sikorsky used to combat torque, were under full power. Cierva, by contrast, saw that rotor blades designed as airfoils needed no power source to make them whirl other than the relative air that flowed above and below them; a conventional engine with a propeller could provide thrust. Of the two approaches, Cierva's was simpler and safer; Sikorsky's was more complicated and capable of heavy lifting.[4]

Cierva patented the autogiro in Madrid in August 1920. He controlled pitch and yaw by adjusting the angle of the rotor blades and using conventional wing and tail surfaces. His creation achieved its first successful flight, demonstrating its ability to move in a circle, in January 1923. In 1925, he went to England and worked with British engineers to refine his concepts. He scored a notable triumph in September 1928 by flying his novel aircraft from London to Paris, where he impressed onlookers by demonstrating its ability to land in small spaces. In 1929, he licensed Harold F. Pitcairn, an American entrepreneur with a zest for new ideas, to use his patents. Pitcairn was so convinced about their potential that he sold Eastern Air Lines, which operated an airmail route from New York to Miami, and began to manufacture autogiros at Bryn Athyn, Pennsylvania, a Philadelphia suburb. Considering that Pitcairn had founded and once owned Eastern, it was fitting that the same enterprise ultimately adopted autogiros under Rickenbacker's leadership.

4 Among other sources on the history of autogiros, see Peter W. Brooks, *Cierva Autogiros: The Development of Rotary-Wing Flight* (Washington, DC: Smithsonian Institution Press, 1988); and Smith, *Legacy of Wings*, pp. 161–302. For other accounts, see Charles Gabelhouse, *Helicopters and Autogiros: A Chronicle of Rotating-Wing Aircraft* (Philadelphia and New York: J. B. Lippincott Company, 1967), pp. 33–67; Richard G. Hubler, *Straight Up: The Story of Vertical Flight* (New York: Duell, Sloan, and Pearce, 1961), pp. 43–48; Kenneth Munson, *Helicopters and Other Rotorcraft Since 1907* (London: Blandford Press, 1968), pp. 19–20; Gorge Townson, *Autogiro: The Story of "the Windmill Plane"* (Fallbrook, CA: Aero Publishers, Inc., 1985); and Young, *The Helicopters*, pp. 15–67. For various milestones in the development of Cierva's autogiros, see Bill Gunston, *Chronicle of Aviation* (Liberty, MO: JL International Publishing, Inc., 1992), pp. 176, 202, 257, 260, 278.

Cierva helped Pitcairn build his first autogiros, the PC-1 and PCA-1. They had duralumin-fuselage frames and stubby fabric-covered wings braced with struts and angled upward at the tips to increase their stability. They also included wire-braced box tail surfaces; a four-bladed laminated 45-foot mahogany rotor mounted on a four-strutted pylon; seven- or nine-cylinder Wright Whirlwind engines with two-bladed propellers; wheels set widely apart with angled bracing to cut down forward roll; and other features that demonstrated Cierva's ingenuity. The first aircraft manufactured by the Pitcairn-Cierva Autogiro Company of America flew in October 1929. Modifications ensued after a fire destroyed the original Pitcairn factory and a new plant was built in Willow Grove, Pennsylvania. The resulting PCA-1B had a stiffer rotor pylon, better-balanced rotor blades, and a simpler tail structure than the previous two models.

Four autogiros, two of which were built in the United States at Pitcairn's factory, made a dramatic spectacle flying over Manhattan in November 1930, looking like overgrown dragonflies. Further development and rigorous testing in cross-country flying led to a superb autogiro, the Pitcairn PCA-2, which had a reduced wingspan, a front cockpit placed to provide a new center of gravity and better load distribution, a more flexibly mounted rotor, and a more efficient tail structure. Test pilot James Ray proved the ruggedness of the PCA-2 by making a 2,500-mile round trip between Willow Grove and Miami, flying through terrible weather and making frequent landings on rough fields. In 1931, President Hoover awarded the National Aeronautic Association's Collier Trophy to Pitcairn after Ray had set down a PCA-2 on the White House lawn to demonstrate its ability to land in a restricted area. Miller later described the PCA-2 as "an absolutely fabulous aircraft," calling it and three other Pitcairn models "the only inherently safe aircraft ever built."[5]

Miller came from Poughkeepsie, New York, where he continues to reside in his late nineties, flying airplanes and leading a vigorous life. He was inspired as a five-year-old to launch a career in aviation when he looked skyward and saw Glenn Curtiss soaring above the Hudson River, making a 152-mile prize-winning flight from Albany to New York City. In 1923, at the age of 18, Miller first soloed in a Curtiss JN-4 Jenny. Learning about Cierva's work with autogiros, he wrote to the Spanish inventor inquiring about them. He was surprised when Cierva responded with two letters, written in excellent English, "explaining his autogiro in detail, including its aerodynamics and its possible development into a future helicopter."

After earning an engineering degree from Pratt Institute, Miller began his career in aviation in 1927 as a mechanic for the Gates Flying Circus, a legendary barnstorming enterprise that thrilled spectators with feats of wing walking, parachute jumps in which

5 The discussion of Miller's early life and career here and in the following paragraphs is taken largely from John M. Miller, *Flying Stories: A Chronicle of Aviation History from Jennys to Jets by the Pilot Who Flew Through It All* (Wichita, KS: American Bonanza Society, 2002), pp. 1–41, supplemented by telephone interviews, e-mail messages, and letters between Miller and Lewis.

Pitcairn PCA-2 flying over docks lining Manhattan in November 1930, with PCA-1 chase plane in the background. (Courtesy of Stephen Pitcairn)

canopies opened only at the last possible moment, and death-defying acts in which aerialists hung with their teeth on trapezes suspended from a plane's landing gear. After the circus went out of business because of restrictions imposed by the newly created Bureau of Aeronautics after the passage of the Air Commerce Act of 1926, Miller acquired a Standard J-1 biplane with a Hispano-Suiza engine. After obtaining a pilot's license from the Commerce Department, he became a racer, winning events in a Travel Air 2000 by secretly loading sandbags in his cockpit to get the weight he needed to maximize speed. He learned instrument flying under Howard Stark, with whom he saw Charles A. Lindbergh take off from Roosevelt Field on 20 May 1927 to begin his epic flight to Paris. Soon thereafter, Miller watched anxiously as an inept pilot nearly crashed while trying to land a Travel Air 6000 on Poughkeepsie's small airfield. After the plane came down safely, Miller drove one of the passengers to her home at Hyde Park. He later mused that her husband, FDR, may never have realized how close his wife Eleanor had come to dying in a crash. The pilot of the Travel Air 6000 was killed the next day attempting the same type of landing in North Carolina.

Miller found a job in 1929 with a company that had acquired a Standard D-25, specifically designed to meet new federal specifications and for barnstorming. Taking passengers on extremely short hops in an uninterrupted sequence to boost his earnings, he made 250 to 350 flights a day. He also operated an airport and a shop in which he repaired planes used by bootleggers. Hearing about the remarkable features of the Pitcairn PCA-2 autogiro, however, led him on a new path that made him America's foremost autogiro pilot. The idea of flying a PCA-2 across the United States and back appealed to him.

Business enterprises including the *Detroit News* were already acquiring autogiros from Pitcairn, but Miller was the first private citizen to order one. He found, however, that Pitcairn was more interested in meeting the needs of Amelia Earhart, who had ordered an autogiro later than he did, but received priority because of her celebrity status. Earhart, who had set an altitude record of 18,450 feet in a production model PCA-2 on 8 April 1931, planned to fly a specially modified autogiro across the United States and back under the sponsorship of the Beech Nut Company. When Miller cried foul, Pitcairn continued to prioritize Earhart's order but agreed to speed production of Miller's autogiro so he could enter it in the upcoming Omaha Air Races.

Miller, a fast learner, soloed in his new autogiro after only 1 hour of instruction. Meanwhile, Earhart's cross-country attempt was delayed by throat surgery, leaving her in a shaky condition. Miller knew that she was having problems learning how to fly autogiros on long trips. As the Omaha Air Races drew near, Miller became incensed about Earhart's delay in carrying out her plans and decided to fly to the Pacific Coast. Taking off at Willow Grove on 14 May 1931, he reached the San Diego Naval Air Station on 28 May, pausing only long enough at Omaha to give demonstration flights. At San Diego he impressed Admirals Joseph M. Reeves and William H. Standley by taking them aloft. Oil companies and film star Mary Pickford lavished attention on Miller, delaying his departure eastward until 21 June. He reached Willow Grove in 10 days, flying by way of El Paso and Kansas City.

Meanwhile, Earhart had begun her projected transcontinental round trip at Newark, New Jersey, on 28 May, the day Miller landed at San Diego. Flying across the Rockies, she reached Oakland on 6 June and was distressed to learn that Miller had gotten to California a week earlier. Trying to return to the East Coast before he arrived there, Earhart had an accident on 22 June at Abilene, Texas, failing to clear a fence on takeoff and crashing into cars parked near the airfield. She escaped unhurt but wrecked her aircraft and lost all hope of crossing the country eastward ahead of Miller, going the rest of the way by train. The Commerce Department compounded her embarrassment by reprimanding her for "carelessness and poor judgment" at Abilene. Despite her crash, she received more publicity than Miller, who felt unjustly treated because he, not Earhart, had been the first person to fly an autogiro across the continent and back.[6]

---

6 For varying accounts on Earhart's autogiro flights, see Smith, *Legacy of Wings*, pp. 176–189; Donald M. Goldstein and Katherine V. Dillon, *Amelia: A Life of the Aviation Legend* (paperback ed: Washington, DC: Brassey's, 1999), pp. 80–84; Susan Butler, *East to the Dawn: The Life of Amelia Earhart* (Reading, MA: Addison-Wesley, 1997), pp. 258–260; Doris L. Rich, *Amelia Earhart: A Biography* (Washington, DC: Smithsonian Institution Press, 1989), pp. 120–123; and Mary S. Lovell, *The Sound of Wings: The Life of Amelia Earhart* (New York: St. Martin's Press, 1989), pp. 169–172. Smith, Lovell, and Rich mention Miller by name, but the other works do not.

Miller continued to fly his autogiro at national and international air races. He became the first person to loop the loop with a rotorcraft and did so regularly, often adding a roll at the top of the loop to demonstrate the ruggedness of the PCA-2. In 1934, his engine stopped in midair, but his whirling rotor enabled him to glide to a safe landing in a cemetery. After the engine was repaired, he took off between rows of tombstones, demonstrating that autogiros could exceed the capabilities of other aircraft.

Miller became a pilot with United Air Lines in the mid-1930s, flying the world's first modern airliner, the Boeing 247. One of the planes he flew, a 247-D, now hangs in a gallery at the National Air and Space Museum, along with a DC-3 that he flew for Eastern Air Lines. In 1937, however, his love of rotorcraft led him to accept an offer from the Kellett Autogiro Company (KAC), which made autogiros under license from Pitcairn. Kellett designed a new model, the KD-1B, with no wings, ailerons, or elevators, but pilots were dubious about testing it for federal certification. Miller had no doubt that he could pass the tests, and he did so.

After Miller proved the airworthiness of the KD-1B, W. Wallace Kellett, President of KAC, started a campaign in Washington for a contract to use it carrying mail for short distances. A likely route would link Camden Central Airport and Philadelphia's 30th Street Post Office, which had been built in 1931. Its 100,000-square-foot roof had been specifically designed for rotorcraft landings. Postmaster General James A. Farley had already staged a demonstration in May 1935 at which an autogiro delivered a mailbag to Farley on the rooftop while a large crowd looked on, but special legislation was needed to implement regular operations. Early in 1938, while Kellett lobbied at the nation's capital for an enabling act, Miller gave senators and representatives demonstration flights to curry favor with them. Frank Dorsey, a Philadelphia congressman supported by George W. Lewis of the National Advisory Committee for Aeronautics, sponsored a bill providing that "the Postmaster General is authorized, under such appropriate rules and regulations as he may prescribe, to provide for and supervise experimental services in connection with the . . . transportation of mail by autogiro shuttle service between outlying airports and central city areas."[7] Congress passed it on 15 April 1938, and President Roosevelt signed it into law. Knowing about progress in autogiro and helicopter development in Nazi Germany, FDR secured an appropriation of $1,250,000 to implement the act and promote research in rotorcraft. Newspapers stressed the military potential of the KD-1B, which could climb 1,000 feet per minute, reach 100 miles per hour, take off in 50 feet, and land at nil speed.[8]

7 *United States Statutes at Large, LII* (1938), chapter 157, H. R. 7448 (15 April 1938), pp. 328–220.

8 Carroll V. Glines, *Airmail: How It All Began* (Blue Ridge Summit, PA: Tab Aero Books, 1990), p. 136; Smith, *Legacy of Wings*, pp. 238–240; "Autogiro as Auxiliary," *Aviation* 38, no. 2 (February 1939): 104; "Helicopters Stir Study," *New York Times* (5 March 1939): section 11, p. 12; and "U.S. to Pioneer in 'Giro Roof-Top Mail Service," *Aero Digest* 25 (December 1939): 32.

Kellett followed up on his victory by trying to persuade airlines to bid on mail contracts under the new law. Every carrier he approached, however, turned him down except TWA, which demanded demonstration flights from Midway Airport to the rooftop of Chicago's main post office before considering even a one-year agreement. Miller thought that TWA merely wanted publicity and warned Kellett that its offer was bogus. Nevertheless, Miller flew a KD-1B to the Windy City and made two flights between Midway and the postal facility. As he had expected, TWA rejected Kellett's proposal, leading Miller to urge approaching Eastern.[9]

The timing was excellent for the move because Eastern was undergoing a transition in ownership after being a subsidiary of GM. Rickenbacker had become general manager of Eastern in 1935 when the post was offered to him by one of GM's rising officials, Ernest Breech, but relations between the two men had soured. Earlier in 1938, Rickenbacker had won a battle to prevent a takeover of Eastern by one of Breech's allies, rental car magnate John Hertz. Investors including Laurance Rockefeller, who admired Rickenbacker as a national hero, acquired Eastern from GM and made Rickenbacker its Chief Executive Officer, with even more power than he had wielded before.

During the mid-1920s, when Rickenbacker was an automobile manufacturer, he had tried to create a small roadable airplane for the masses.[10] The autogiro probably reminded him of his former dream. Also working in favor of Kellett and Miller was finding an ally in Eastern's public relations director, Beverly Griffith, who, like Rickenbacker, was perpetually alert for new ideas. Griffith realized that carrying mail by ground transport 6 miles between Camden's airport and Philadelphia's 30th Street Post Office took about 45 minutes, whereas an autogiro could cover the distance in less than 10. Griffith arranged for Kellett and Miller to meet Rickenbacker, who knew about Miller's piloting skills and readily endorsed their proposal. When Rickenbacker asked Miller how much pay he wanted, Miller shocked Kellett by asking for twice the figure he was currently earning. Rickenbacker accepted without a murmur and even agreed to hire Miller as a pilot with Eastern if the shuttle service ended. Soon Eastern bid $3.86 per mile to the Post Office Department to approve a one-year experimental service creating the world's shortest airmail route.[11]

---

9 Miller, *Flying Stories*, pp. 71–73.

10 "Ace Promises Air Flivver," undated clipping from *New York American*; "New Plane Presages Ship for Home Use," clipping from *Los Angeles Sunday Times* (26 September 1926); and other clippings in Rickenbacker Scrapbook No. 2 (1920–1935), Eddie Rickenbacker Papers (hereafter cited as ERP), Auburn University Library, Department of Archives and Special Collections.

11 Rickenbacker, *Rickenbacker*, pp. 182–195; W. David Lewis, *A Hero in His Prime: Edward V. Rickenbacker and Eastern Air Lines, 1934–1941* (Dallas, TX: University of Texas at Dallas, 2002). On Griffith's background, see Robert J. Serling, *From the Captain to the Colonel: An Informal History of Eastern Airlines* (New York: Dial Press, 1988), pp. 116–156.

Eastern used National Air Mail Week, held in May 1938 to honor the 20th anniversary of the federal airmail system, to demonstrate the advantages of carrying mail with autogiros. On 19 May, Miller, piloting one of Kellett's ships, swooped down in a field adjacent to Washington's main post office with mail he had picked up at the Bethesda postal station. The 9-mile flight took only 7 minutes. Miller also carried mail between a temporary post office, a tent across the street from the Commerce Building, and Washington's Hoover Airport, an antiquated facility on the other side of the Potomac where the Pentagon building now stands. Crowds lining the street admired the ease with which Miller used it for takeoffs and landings. Prominent officials, including Harllee Branch, Vice Chairman of the newly created Civil Aeronautics Administration, witnessed the flights.[12]

Before regular service could begin, the rooftop of the 30th Street Post Office had to be converted into a small airport with a radio station, facilities for weather reporting, and maintenance equipment. The roof, 365 feet long by 285 feet wide, had an asphalt pavement leading to takeoff ramps on the north and south sides of the building. A heating system of pipes beneath the surface kept the pavement free of ice and snow in winter. To provide optimum service, Eastern ordered a modified Kellett KD-1B that had a radio, instrument flying ability, a compartment for mail bags, and a sliding cockpit canopy. Other features included a stationary tail with a hinged rudder, three rotor blades of 20-foot radius and a 1-foot chord, and a 220-horsepower Jacobs engine manufactured in nearby Pottsville, Pennsylvania, equipped with a fixed-pitch propeller. After these changes had been made, government regulations required Miller to put the ship through grueling tests including a dive that tore plywood surfaces and tips from its rotor blades and threw sandbags around the mail compartment so violently that its door burst open and spewed their contents into the air, causing the autogiro to seem to be trailing smoke. Otherwise, it came through the ordeal in good condition. Though badly shaken by gravity forces, Miller persuaded the federal inspector who witnessed the test to approve the aircraft for airmail operations. The way was now clear for Postmaster General Farley to formalize a contract with Eastern, on 21 March 1939, to operate Experimental Airmail Route 2001. Service was scheduled to begin on 7 July 1939 for a one-year period, after which Eastern could apply for permanent certification.[13]

---

12 "Miller Carries Mail Over D.C. By Autogyro," *Washington Post* (19 May 1939): 6; "EAL To Operate World's First Autogiro Air Mail Service," *The Great Silver Fleet News* (Eastern Air Lines company magazine, hereafter cited as *TGSFN*) 4, no. 3 (February–March 1939): 2–3.

13 *New York Times* (7 July 1940): 1–2; Miller, *Flying Stories*, pp. 71–74; "Eastern Airlines, Inc.—Autogiro Service—Philadelphia," *Decisions of the Civil Aeronautics Authority, February 1939 to July 1940*, vol. 1 (Washington, DC: Government Printing Office, 1941), Docket no. 403 (16 July 1940), p. 56; and "EAL To Operate World's First Autogiro Air Mail Service," *TGSFN*, cited above.

Test flights between the post office and airport a few days before the experiment officially began were hopeful, covering the distance between the two points in 6½ minutes.[14] On 6 July 1939, a ceremonial luncheon at the Bellevue-Stratford Hotel attracted almost 600 dignitaries. That evening Miller brought orchestra leader Andre Kostelanetz to the roof of the post office in the mail compartment of Eastern's new autogiro, marking the first passenger flight of its type. On the next day, 7 July, about 400 people (including Harold Pitcairn, Laurance Rockefeller, General Henry H. "Hap" Arnold, and Jacqueline Cochran) watched as First Assistant Postmaster General W. W. Howes helped load 42,000 pieces of mail into the autogiro. Just before it took off "a midget automobile with a 50-gallon gasoline tank shot across the roof to the side of the craft and began fueling it." At 3:16 p.m., Miller took to the sky with his cargo in the official start of operations, which were scheduled to cover five round trips per day. Fourteen minutes later, he was back from the Camden airport with return mailbags, showing the dispatch with which the transfer process had been accomplished. "Eastern Air Lines is proud to be in the forefront of this progressive, pioneering movement," Rickenbacker declared. "We firmly believe the success of this experimental service will mean the inauguration on a large scale of similar services throughout the United States." Receipts for the opening day amounted to more than $3,000 as 52,128 letters passed back and forth, bearing a first-day cover with an autogiro superimposed on a keystone, emblematic of Pennsylvania. Post office officials said that most of them were addressed to locations in the eastern parts of the United States, but that some were scheduled to cross the Atlantic and Pacific Oceans by air.[15]

Beaming, Eddie Rickenbacker holds a mailbag as Johnny Miller smiles from the cockpit of the Kellett KD-1B autogiro on 7 July 1939, the opening day of mail service between Camden Central Airport and Philadelphia's Market Street Post Office. (Courtesy of John M. Miller)

14 *New York Times* (4 July 1930): 30.

15 "'Gyro on Philadelphia-Camden Air Run; Opens City to Airport Taxi Flight Service," *New York Times* (8 July 1939): 17; "Airmail Via 'Windmill,'" *TGSFN* 4, no. 6 (September 1939): 13–15; Lawrence Davies, "'Giro Flies Mail a Year," *New York Times* (14 July 1940) section 10: 9; and Miller, *Flying Stories*, p. 75.

Johnny Miller takes off from the rooftop of the Market Street Post Office in the Kellett KD-1B autogiro carrying mail bound for Camden Central Airport, with the Philadelphia skyline in the background. (Courtesy of John M. Miller)

The takeoff and landing area on the roof of the post office was oriented in a north-south direction between two "penthouse structures" on the east and west sides of the building, creating what Miller called "the potential for an aerodynamic disaster with severe turbulence of varying kinds in winds blowing in different directions." Kostelanetz had landed with such a jolt the evening before that Miller tested wind conditions on the rooftop by dropping hundreds of sheets of toilet paper and noting how they swirled so that he could identify the best takeoff and landing routes. Although at times he encountered gusts of up to 60 miles per hour, his methodical approach enabled him to complete 2,634 round trips without accident. Mail from Camden reached the post office in an average of 8 minutes, a figure that remained about the same after a new airport that had been built on the south side of Philadelphia opened on 15 June 1940. Elevators took incoming mail to the lower floors of the post office for sorting and dispatching by ground transportation to destinations that were mostly in Pennsylvania but also in Maryland, Delaware, New Jersey, and Virginia.

To ensure predictable service, Eastern leased a second autogiro, and Miller trained John Lukens, a test pilot for Pitcairn, to fly it. Accounts of a crash that assertedly took place when an autogiro fell from the rooftop of the post office during a windstorm were false, stemming from an incident in which Lukens's aircraft tipped over on its side and part of its plywood rotor covering blew over the parapet.[16]

Gross revenue during the year ending 6 July 1940 was $60,633. Operating costs were estimated at $56,838 for 15,708 revenue miles, yielding a net profit of $3,795. Despite the small return on its investment, Eastern considered the service sufficiently promising to apply to

16 Miller, *Flying Stories*, p. 75; telephone interview by Lewis with Miller and subsequent e-mail correspondence; and "Eastern Air Lines, Inc.—Autogiro Service."

the Civil Aeronautics Authority (CAA) for permanent certification on 13 April 1940, leading to a public hearing on 22 May. Supporting Eastern's application, the Post Office Department urged that further experience operating autogiros and other types of rotorcraft would benefit not only the postal service but also contribute to national defense. The department's superintendent of airmail service strongly advocated granting Eastern's petition, as did the Postmaster General. A banker stated that Eastern's autogiro operations had greatly expedited delivery of funds to various destinations and pointed out that night operations for which the 30th Street Post Office had been equipped, and which Eastern was ready to deliver, would further increase the utility of the service. Eastern sought publicity by having Miller give demonstration flights in one of its autogiros at the New York World's Fair.

On 14 July 1940, eight days after operations had ended and service was in abeyance pending the CAA's decision, the *New York Times* stated that Miller and Lukens had completed 2,634 flights and carried out 85.8 percent of all scheduled trips, exceeding expectations by 10 percent. It pointed out that high winds had not kept the autogiros from operating and quoted Miller that the service could "be maintained at any time the regular airlines are able to operate." Miller advocated testing improvements that would cut rotor-blade vibration to "about 25 percent of the present amount" and pitch controls that would permit an autogiro to take off and land vertically.[17]

On 16 July, however, the CAA turned down Eastern's application, stating that it was "not clear . . . that the development of rotor aircraft operations will be best and most efficiently fostered by the present establishment on a permanent basis of the privilege of carrying mail by autogiro." Though willing to give further temporary permission to Eastern to continue its autogiro service, the CAA denied the permanent certification it sought. Efforts in Congress to pass a law permitting the extension of operations for two years were fruitless, and the experiment ended.[18]

Kellett, who was now developing a twin-rotor helicopter, asked Miller to remain as a test pilot, but Miller refused because Kellett would not modify a feature he considered unsafe. True to his word, Rickenbacker hired Miller as a pilot for Eastern, and he remained with the company for 25 years. On several occasions his excellent airmanship prevented serious and potentially fatal accidents. He was not afraid to confront Rickenbacker when he thought he was wrong and told him bluntly that acquiring the Lockheed L-88 Electra turboprop in 1955 was a mistake. Events proved that Miller was correct. After Eastern began service with the L-88 on 12 January 1959, the plane had a series of crashes with other carriers, and federal regulators imposed severe speed restrictions that hurt Eastern badly in trying to compete with Douglas DC-8 and Convair CV-880 jetliners flown by its archrival, Delta Air Lines.[19]

17 *New York Times* (14 July 1940), section 10: 9.

18 "Eastern Air Lines, Inc.—Autogiro Service"; and Davies, "'Giro Flies Mail a Year."

19 W. David Lewis and Wesley Phillips Newton, *Delta: The History of an Airline* (Athens: University of Georgia Press, 1979), pp. 271, 280.

Rickenbacker died during a visit to Switzerland in 1973, but Miller had many years to live. After he retired from Eastern in 1963, with 35,000 hours to his credit, he remained active in aviation and became a charter member and President of a newly founded organization, the United Flying Octogenarians, in 1987. Still honing his piloting skills in his 90s, and particularly devoted to Beech Bonanzas and Barons, he became a "Nonagenarian Nomad." In 2001, taking off at Poughkeepsie, he observed the 70th anniversary of his round-trip transcontinental autogiro venture by flying his V35A Bonanza across the country and back. He continues to believe in the future of the autogiro because of its inherent safety. He keeps his weight down; takes frequent "brisk walks up and down hills"; avoids alcohol, tobacco, coffee, and drugs; stays in excellent physical condition; and intends to continue living a long time. As he says, "Flying is a youth preservative—if you live through it."[20]

Miller remains an outspoken champion of the autogiro, believing that such aircraft have a potentially important role to play in the future of aviation. The merits of the aircraft hit home to an engineer who heard the paper on which this essay is based at a symposium sponsored by NASA on 5 November 2003 in observance of the Centennial of Powered Heavier-Than-Air Flight. Being previously unacquainted with the autogiro and its history, he spoke to me excitedly about the continuing potential of Cierva's remarkable invention to meet current needs for short-range aircraft that cost relatively little to build and could serve a number of useful functions that do not require the power and size of a helicopter.

I have long since lost my boyhood faith in the future of the roadable autogiro. It exemplified a democratic, egalitarian dream, widely prevalent in an age of technological exuberance, of mass aircraft ownership among citizens possessing their own flying machines and using them as freely and regularly as automobiles. Henry Ford, Eddie Rickenbacker, Fred Weick (designer of an easy-to-fly plane called the Ercoupe), and Eugene Vidal (head of the Bureau of Air Commerce under FDR) were only a few of the many persons who subscribed to that idea in an era when intrepid aviators shattered speed and distance records, radio waves spanned vast distances without the benefit of wires, skyscrapers reached unprecedented heights with hidden steel frames, and television sets in every home were just around the corner.[21]

---

20 Miller, *Flying Stories*, pp. 76–117; and telephone interview of Miller by Lewis.

21 William M. Leary, "Henry Ford and Aeronautics During the 1920s," in *Aviation's Golden Age: Portraits from the 1920s and 1930s*, ed. William M. Leary (Iowa City: University of Iowa Press, 1989), p. 13; Fred E. Weick and James R. Hansen, *From the Ground Up: The Autobiography of an Aeronautical Engineer* (Washington, DC: Smithsonian Institution Press, 1988), pp. 159–191, 215–238; and Joseph J. Corn, *Winged Gospel: America's Romance with Aviation, 1900–1950* (New York: Oxford University Press, 1983), pp. 96–97, 110. On the technological exuberance of the 1920s and 1930s, see also Ann Douglas, *Terrible Honesty: Mongrel Manhattan in the 1920s* (New York: Farrar, Straus, and Giroux, 1995), p. 434; and Carol Willis, "Skyscraper Utopias: Visionary Urbanism in the 1920s," in *Imagining Tomorrow: History, Technology, and the American Future*, ed. Joseph J. Corn (Cambridge, MA: MIT Press, 1986), pp. 164–187.

But skeptics abounded. Among them was Donald Douglas, who argued that visions of "aircraft for everyone" were impractical because it was hard enough for many people simply to drive a car, let alone become qualified as pilots. Among other obstacles, Douglas pointed out that having untold numbers of privately owned aircraft would pose insurmountable problems of regulation and control.[22] It seems fortunate today that the skies above densely populated areas are not darkened by swarms of airborne vehicles that would tax air traffic controllers beyond human endurance trying to keep up with the blips on their computer screens. It is not hard to imagine the consequences that could ensue if roadable autogiros suddenly swooped down and landed with little or no warning on interstate highways amid automobiles, buses, and 16-wheelers, or spread their wings and took off at the whim of their operators. It is still reasonable to believe, however, that autogiros can play a more limited but nevertheless valuable role, not merely as aircraft for enthusiasts but as flying machines fulfilling many more practical purposes. Should this vision become a reality, Miller and Rickenbacker will deserve prominent places among the prophets of a mode of air transport that many people know all too little about, thinking that helicopters are the only rotorcraft that can play a meaningful role in everyday life.

ENDNOTES

Preparation of this essay was greatly aided by Mr. John M. Miller, from Poughkeepsie, New York, who generously contributed material in letters, e-mail messages, and telephone calls, and made corrections to a final draft to ensure its accuracy. The author wishes to thank Mr. Miller for his cooperation and help. He also thanks his colleague and department chairman, Dr. William F. Trimble, for providing information, perspective, and insight about Pitcairn, Kellett, and autogiros based on his authorship of *High Frontier: A History of Aeronautics in Pennsylvania* (Pittsburgh: University of Pittsburgh Press, 1982). See particularly pp. 173–181 of Trimble's work, from which this essay has benefited at various points.

22 Douglas's views are quoted in an undated article by Chester Hanson in an undated newspaper clipping in Rickenbacker Scrapbook no. 2 (1920–1935), ERP. For a valuable work about the dependence of technological innovations on the development of controls in evolving infrastructures, see Miriam R. Levin, ed., *Cultures of Control* (Amsterdam: Harwood Academic Publishers, 2000).

5

Donald Douglas

From Aeronautics to Aerospace

ROGER BILSTEIN

ON 23 MAY 1938, DONALD WILLS DOUGLAS MADE THE COVER OF *TIME* MAGAZINE, WHICH DESCRIBED HIM AS ONE OF THE LEADING YOUNG INDUSTRIALISTS OF HIS DAY AND BUILDER OF A NEW generation of modern, four-engine commercial transports. The company had already made its mark as supplier of military designs and commercial aircraft like the outstanding DC-3, but the new airliner substantially advanced the DC (for Douglas Commercial) family of transports as globally recognized icons of modern air transportation. The following decades of World War and Cold War transformed the Douglas Aircraft Company into an industrial leader. In 1961, the noted British aviation expert, Peter Brooks, wrote a classic history of air transport development, published as *The Modern Airliner.* He titled one chapter as "The DC-4 Generation," a summary of technical and operational developments of the post-WWII revolution in air travel. The choice of the DC-4 as exemplifying this era further

Donald W. Douglas, c. 1948–1950. (McDonnell Douglas photo number C62-11)

cemented the reputation of Douglas Commercial transports as the ultimate symbols of air transportation.[1]

Donald Douglas's career demonstrated admirable business acumen, but the fortuitous role of government funding represented a continuing thread in the history of his company. Clearly, military contracts stemming from cold and hot wars represented a major element in the success of Douglas Aircraft, and the factor of federal funds played a crucial role in the evolution of the DC-3. Success in highly publicized competitive flights as well as equally publicized world events like the Berlin Airlift gave Douglas products worldwide recognition. In an era of intense competition, his company also delivered high-quality aircraft to airlines during several decades of rapid growth of air travel and, in the 1950s, essentially dominated the global market for large airliners. As an aerospace giant, Douglas also shared in the billion-dollar space program. But Douglas Aircraft

1 *Time* magazine cover and article, "DC-4," *Time* 31 (23 May 1938): 33–38; Peter Brooks, *The Modern Airliner: Its Origins and Development* (London: Putnam, 1961), pp. 67–68, 89–111.

became overwhelmed by its success; Douglas's personal life and family loyalties also contributed to the decline of the company he founded.

Donald Wills Douglas was born in Brooklyn, New York, in 1892. His father's job in the banking business afforded a comfortable, secure upbringing. In 1908, while visiting Washington, DC as a petitioner to enter the U.S. Naval Academy, he observed flight trials of the Wright biplane at nearby Ft. Myer, Virginia. Although Douglas entered the academy with the class of 1909, his growing interest in aviation led to his departure in 1912 in order to enroll at the Massachusetts Institute of Technology (MIT), where Jerome Hunsaker had begun to offer coursework embracing aeronautics. Douglas graduated with a degree in mechanical engineering in 1914 and spent the next year as a research assistant for Hunsaker, one of the leading aeronautical experts of the era. Douglas's degree and research in advanced aeronautics at MIT made him one of the few academically trained professionals in the young aviation manufacturing industry.

In 1915, Hunsaker helped him obtain a job with the Glenn Martin Company in California, followed by a challenging position with the aviation office of the U.S. Signal Corps from 1916 to 1918. Douglas then returned to a reorganized Martin company in Cleveland, Ohio, where he became chief engineer for the MB-2 bomber, a twin-engine biplane recognized as an outstanding design of the early postwar era. In 1920, with five years of experience, Douglas headed back to California to establish his own aircraft company. He was 28 years old.[2]

THE DOUGLAS WORLD CRUISER

Shrewdly, Douglas chose the West Coast to take advantage of favorable geographic factors and sources of investment. A robust economy and burgeoning petroleum industry provided sources of risk capital at several points in his young company's evolution. Nonetheless, its beginnings were modest enough, with offices at the rear of a building housing a barber shop. On occasion, employees' wives trekked to a nearby shed where they helped stitch fabric to the structural skeletons of airplanes under construction.[3] Douglas also proved to be a shrewd judge of aviation engineering talent, hiring a number

2 Wayne Biddle, *Barons of the Sky* (New York: Simon & Schuster, 1991), pp. 81–86, 96, 102–03; and Wilbur H. Morrison, *Donald W. Douglas: A Heart with Wings* (Ames: Iowa State University Press, 1991), pp. 3–16. A former public relations manager at Douglas Aircraft, Morrison interviewed Douglas and other corporate personnel. Douglas was married in 1916; his family grew to include four sons and one daughter.

3 Morrison, *Douglas*, pp. 44, 130; Douglas J. Ingells, *The McDonnell Douglas Story* (Fallbrook, CA: Aero Publishers, 1979), pp. 18–19; and William Glen Cunningham, *The Aircraft Industry: A Study in Industrial Location* (Los Angeles: Lorrin L. Morrison, 1951), pp. 3–42. Ingells was a veteran aviation journalist whose career began in WWII; during his career he interviewed Douglas and many other industry leaders.

of technically apt individuals even though they often lacked the college credentials that he himself possessed. College trained or not, aviation professionals tended to be a colorful lot. "In those days," Douglas recalled, "the most conservative type of engineers wouldn't consider going into aviation. They thought it was rather poor affair. I suppose a person had to be a little out of the ordinary to go into it . . . ." One of these professionals was J. H. Kindelberger, who, typical of many nonacademic professionals, had literally grown up with the aviation industry. Kindelberger had picked up one year of college prior to wartime service with the Corps of Engineers before joining the Martin Company's operations in Cleveland, where he became a close friend of Donald Douglas. Five years after leaving Martin, Douglas contacted Kindelberger and hired him as chief engineer. Kindelberger represented a bridge between individuals like himself, who achieved success through an apprentice-like work career, and cadres of younger, professionally trained individuals with degrees from formal college curricula. When Kindelberger joined Douglas in 1925, he went in with a newly minted engineer from MIT, Arthur Raymond. The latter's starting salary, Kindelberger recalled, amounted to a grand total of 25 cents per hour.[4]

Considering their collective achievements, many of the personnel who worked for Donald Douglas at one time or another were certainly out of the ordinary. Art Raymond, for example, replaced Donald Hall, a stress analyst, who went on to the Ryan Corporation in San Diego, where he presided over the design of the *Spirit of St. Louis* flown across the Atlantic by Lindbergh in 1927. Jack Northrop and Jerry Vultee also worked at Douglas Aircraft before moving on to organize their own aviation companies. Kindelberger departed in 1934 to head up North American Aviation. Another early stalwart, Leland Atwood, played a key role in the evolution of the original DC-1 series before he joined Kindelberger at North American and eventually became its President.[5]

During the early 1920s, the experiences of the Douglas Aircraft Company often read like cliffhanger episodes from movie serials like *The Perils of Pauline*, surviving from one crisis to the next. Even though the *Cloudster*, a custom-designed biplane built to set a transcontinental record, turned back due to mechanical problems, its ability to lift its own weight won favorable notoriety for Douglas. It became the basis for Navy contracts to build torpedo bombers, the DT (Douglas torpedo) series. More importantly, the DT design became the first airplane to fly around the world.

After WWI, declining military budgets often led fliers to promote long-distance flights in an effort to generate headlines and garner funding for military air arms. During 1923, a coterie of aviation officers in the U.S. Army convinced the Secretary of War to

---

4 Interview of Donald Douglas (1959) and interview of J. H. Kindelberger in the Arnold Collection, Oral History Collection of Columbia University, NY (cited hereafter as Columbia-OHC).

5 Interview of Arthur Raymond (1964) in the Special Collections Room, Honnold Library, Oral History Collection, Claremont Graduate School, CA (cited hereafter as Claremont-OHC). I wish to acknowledge the late John B. Rae, who alerted me to this collection and supplied transcripts.

The Douglas World Cruiser made headline news. (McDonnell Douglas photo number SM-7574)

endorse an ambitious bid to fly the first airplanes around the world. The project stirred a flurry of national interest, intensified by similar announcements from Britain, France, and others. In 1924, aviators in a quartet of United States Air Service biplanes set out; two of the original planes (along with a replacement for a third) made it. As competing nations dropped out, the Americans became the focus of international attention as they progressed across oceans, mountain ranges, and deserts. Their journey became an impressive demonstration of gritty determination as well as considerable planning and organization by federal entities—a legacy of federal largesse that continued to benefit Douglas Aircraft and the young aviation industry.

In the case of the world flight, a selection board with special funding from the War Department fortunately settled on the Douglas DT-2, a two-seat, open-cockpit, biplane torpedo bomber in production for the U.S. Navy. The plane received numerous modifications including beefed-up landing gear struts that allowed the use of wheels from airstrips or for attachment of pontoons for operations over water. With all the changes, the planes received a special appellation—DWC, for Douglas World Cruiser. The success of the Douglas planes said a great deal about their inherent durability. But a great deal also rested on the extraordinary diplomatic efforts required for clearances to use foreign airfields and harbors. There was considerable liaison with the Secretary of the Navy, since various vessels (as well as Coast Guard ships) were strung out along the path of the flight to support maintenance and repair as well as provide weather forecasting. Nonetheless, the airplanes themselves became headline news, launching Douglas Aircraft as an internationally recognized designer and builder. As Donald Douglas reminisced from the vantage point of 1959, the odyssey of the Douglas World Cruisers was "the best thing that occurred for us."[6]

---

6 For a detailed summary of military projects, see Rene Francillon, *McDonnell Douglas Aircraft Since 1920* (London: Putnam, 1979); on the Douglas World Cruiser flight, see Carroll V. Glines, *Around the World in 175 Days: The First Round-the-World Flight* (Washington, DC: Smithsonian Press, 2001); and the Douglas quote was cited in Biddle, *Barons*, p. 136.

AN ICON OF MODERNITY

Through the 1920s, Douglas and other manufacturers characteristically built biplanes that continued to utilize wooden spars and fabric covering. Even monoplanes used these materials, including the Fokker Company, whose trimotor designs with fixed-gear and dated appearance equipped many airlines in the United States and overseas. The move to modern, metal construction and monoplane airliners occurred rather suddenly. During 1931, Boeing's development of the twin-engine 247, with metal construction, retractable landing gear, and other advanced features, made competing airlines nervous. Moreover, United Air Lines, as part of a holding company that also controlled Boeing, enjoyed a monopoly on existing deliveries for the new Boeing transport. The same year, TWA sent out a two-page prospectus to manufacturers, specifying a new plane to exceed the Boeing's expected performance. Donald Douglas and his design group decided to go for the TWA job and crafted a proposal for the DC-1 (Douglas Commercial No. 1).

During the 1920s, Arthur Raymond recalled that Donald Douglas took an active role in stress analysis as well as overall design. By the time work began on the DC-1, the final engineering and design decisions were handled by a project team; Donald Douglas spent most of his time dealing with overall corporate issues. Company operations at Santa Monica focused mostly on civil aviation; a second facility at El Segundo primarily involved military contracts. The DC-1 took shape in Santa Monica, where Ed Burton, a long-time Douglas employee, did the drafting work, assisted by Leland Atwood, who focused on stress analysis. Jack Northrop had been working on military projects at El Segundo, and some of his expertise involving all metal, monocoque structures leavened the design evolution of the new Douglas transport.[7] Arthur Raymond was also one of the key team members. When TWA needed a face-to-face conference in New York City to verify the bold performance estimates by Douglas, Raymond and a colleague boarded a Pullman train to make the trip back east. As Raymond later recalled, the train trip gave them an extra couple of days to work out crucial last-minute details. Moreover, as he admitted years later, "air travel across the continent was then rather rudimentary, and we didn't have much confidence in it." But Raymond decided to fly back to California—at least as far as Kansas City.[8]

By the time the first DC-1 got airborne in 1933, Douglas already realized that a modified version could easily add two more seats (accommodating 14 passengers compared to 10 in the Boeing 247), offer improved range, and utilize newer engines that promised a superior cruising speed of 196 mph. Consequently, the DC-2 went into production,

---

7 For the origins of the DC-1, see Arthur Pearcy, *DC-3* (New York: Ballantine, 1975), pp. 28–51; and Raymond interview, Claremont-OHC.

8 Arthur Raymond, *Who? Me? Autobiography of Arthur E. Raymond* (privately printed: 1974), chapter 2, part 3, pp. 1–2. Copy in the Hatfield Collection, Rare Books, Museum of Flight, Seattle, WA.

entered service in 1934, and became a phenomenal success. International acclaim followed, especially after the Dutch airline, KLM, entered the famous MacRobertson Air Race from London to Australia in the same year. Named after an Australian firm that made a fortune from chocolate sweets and other candies, the international contest captured worldwide headlines. Against some 20 international competitors, Britain intended to win, building a highly streamlined, custom-built racer, the *de Havilland D.H. 88 Comet*, and fielded a trio of them in the contest. A Boeing 247 flown by the flamboyant American flier Roscoe Turner also entered the lists. KLM's DC-2 entered with no special modifications—a standard transport with a two-man crew, three paying passengers, and a consignment of mail. Although a D.H. 88 eventually won, the surprising DC-2 nipped at its heels all the way and nearly took the prize. British newspapers published astonished articles reporting that not even a Royal Air Force plane could have done as well as the American passenger transport. "It is almost incredible," admitted one writer, "but it is true."[9] The DC-2 became the airliner of choice for airlines overseas as well as in the United States, and Douglas Aircraft began its long run of air transport deliveries.

But C. R. Smith, impresario of American Airlines, almost immediately saw the need for an even better DC-2 type, especially a larger one that could offer sleeping berths on American's transcontinental routes. With his chief engineer, William Littlewood, the two men sketched diagrams for wider fuselage with various schemes for additional seats, estimated performance using bigger engines, and developed a list of additional improvements for a plane to be known as the DC-3. In the autumn of 1934, Smith decided to call Douglas Aircraft. During the course of a legendary 2-hour phone conversation between Smith in Chicago and Donald Douglas in Santa Monica, Smith finally convinced a reluctant Douglas to go ahead with the design. In the process, Smith promised Donald Douglas an order of some 20 aircraft worth an estimated $4 million. Hanging up the phone, Smith faced the reality that American Airlines had neither the cash nor a feasible line of credit to pay for the planes he had just ordered. He quickly made arrangements for an emergency trip to Washington, DC in order to corner an old friend and convince him to bestow several million dollars on American Airlines.

Smith's friend, Jesse Jones, headed the Roosevelt administration's Reconstruction Finance Corporation (RFC), a powerful New Deal program to make emergency loans as a means to stimulate qualifying industries and sustain employment. Both men hailed from a cohort of Texans whose tribal rituals embraced dominoes, poker, tall tales, good whiskey, and the art of the deal. Smith persuasively argued the case for his coast-to-coast

9 Terry Gwynn-Jones, *Farther and Faster: Aviation's Adventuring Years, 1909–1939* (Washington, DC: Smithsonian Press, 1991), pp. 252–259.

airline, whose new planes would stimulate the airframe industry on the Pacific coast and the engine builders on the Atlantic coast. When Smith departed Washington, DC, he took a $4.5-million RFC loan with him. Thus, the DC-3 took wing with crucial dollars allocated by the government.[10]

Another key federal legacy involved the Wright Cyclone and Pratt & Whitney Twin Wasp engines that powered subsequent versions of DC-2 planes and legions of DC-3 transports. Both series originated in programs for Navy and Army requirements and also benefited from NACA studies that included refined engineering for the NACA cowling. With the contract from American Airlines in hand, Arthur Raymond recalled that Douglas management debated about acquiring tools and jigs to build 25 or 50 airplanes. "We decided to be bold and to tool up for 50." Eventually, some 350 twin-engine aircraft came off the original jigs; wartime requirements raised production into the thousands.[11]

Although the DC-2 and the DC-3 looked alike, they became very different airplanes. Not only had the DC-3 originated with American Airlines, but the airline also influenced the majority of engineering and design details. Bill Littlewood, American's chief engineer, literally moved into the Douglas design offices in Santa Monica. As the principal engineer at Douglas by this time, Raymond nonetheless recognized Littlewood's expertise. "I gave them [American Airlines] almost a free hand in establishing the dimensions of the cabin and deciding what went in it, and in the cockpit layout," said Raymond. The creative teamwork between the airline and the builder became a major factor in the success of the DC-3.[12] The Boeing 247 transport quickly faded, and Boeing spent the next several decades trying to recapture the technological momentum in civil airliners that now passed to Douglas Aircraft. By the decade's end, the safety of airline travel had improved so much that insurance companies finally began to offer air travel insurance at the same rates as coverage for rail travel. This example of equity in travel reliability and safety prompted the Air Transport Association (ATA) to commission full-page advertisements in such national periodicals as the *Saturday Evening Post*, trumpeting the significance of air travel achievement. Again, because the principal equipment of airlines representing the membership of the ATA consisted of DC-3 transports—the image used in the ads—there could be little argument about the implicitly high level of the plane's design and engineering.[13]

---

10 Pearcy, *DC-3*, p. 70; Roger Bilstein, "C. R. Smith: An American Original," in *Airline Executives and Federal Regulation: Case Studies in American Enterprise from the Airmail Era to the Dawn of the Jet Age*, ed. W. David Lewis (Columbus: Ohio State University Press, 2000), pp. 86–87.

11 Pearcy, *DC-3*, pp. 71, 75; Roger Bilstein, *Testing Aircraft, Exploring Space: An Illustrated History of NACA and NASA* (Washington, DC: Smithsonian Press, 2003), pp. 12, 27–28; Bilstein, *The Enterprise of Flight: The American Aviation and Aerospace Industry* (Washington, DC: Smithsonian Press, 2001), pp. 29–30; and Raymond interview, Claremont-OHC.

12 Raymond, *Who? Me?*, chapter 2, part 4, pp. 1–2.

13 Air Transport Association advertisement, *Saturday Evening Post* 212 (16 November 1940): 83.

Loading the DC-3, the first commercial airliner to make money handling passengers. (McDonnell Douglas photo)

While Douglas produced high-quality, enthusiastic brochures about its military planes and carried on its own publicity campaign for commercial transports,[14] the company invariably benefited from even more visible programs funded by airlines, especially when new planes or major new routes were introduced. American Airlines inaugurated its DC-3 coast-to-coast "Skysleeper" service in the fall of 1936; passengers boarded in New York at 5:10 p.m. and stepped off in Los Angeles the next morning, a trip that consumed 16 to 18 hours with three to four scheduled stops depending on the route. The flight carried the designation of "American Mercury" service in the airline's timetable, and attendant press releases made the most of the new plane and the new flight connections. The idolized child movie actress Shirley Temple appeared to receive the first ticket, and news cameramen snapped dozens of pictures of the diminutive child star, American Airlines personnel, and, of course, the imposing Douglas DC-3. During 1939 and 1940, after Braniff

14 Brochure, "Douglas O-38" (no date). Copy in Hatfield Collection, Rare Books, Museum of Flight, Seattle, WA. Douglas marketed the O-38 biplane (circa 1931) as an observation plane and light bomber, stating that "prices and deliveries will be quoted upon request," and noted its cable address, "Douglasair." Between 1927 and 1933, Douglas reported $15.83 million in sales, with 91 percent to military sources. Boeing reported $17.37 million in sales, with 59 percent in military sales. Bilstein, *Enterprise*, p. 31.

purchased a quartet of new DC-3 transports, the airline bombarded news desks with hyperbolic press releases ("luxurious . . . latest in safety and comfort refinements . . . artistic color schemes . . . most modern and beautiful of their type in use today") and placed full-page spreads in strategic magazines. Nonetheless, the DC-3 essentially sold itself on its inherent qualities. Its performance and passenger capacity made its seat-mile costs as much as one-third to one-half lower than its contemporaries. It became the first commercial airliner to make money for its owners by hauling passengers rather than subsidized cargoes of mail. By 1938, DC-3s carried 95 percent of scheduled air passengers in the United States, and 30 major airlines overseas operated the plane. By 1939, the redoubtable Douglas was transporting an estimated 90 percent of the world's airline traffic.[15]

In the course of its phenomenal success, the DC-3 design became one of the icons of 20th-century progress. Its silhouette appeared with raincoats, automotive products, and other consumer goods as an equation of high quality. It was no accident that advertisements for cigarettes, stressing that modern men and women in a modern world smoked Pall Malls, chose an airport as the locale and sited state-of-the-art DC-3 transports in the background.[16] The airliner not only embodied a synthesis of the best of the prewar era's engineering, but also represented the aerodynamic, streamlined motif that influenced art deco styling and industrial design of later years. Designers like Raymond Loewy, Walter Dorwin Teague, and others paid homage to modern aircraft like the DC-3. Teague declared that "I do not know where in modern design to look for an example of rhythm of line composed more perfectly than in these transport planes." During the era of the 1930s and early 1940s, the form of the DC-3 appeared in lamps, clocks, and other appliances; its aerodynamic lines seemed to surface everywhere in industrial design.[17]

DOUGLAS AIRCRAFT AND THE MILITARY

An even more revolutionary plane followed on the heels of the DC-3, and it is important to remember that its ancestry originated before the war; American airliners forged ahead of foreign counterparts well before the appearance of modern designs from Euro-

15 Pearcy, *DC-3*, p. 85; Kenneth Munson, *Airliners Between the Wars, 1919–1939* (New York: Macmillan, 1972), pp. 164–165; and Braniff press releases from Braniff Files, Box 27, Folder 8, Aviation Collection, Special Collections, McDermott Library, University of Texas—Dallas.

16 Roger Bilstein, "Air Travel and the Traveling Public: The American Experience, 1920–1970," in *From Airships to Airbus: The History of Civil and Commercial Aviation*, vol. 2, ed. William Trimble (Washington, DC: Smithsonian Press, 1995), pp. 97–98; and Pall Mall cigarette ad from the *Saturday Evening Post* 212 (16 September 1940).

17 Phil Patton, *Made in USA: The Secret Histories of the Things That Made America* (New York: Grove Weidenfeld, 1992), pp. 218–222, 234–242 (Teague quoted on p. 220).

pean manufacturers after 1945. In 1935, a consortium of major carriers, led by United Air Lines, collated their requirements for an advanced airliner and negotiated a contract with Douglas Aircraft to build a very large, long-range transport with four engines. It carried the designation of DC-4E and made big headlines during an inaugural tour in 1938–1939. But, as Arthur Raymond dryly observed, "It was an example of design by committee [and] took so long to build it that technology had advanced in the meantime." Consequently, Douglas engineers "took a fresh sheet of paper" and came up with the classic DC-4. Tricycle landing gear had appeared on some planes before WWII, but the DC-4E, the DC-4, and their successors made this feature standard on postwar American airliners. Boarding various "tail draggers" like the DC-3 required passengers to scramble up a steep aisle and necessitated some alacrity to settle into their seats; planes with tricycle gear allowed customers to board and stow their gear with everything comfortably horizontal. For ground crews, tasks of handling luggage, cargo, and various service requirements also went more smoothly. Pilots appreciated a steerable nose gear that enhanced taxiing around on the ground, the cockpit position afforded pilots a far better view of busy airport activities, and tricycle gear automatically positioned wings for the big airliners at a more efficient attitude to enhance takeoff performance. The Douglas DC-4 also introduced power-boosted controls, along with advanced engineering details like flush-riveted skin. With Pratt & Whitney R-2000 Twin Wasp engines, each generating some 1,450 horsepower, the DC-4 carried an impressive load of 42 (or more) passengers and could cruise at more than 200 mph over ranges of more than 2,000 miles. And its tubular fuselage and stoutly engineered wings lent themselves to stretching the basic DC-4 into bigger and better airplanes.[18] But when the DC-4 reached flying status early in 1942, military requirements took precedence, and they went directly into service with the USAAF as C-54 transports.

Meanwhile, military aviation technology forged ahead in the 1930s, carried along by the tide of advances in engines, structures, and ancillary technologies. While fabric-covered biplanes remained in Air Corps service, the Soviet Union produced several large, all-metal bombers. For some American officers and military planners, the surge in Soviet progress came as a clarion call to catch up—similar to reactions after the Soviet Sputnik went into orbit in 1957. During the 1930s, the Army Air Corps began to formulate strategics that called for long-range bombers for coastal defense as well as a means to strike at enemy industries needed to wage war. All of this presumed that there would be a large intercontinental bomber to serve as a strategic deterrent; as early as 1934, the general staff authorized secret feasibility studies. These in turn led to authorization to build prototypes from designs submitted by Boeing, Douglas, and Sikorsky. The latter's candidate was later canceled; the Boeing XB-15 became airborne in 1937; and the Douglas entry—the XB-19—dwarfed it.

18 Carl Solberg, *Conquest of the Skies: A History of Commercial Aviation in America* (Boston: Little, Brown, and Company, 1979), pp. 313–314; and Raymond interview, Claremont-OHC.

The Air Corps wanted to learn about manufacturing techniques—oversized tooling, assembly hardware, a dramatic multiplication of parts numbers, and management—as well as operational requirements for such large-scale aircraft. In addition to its size, the XB-19 became the first large bomber to employ a tricycle gear (in contrast to the tail-wheeled Boeing design) requiring Douglas to modify smaller planes to understand its dynamics. After taking to the air in June 1941, the XB-19 fortuitously alerted planners to problems in the Wright Cyclone twin-row engines destined for installation on the B-29.[19]

Despite headaches with the XB-19, Douglas Aircraft made the most of it. A lone example built expressly for the Air Corps with federal funding, the XB-19 became skillfully exploited by the Douglas public relations department for upgrading the company's image as a major military contractor. Hoping to avoid the European war, many viewed the huge plane as a major deterrent. Douglas Aircraft touted its role as producer of the world's largest bomber, taking out full-page advertisements in national magazines such as *Collier's.* The hulking aircraft itself became a popular news subject for tabloids as well as mainstream periodicals. With cooperation from the Douglas Company's corporate offices, approving images of the plane even showed up in movie cartoons produced by Warner Brothers in 1942 and in MGM releases as late as 1943 and 1944.[20]

The advent of WWII made Douglas Aircraft into an industrial behemoth. During the 1930s, the company began delivering monoplane dive-bombers to the U.S. Navy, leading to the SBD Dauntless, the mainstay of the Navy's bomb squadrons. During the spring of 1942, its performance during the Battle of the Coral Sea and the Battle of Midway—inflicting crippling blows against Japanese naval air power and blunting the momentum of Japanese aggression in the Pacific—made it a wartime classic. Its ongoing role in the sprawling Pacific theater of operations kept the name of Douglas Aircraft in the national news. A different plane, the twin-engine A-20 light bomber, also played a significant role in both Pacific and European theaters; although the "Havoc," as the Air Force called it, is often overlooked. Early versions of the plane went into service with the air forces of France and Britain during the late 1930s—examples of an export business that constituted an important element of Douglas Aircraft sales before and after WWII. During the

---

19 Bill Yenne, "Experimental Bomber, Long Range: Boeing XB-15 and Douglas XB-19," *International Air Power Review* 5 (summer 2002): 169–170; Douglas Ingells, *They Tamed the Sky* (New York: Appleton-Century, 1946), p. 138; Jacob Vander Meulen, *Building the B-29* (Washington, DC: 1995), pp. 96–97. As a matter of reference, the wingspan for the XB-15: 150 ft.; the B-17: 103 ft.; the XB-19: 212 ft.; and the B-29: 141 ft. Eventually, four-engine bombers like the B-17, B-24, and B-29 entered mass production. The sole XB-19 prototype was used for wartime public relations and occasional transport duties; it was broken up in June 1949. Intensive engineering studies and modification of the Wright Cyclone R-3350 engines for the B-29 continued, but they remained troublesome engines throughout the war. See Lloyd Jones, *U.S. Bombers* (Fallbrook, CA: Aero Publishers, 1974), pp. 55–58.

20 The Douglas B-19 advertisement appeared in *Collier's* 107 (15 March 1941). Commentary on cartoons from E. O. Costello, comp., "Warner Brothers Cartoon Companion," *http://members.aol.com/EOCostello/b.html.*

war itself, the A-20 operated in many variants, including that of night fighter equipped with radar.[21] In this respect, the integration of electronic warfare into operational combat planes put Douglas in the vanguard of advanced technologies that characterized postwar military designs.

Nonetheless, by far the most ubiquitous Douglas military plane of WWII, and one of the most significant weapons of the war, was the C-47, the armed forces version of the DC-3. With some 10,000 models of the C-47 delivered during the war, Douglas turned to a network of subcontractors, including McDonnell Aircraft, recently incorporated in St. Louis, Missouri.[22] The planes airlifted crucial supplies, carried paratroopers into combat, evacuated the wounded, and generally performed beyond all expectations for American and foreign air forces. As General Eisenhower allegedly said, the key equipment for Allied success in the war included five examples: the 2½-ton truck, the bulldozer, the amphibious DUKW vehicle, the Jeep, and the C-47.[23] In American service, the C-47 carried thousands of tons of supplies from India to China over the "Hump," as aircrews called the formidable Himalayas. Of all the supplies delivered to China from 1942 through 1945, 81 percent came by air over the Hump, and Douglas transports carried the bulk of this cargo. Moreover, in the service of the Air Transport Command (ATC), the C-47 equipped a global network for cargo and passengers. The Douglas C-54 (military designation for the DC-4), after becoming operational in 1942, accelerated the revolution in air travel inaugurated by ATC operations. By 1945, the ATC carried 275,000 passengers per month, crisscrossing the globe with the regularity of passenger trains and transforming intercontinental air travel from a state of exciting adventure to a matter of daily routine. At its peak of operations, ATC aircraft crossed the Atlantic at an average rate of one plane every 13 minutes; the long-legged C-54 played a central role in these flights.[24]

Due to extensive use by Allied forces around the world as well as U.S. forces, the Douglas planes seemed to be everywhere. By the end of hostilities, even though other cargo aircraft made important contributions to the war effort, mention of a transport plane mostly conjured up images of the ubiquitous Douglas C-47s and C-54s in olive drab. Global headlines continued as the result of dramatic Cold War confrontations like the Berlin Blockade of 1948–1949, when the Soviet Union attempted to cut off the flow of land-based shipments of food and coal to the Allied sector of the city. Cynically timing

---

21 The SBD Dauntless and its combat role are assessed in Clark G. Reynolds, *The Carrier War* (Alexandria, VA: Time-Life Books, 1982). For this plane and other Douglas wartime aircraft, see Francillon, *McDonnell Douglas Aircraft Since 1920*.

22 Ingells, *McDonnell Douglas Story*, p. 43.

23 Pearcy, *DC-3*, p. 159.

24 Roger Launius, *Anything, Anywhere, Anytime: An Illustrated History of the Military Airlift Command, 1941–1991* (Scott AFB, IL: Headquarters Military Airlift Command, 1991), pp. 23–42; and Roger Bilstein, *Airlift and Airborne Operations in WWII* (Washington, DC: Office of Air Force History, 1998), pp. 38–43, 46.

their actions just as winter set in, the Soviets left narrow air corridors open, clearly assuming that any attempt to airlift crucial supplies would inevitably fail. As a diplomatic gambit largely sustained by C-47 and C-54 transports, the Berlin Airlift brilliantly succeeded, including the heartwarming "Operation Little Vittles" that captivated Berlin's children (and the world press) when planes dropped strings of small parachutes dangling candy bars underneath.[25] Marketing Douglas airliners in the postwar era became immeasurably easier as a result of the global exposure of Douglas products.

Given the success of the C-47 and the C-54 cargo planes, Douglas inevitably became involved in postwar projects for a new generation of military transports. The origins of this new line of aircraft emerged during WWII; although, actual planes did not take to the air until the war had ended. Douglas built the C-74 Globemaster—a name that aggressively advertised its intended postwar role. The worldwide deployment of American military assets during the era of Cold War led to an improved version, the C-124 Globemaster II, of which 448 planes rolled Douglas assembly lines. Conceived from the start as a dedicated military transport, unlike the conversions of the DC-3 and DC-4 to military service, the Globemaster II represented a major step ahead in terms of long-range airlift. Moreover, in the context of America's new intercontinental posture in the Cold War era, the designation of Globemaster II took on added significance as an aerial extension of strategic and national security intentions. It was followed by a fast, turboprop design, the C-133.[26]

Postwar combat planes of the 1940s and 1950s included the burly, radial-engine Douglas AD-1 Skyraider attack plane for the U.S. Navy, with several thousand produced in over 20 variations. The F3D Skynight, a two-place, twin-engine jet, became one of the first radar-equipped night fighters of the postwar era and helped make Douglas Aircraft one of the leaders in increasingly arcane electronic warfare scenarios.[27] Moreover, Douglas Aircraft became involved in one of the most sophisticated entities that tried to grapple with equally arcane dimensions of Cold War strategy—the RAND organization.

25 Launius, *Anything*, p. 87.

26 Ibid, pp. 98–115; and Andrew W. Waters, *All the U.S. Air Force Airplanes, 1907–1983* (New York: Hippocrene Books, 1983), pp. 154–155. Delivery of production Globemasters began in 1950 and ended in 1955; they continued to serve with USAF Reserve units until the middle of 1961.

27 Francillon, *McDonnell Douglas Aircraft Since 1920*, covers postwar military types including the B-66 twin-net bomber and others. On the Skynight in particular, see Richard Hallion, *The Naval Air War in Korea* (Baltimore, MD: Nautical & Aviation Publishing Company of America, 1986), pp. 174–187, which includes an account of the first radar-guided night victory for a jet fighter recorded by a Skynight crew in 1952.

AERONAUTIC RESEARCH

During WWII, General H. H. Arnold and a number of high-level Air Force officers, government executives, and industrial leaders in the aircraft industry concluded that cutting-edge technologies and the formulation of policies to utilize them needed to be anticipated and acted upon more adroitly. This was especially true as the scope and complexities of global military power increased through the acquisition of nuclear weapons along with delivery systems like jets and rockets. All this took on new urgency not long after the end of hostilities, as friction with the Soviet Union escalated. Consequently, Edward Bowles of MIT, a consultant to the Secretary of War, joined Arnold and a small group of key individuals to organize a research and development organization. The acronymic shorthand for this function—R&D—morphed into a formal name, Project RAND, a special contract assigned to the Douglas Aircraft Company. Activities began in the spring of 1946, operating from offices located in a separate area of the Douglas Aircraft plant at Santa Monica's municipal airport.

A prescient research team quickly got to work on RAND's first position report, which dealt with rocketry and space exploration. As additional work followed, RAND recognized the need for added space as well as a venue that offered insulation from the more public offices of the busy Douglas operations. Meanwhile, Donald Douglas became lukewarm about the whole idea of RAND as a part of Douglas Aircraft. Arthur Raymond felt that original endorsement had more to do with his long friendship with General Hap Arnold, and he continued to harbor concerns about a conflict of interest between his company and the Pentagon think tank within it. Formal separation began in 1947 when RAND relocated to new offices in downtown Santa Monica and also added ranks of social scientists to its staff. The Air Force also concluded that RAND should be independent of Douglas Aircraft or any other company. By the spring of 1948, RAND had been reinvented as a nonprofit entity.[28]

In the meantime, Douglas Aircraft moved rapidly into the postwar era of jet technology, supersonic speeds, and exotic aircraft that persistently generated headlines in the aviation press as well as the popular media.[29]

---

28 Raymond, *Who? Me?*, chapter 3, part 2, pp. 2–3. For a detailed history of the RAND Corporation, its leading personalities, and an analysis of its catholic range of studies, see Martin Collins, *Cold War Laboratory: RAND, the Air Force, and the American State, 1945–1950* (Washington, DC: Smithsonian Institution Press, 2002).

29 A review of articles under the name of Donald Douglas, Sr., and Douglas Aircraft as cited in the *Reader's Guide to Periodical Literature* reveals a plethora of references. See also Bill Gunston, ed., *Chronicle of Aviation* (London: Chronicle Communications, 1992), a massive compendium of myriad events in the history of flight, which invariably includes many references to Douglas products as they appeared in news reports.

The Douglas F4D-1 Skyray was capable of supersonic speeds. (McDonnell Douglas photo number ES-145103)

As the advent of gas turbine engines drove the top speed of new military designs into the realm of supersonic speeds, American military planners began to fund a round of experimental aircraft in order to understand the dynamics of jet-powered bombers and fighters with very high operational speeds. Donald Douglas adroitly positioned his company to benefit from several experimental projects. Soon after the war, Douglas Aircraft signed NACA and U.S. Navy contracts to design and build two types of high-speed experimental planes. The first, the D-558-1 Skystreak, was a jet-powered conventional design with straight but thin wings intended to investigate aerodynamic phenomena in the transonic range. The second, the D-558-2 Skyrocket, had a far more radical appearance with swept wings and tail surfaces. Early versions employed a jet engine; later versions mounted a rocket in the tail. The jet/rocket combination allowed it to take off and land under its own power and provided valuable flexibility in various test-flight scenarios. Both planes generated invaluable data that influenced the design of a new generation of postwar fighter aircraft and kept Douglas Aircraft in the forefront of exotic, high-speed aerodynamic research. In 1953, a modified, rocket-powered Skyrocket became the first aircraft to fly faster than twice the speed of sound.[30]

With growing ties to the Navy for designing high-speed jets appropriate for operations from the confined spaces of aircraft carriers, Douglas Aircraft developed a truly radical combat jet known as the F-4D Skyray. Its chief designer, Edward Heinemann, had joined Douglas Aircraft in the 1930s, working on early transport designs as well as the wartime projects and the Skyrocket/Skystreak series. The distinctive Skyray featured a

30 Richard Hallion, *Supersonic Flight: Breaking the Sound Barrier and Beyond* (New York: Macmillan, 1972), pp. 56–81 on the Skystreak, and 129–159 on the Skyrocket. Three Skystreak and three Skyrocket research planes were eventually built. Later, Douglas also built the X-3, with twin jet engines and a supersonic design that yielded pertinent information about exotic alloys and fabrication techniques for a new generation of supersonic fighters.

delta-shaped wing and a single vertical stabilizer, and was one of the first operational jet fighters capable of attaining supersonic speeds. In a unique ceremony in 1954, Heinemann and "Dutch" Kindelberger (former Douglas employee turned CEO at North American) shared the prestigious Collier Award, given annually for great achievement in American aeronautics—Heinemann for the Skyray and Kindelberger for the F-100 Super Sabre. Both had worked on the original DC-1 project. One of Heinemann's most enduring engineering legacies involved the A-4 Skyhawk. In the early 1950s, the U.S. Navy released requirements for a multicrew attack jet to weigh some 45,000 pounds. Heinemann took the weight and complexity as a challenge, and his successful design, the remarkably compact A-4 Skyhawk, weighed 10,000 pounds empty and 17,000 pounds loaded. Advanced models flew at 670 mph and could carry as much as 9,000 pounds of weaponry. Production began in 1954 and continued until 1979, with nearly 3,000 aircraft delivered to the U.S. Navy, Marines, and several air arms around the world. It remained in service into the 21st century.[31]

COMMERCIAL AIR TRANSPORTATION

In parallel with outstanding progress in aeronautical research and production of acclaimed products for military service, Douglas Aircraft achieved remarkable success in building commercial planes. Once the war ended, Douglas sold several dozen new DC-4s to the airlines, but the number of war surplus C-54 transports stifled the market for substantial orders of this type. Moreover, the Lockheed Constellation, which went into military service late in WWII, boasted cabin pressurization and superior performance in comparison to the DC-4. Donald Douglas realized that his company needed a competitive model, and it needed one in a hurry. Although the DC-4 did not have pressurization, it had been designed with that prospect in mind, and this legacy quickly became part of plans for a postwar successor, designated the DC-6. Even before the end of the war, Donald Douglas signed off on the start of design studies for the new plane, which had the same wings as the DC-4 but incorporated a stretched fuselage, new engines, and other embellishments.[32]

31 David Donald, ed., *The Complete Encyclopedia of World Aircraft* (New York: Barnes and Noble, 1997), p. 609; and Ingells, *McDonnell Douglas Story*, pp. 97, 100. Edward H. Heinemann, with Rosario Rausa, *Ed Heinemann: Combat Aircraft Designer* (Annapolis, MD: Naval Institute Press, 1980). When Heinemann arrived at Douglas in the mid-1930s, he had a high school degree along with experience in mechanical drawing and designing for a Los Angeles company that specialized in bodies for fire trucks. Experience, engineering skill, and intuition made him a master airplane designer. He played a major role in designing the SBD Dauntless, the AD Skyraider series, as well as other major military projects. The Skyhawk operated in several overseas air forces, including Argentina, Israel, and others. The Republic of Singapore still had active Skyhawk squadrons in 2003, and a private company in Arizona operated over a dozen for contract training and research duties.

32 Raymond interview, Claremont-OHC.

The first DC-6 became airborne in 1946 and entered service with American Airlines only a year later. The well-appointed, pressurized cabin made passengers happy because the airliner's ceiling of 28,000 feet allowed it to fly above storms and avoid choppy weather en route. Moreover, at higher altitudes, the "thinner" air actually caused less aerodynamic friction, and the plane's radial engines drove it along at speeds of 300 mph. Various seating arrangements accommodated 48 to 86 passengers. The first planes were soon followed by an improved DC-6B that featured seating for 54 to 102 passengers, higher cruise speed, and generally better performance—often called one of the most economical and efficient airliners ever built. Operators in the United States and all over the world tended to agree; including several dozen military versions, over 700 of the DC-6 types were delivered. One of them, called *The Independence* by President Harry Truman, began the tradition of an official presidential transport, eventually referred to as Air Force One. In 1952, Pan American used the Douglas airliner to launch the first scheduled around-the-world flight, a journey of 83 hours.[33]

The decade of the 1950s defined the pinnacle of Douglas Aircraft success for air travel. During this period, Donald Douglas maintained a personal touch with his company with visits to R&D shops working on advanced projects like the Skyrocket series and in occasional forays along production lines. He continued to negotiate major airline deals through personal telephone calls, leaving the paperwork to corporate vice presidents, and invariably appeared to preside over the formalities and photographs that accompanied the delivery of new models to the airlines. He remained a busy executive. In 1950, approximately two-thirds of all the aircraft in the airline fleets of the noncommunist world came from Douglas assembly lines. During the next several years, many DC-3 aircraft were retired, but, as late as 1955, Douglas products still accounted for some 50 percent of airliners operated outside the communist bloc.[34]

On premier routes across the United States and over the North Atlantic, airlines vigorously competed for leadership. By the late 1950s, American Airlines and Pan Am both wanted bigger, faster transports to exceed the performance of late-model Lockheed Constellations. Reports of a new Boeing jet airliner had already surfaced. The Douglas executive leadership temporized, torn over the decision of whether to build a new piston-engine airliner or take a leap into the future with a new turbojet. An informal group at Douglas, led by Art Raymond, "kicked around a number of airframe-engine combinations, but all the while in the back of minds was the big question . . . could the engine people stifle the turbo-

33 Arthur Pearcy, *Douglas Propliners: DC-1–DC-7* (Shrewsbury, England: Airlife Publishing, 1995), pp. 137–146.

34 Morrison, *Donald W. Douglas*, pp. 140, 204–205; John Newhouse, *The Sporty Game: The High-Risk Competitive Business of Making and Selling Commercial Airliners* (New York: Knopf), pp. 133–134; and Paul Eddy, et al., *Destination Disaster: From the Tri-Motor to the DC-10* (New York: Quadrangle, 1976), p. 41.

jet's fuel-guzzling appetite?"[35] Eventually, influenced by major customers like American and United, who questioned the economics of fuel-hungry jets, Donald Douglas committed his company to a new piston-engine transport. A key factor involved the extremely promising performance of a new Wright turbo-compound engine, which used exhaust-driven turbochargers that yielded 20 percent more power from each engine. The new DC-7 (1953) and DC-7B were soon followed by the DC-7C, the ultimate evolution of the piston-engine airliner designed for nonstop transcontinental service in the United States or to thunder across the Atlantic without having to land in Newfoundland or Iceland to refuel.

Joining airlines in 1956, the elegant transport with matchless service became internationally renowned as the "Seven Seas," with a speed of 355–400 mph and a range of over 4,000 miles. For Pan Am, British Overseas Airways Corporation, and other international carriers, it handled the North Atlantic with aplomb and also equipped new, long-distance routes across the Pacific. The Scandinavian airline, SAS, launched a sensational Europe-to-the-Far-East route that crossed over the North Pole.[36] Because Douglas Aircraft Company continued to field various advertising campaigns, the launch of new service by customer airlines clearly multiplied the awareness of Douglas as a leading brand name in the airliner business. During the years that the postwar DC-4/-6/-7 series won such a wide customer base, the Douglas name constantly cropped up in regional as well as national venues. For example, when Braniff took delivery of its first DC-7C, its own publicity department trumpeted the virtues of the new equipment and disbursed a flurry of press releases that featured Texas native Ginger Rogers posed with the new plane. Although Douglas Aircraft delivered 127 models of the big plane, production ended in 1958, the inaugural year of jet service offered by the Boeing 707.[37]

Suddenly, Donald Douglas had to lead his well-known company into the age of jet airliners or close the books on its role as a premier force in the manufacture of passenger transports. The company forged ahead with jets, belatedly committing tens of millions of dollars, but consistently found itself in the wake of archrival Boeing. Ironically, Ivar Shogran, one of the Douglas leadership who opted for the DC-7, now found himself in charge of a study group for what became the jet-powered Douglas DC-8. "We should have been in the race much earlier," he admitted later. Corporate indebtedness accelerated with the development of a new twin-engine jet, the DC-9, followed by a wide-body tri-jet, the DC-10.[38]

---

35 Quoted in Ingells, *McDonnell Douglas Story*, pp. 124–126.

36 Ibid., pp. 125–127; Donald, *World Aircraft*, p. 362; Newhouse, *Sporty Game*, p. 133; and Mike Machat, "Too Much, Too Late: Battle of the Sky Giants," *Airpower* 32 (November 2002): 43–52.

37 Braniff press releases from Braniff Files, Box 43, Folder 5, Aviation Collection, Special Collections, McDermott Library, University of Texas—Dallas.

38 Quoted in Ingells, *McDonnell Douglas Story*, p. 131; and Newhouse, *Sporty Game*, p. 133.

THE AEROSPACE INDUSTRY AND ADMINISTRATIVE TURMOIL

In the meantime, Douglas Aircraft had become a major player in a new enterprise involving rocketry and space research, known by pundits in the late 1950s as the aerospace industry. In a project endorsed by Donald Douglas during WWII, engineers had actually begun experiments with an air-to-ground rocket, the ROC, and late in the war joined Bell Laboratories in developing a rocket that might intercept German V-1 flying bombs. The ground-to-air weapon led to a series of rocket-propelled air defense systems in the post-war era—the Nike family of missiles. By the mid-1950s, Douglas Aircraft won an Air Force project as prime contractor to build the Thor Intermediate Range Ballistic Missile. The 65-foot-high missile became operational in 1957, with nuclear-tipped versions based in England during 1958. This military technology formed the basis for a commercial launch vehicle, called the Delta, which carried out its first mission in 1960. In various multistage launch configurations, it continued to carry payloads into the 21st century. Thus, in the early 1960s, Douglas Aircraft had a solid background in rocketry when it successfully competed for NASA contracts in the Apollo program, intended to carry the first humans on a journey to the Moon. Douglas acquired responsibility to build the S-IV and S-IVB upper stages of the Saturn launch vehicles. The S-IV/IVB hydrogen-fuelled stages boosted the Apollo spacecraft out of Earth's orbit on a trajectory to the Moon.[39]

In 1924, the Douglas World Cruisers had carried intrepid crews around Earth. In 1968, a Saturn S-IVB upper-stage rocket fueled by liquid hydrogen carried a trio of brave astronauts into orbit around Earth and then set them on course as the first humans to circumnavigate the Moon. A year later, another S-IVB repeated these maneuvers during the first human exploration to set foot on the lunar surface. In his career as a pioneer in aviation and as a leader in the aerospace industry, Donald Douglas had certainly left his mark. But the vagaries of business and finance had required him to surrender control of the company that bore his name. In 1967, a merger with one of his old suppliers in St. Louis resulted in the McDonnell Douglas Corporation.

The unraveling of Douglas Aircraft Company arguably began somewhere around the height of its strength and reputation, about 1957. That year, Donald Douglas, Sr., at the age of 65, assumed the title of Chairman of the Board and decided to bestow the mantle of company President on his son, Donald Douglas, Jr. The elder Douglas still continued

---

39 For a reprise of early rocketry work, see Ingells, *McDonnell Douglas Story*, pp. 217–229. The evolution of Thor, Delta, and their significance, is covered by Kenneth Forsyth, "Delta: The Ultimate Thor," in *To Reach the High Frontier: A History of U.S. Launch Vehicles*, ed. Roger Launius and Dennis Jenkins (Lexington: University Press of Kentucky, 2002), pp. 103–146. On the S-IV/S-IVB, see Roger Bilstein, *Stages to Saturn: A Technological History of the Apollo/Saturn Launch Vehicles* (Washington, DC: Government Printing Office, 1980; revised edition, University of Florida Press, 2003), pp. 155–190.

to take a strong interest in the affairs of his company, but his son began a campaign to unseat his father's senior executives, replacing them with his own choices. In the space of a few years, stalwarts like Arthur Raymond and others had resigned, either due to indignation or to the stresses that attended the bitter administrative turmoil that began to characterize the company. Ed Heinemann, the brilliant designer and engineer, suddenly found himself shunted into a lateral slot in engineering, then assigned as industrial liaison for the Douglas Aircraft Company sales team in Europe. Although infuriated, a dutiful Heinemann endured what amounted to an entertainment and cocktail circuit overseas for a few months, and then he returned to confront Donald Douglas, Sr., who backed his son. Heinemann left the company to join General Dynamics. Some observers interpreted the loyalty of Douglas, Sr., to his son as a consequence of a bitter divorce in 1953, in which Donald Douglas, Jr., had stood by his father. After 37 years of marriage, Mrs. Douglas belatedly discovered her husband's 20-year liaison with Marguerite Tucker. The situation became a major news story at a time when public revelations of infidelity in American corporate culture were not as commonplace as they came to be in subsequent decades. Details of the affair and the highly publicized divorce proceedings in 1953 rattled the company. Donald Douglas, Sr., married Tucker in 1954 and made her a high-level corporate special assistant; her position on organization charts appeared in a box at the same level as her new husband's. Subsequently, many Douglas Company executives found that she made it difficult to gain access to the office of Donald Douglas, Sr. This isolation became especially troublesome after 1957, when Donald Douglas, Jr., generated so much turmoil and animosity within the management structure.[40]

During the ensuing decade, despite occasional upturns in corporate fortunes and its prominent role in the lunar-landing program, Douglas Aircraft Company began to disintegrate from the inside out. In contrast to past successes, the company failed to land several defense contracts. The company's market offensive to sell jet airliners in competition with Boeing brought a flood of orders that exceeded production capacity, and the company wound up paying penalties for late delivery to customers. Efforts to hire and train new workers on production lines raised costs and often led to even slower delivery rates. The Vietnam conflict created shortages and deferred deliveries from subcontractors. Moreover, the company's accounting procedures had become antiquated, and the true financial picture eluded Douglas's upper management. As evidence of the company's widespread financial disarray became public, lenders and banks eventually refused to extend credit. When Donald Douglas, Sr., made a desperate trip to New York City in an effort to secure sources of credit, he took along his son. Laurance Rockefeller

40 Morrison, *Douglas*, pp. 195–197, 214–215, 231–234, 251–252; Biddle, *Barons*, pp. 184–45; and Newhouse, *Sporty Game*, pp. 132–137. For contemporaneous business analysis, see T. A. Wise, "How McDonnell Won Douglas," *Fortune* 75 (March 1967): 156 ff. See also, Donald Pattillo, *Pushing the Envelope: The American Aircraft Industry* (Ann Arbor: University of Michigan Press, 1998), pp. 233–235.

finally agreed to see Douglas, but when he realized that Donald Douglas, Jr., was along, Rockefeller summarily canceled the meeting.

Several aerospace companies made serious bids for a merger with Douglas Aircraft Company, but, in the end, the successful bid of $69 million came from McDonnell Corporation in St. Louis, Missouri. The latter employed 45,000 people, compared to the Douglas payroll of 80,000. Although McDonnell had no experience in commercial airliners, many analysts felt that the new entity, McDonnell Douglas Corporation, would do well because the St. Louis organization's large, lucrative military contracts and specialized astronautics projects would complement the California company's commercial airliner production and rocket booster programs. Despite use of the term merger, the official title of the new entity, McDonnell Douglas, clearly defined the pecking order within the new organization. Following the acquisition, Donald Douglas, Sr., held a seat on the board of directors but rarely attended its meetings. He died in 1981 at the age of 89.[41]

The output of RAND, as an invaluable think tank grappling with arcane national issues as well as domestic programs, has made it one of the most pervasive legacies of the Douglas era. There is an irony here because Donald Douglas did not embrace it as warmly as some of his cohorts wished. But irony is pervasive in the glory years of Donald Douglas when his company achieved its pinnacle and at the same time failed to make decisions to ensure its ongoing success. The company eventually had to bow to a takeover by one of its early subcontractors and ultimately disappeared when its perennial rival, Boeing, bought out McDonnell Douglas in 1997.[42] As part of a company title, the word "Douglas" vanished from the roster of leading aerospace corporations. The name itself, however, remains legendary.

ENDNOTES

The author wishes to acknowledge the archivists for their knowledgeable assistance during visits to the McDermott Library, University of Texas—Dallas; the C. R. Smith Museum, American Airlines, DFW Airport, Texas; and the Museum of Flight, Seattle, Washington.

---

41 Morrison, *Douglas*, p. 243; and Biddle, *Barons*, pp. 315–319, 323. Historically, the transition from Douglas, Sr., to Douglas, Jr., did not have high odds for success. According to the Family Business Resource Center at Rutgers University, only one-third of "family businesses" survive to the second generation; just 13 percent survive to the third. Peter Fleming, "Planning," *Merrill Lynch Advisor* (Summer 2003): 22.

42 Until 1997, the new combination appeared to work well until McDonnell Douglas succumbed to a combination of debt load, management issues, and financial woes that seemed to echo the problems of Douglas Aircraft of the 1960s. Additionally, its airline transport division was buffeted by stiff competition from the European consortium, Airbus Industrie. Roger Bilstein, *The Enterprise of Flight*, pp. x–xi.

6

Benjamin O. Davis, Jr.

American Hero

ALAN L. GROPMAN

GENERAL BENJAMIN O. DAVIS, JR., IS AN AMERICAN HERO, A CHAMPION WHO ABUNDANTLY DEMONSTRATED BOTH PHYSICAL AND MORAL COURAGE.[1] WE CALL THOSE who display physical courage heroes because they risk their lives for something bigger than themselves—for the greater good of their nation or country. General Davis often risked his life for his nation, his people, and also for his country.

He believed in racial integration all his life and was convinced he could aid in promoting this essential reform for America if he could help bring victory to the United States in WWII. His chosen battlefield was the skies of Europe; his weapon was the airplane. Ben Davis thought if he established that blacks could fly, fight, and lead with the same courage, dedication, discipline, and skill as whites—a notion utterly foreign to almost all

1 The best source on Benjamin O. Davis, Jr., is *Benjamin O. Davis, Jr.,: American* (Washington, DC: Smithsonian Institution Press, 1991). Alan L. Gropman refereed the autobiography for the Smithsonian Press and did

continued on next page

whites in America in 1941—he would help destroy the myth of racial inferiority. The legend of racial inferiority was the foundation for race segregation in this country, and it had to be demolished.

To do so he had to risk his life above foreign fields in distant skies against some of the most skilled and well-equipped fighter pilots in the world, the Luftwaffe. Certainly as importantly, he also had to stand up to, confront, and openly disagree with his military superiors when they tried to inhibit or destroy his Tuskegee Airmen.

It is important here to note, the Tuskegee Airmen—pilots and their ground crews trained to fly and fight and maintain aircraft at Tuskegee Army Air Field in the early and mid-1940s—shared his vision and courage, and General Davis succeeded not only through his genius for command, his courage and vision, but also through the devotion to the mission of the other Tuskegee Airmen.

We, therefore, honor General Davis partly for his physical courage signified by his 60 combat missions during WWII and the decorations he earned, including the Distinguished Flying Cross and the Silver Star, and partly for his leadership of the competent Tuskegee Airmen pilots and ground crew. We also pay tribute to him for his open display of moral courage. His entire professional life he held to the West Point creed of "duty, honor, country." General Davis devoted 43 of his 89 years to service to his country, every part of it engaged with aviation. He loved his country, and he loved to fly.

What about the two other parts of the West Point trinity, duty and honor? Regarding the former, duty was equally paramount to honor and country to General Davis—duty in the face of bigotry and discrimination, duty in front of a highly skilled enemy, and duty to his people and to his country when he could have chosen a much less arduous and certainly infinitely less dangerous career.

continued from previous page

the first editing of the manuscript. Gropman wrote a 15,000-word biography of Davis for the Air Force Historical Foundation called "History on Two Fronts" in *Makers of the United States Air Force*, ed. John Frisbee (Washington, DC: Air Force Office of History, 1987). Gropman also wrote the cover story for the summer of 1999 issue of *Air Power History* on Davis's promotion to a four-star general. The best account of the achievements of the Tuskegee Airmen is Stanley Sandler, *Segregated Skies: All-Black Combat Squadrons of WWII* (Washington, DC: Smithsonian Institution Press, 1992). Another scholarly account is Alan Osur, *Blacks in the Army Air Forces in WWII* (Washington, DC: Air Force Office of History, 1977). One also can find numerous mentions of Ben Davis and his father in the monumental, scholarly, and objective account by Ulysses Lee, *The Employment of Negro Troops* (Washington, DC: Chief of Military History, 1966). Also, see Morris MacGregor's definitive account of the racial integration of the armed forces in *Integration of the Armed Forces, 1940–1965* (Washington, DC: Center of Military History, 1981). Less scholarly sources include Charles Francis, *The Tuskegee Airmen* (Boston: Bruce Humphries, 1955); and Robert A. Rose, *The Lonely Eagles: The Story of America's Black Air Force in WWII* (Los Angeles: Tuskegee Airmen, Incorporated, Western Region, 1983). For Davis's contribution to racial integration, see Alan L. Gropman, *The Air Force Integrates, 1945–1964*, 2d ed. (Washington, DC: Smithsonian Institution Press, 1998). For the best one-volume account of blacks in American military history, read Bernard Nalty, *Strength for the Fight: A History of Black Americans in the Military* (New York: The Free Press, 1986).

United States Military Academy photo of Benjamin O. Davis, Jr., taken from The Howitzer *United States Military Academy yearbook in 1936. (Supplied by the United States Air Force)*

Though the cadets at West Point shunned him completely because of his race, he adopted the credo of those who tried to drive him out and stood defiantly against their bigotry. No cadet talked to Ben Davis except for official reasons while at West Point, and the silencing even followed him into the Army for several years after graduation. His lonely four years at West Point are symbolic of his determination, discipline, resolve, and sense of duty—his moral courage. He knew the bigots wanted him to fail, and he was determined to succeed. He graduated in the top 15 percent of the class of 1936 at West Point.

And what about honor? The cadets at West Point between 1932 and 1936 acted dishonorably, and so too did the leadership of the military academy. West Point violated its own code, and nobody at the academy or in the Army intervened. General Davis knew he was fighting something bigger than the racism of young men in their teens and early twenties, but he was undaunted. He stood up to intolerance with dignity and never relented. His honor is unquestioned.

After graduating from flying school at Tuskegee Army Airfield, Alabama, he took the 99th Fighter Squadron (the first of the Tuskegee Airmen) to North Africa and suffered discrimination at the hands of the commander of the 33rd Fighter Group. When that colonel tried to exile the 99th from combat and terminate the new Tuskegee Airmen units being formed—the 332nd Fighter Group and the 477th Medium Bombardment Group (the rest of the Tuskegee Airmen)—General Davis fought his commander and also the leadership of the entire Army Air Forces, all of whom had endorsed the 33rd Group commander's bigotry.

General Davis battled a four-star general trying to destroy the reputation of the 99th and marginalize the entire race. General Davis won that battle in the Pentagon. It took moral courage for a lieutenant colonel to fight the commanding general of the Air Corps.

In Italy, with the 332nd Fighter Group flying obsolete P-39 Airacobras, he was given the opportunity to change missions from ground attack to bomber escort and saw a great opportunity. Success in this mission would underwrite his goal of exploding myths by demonstrating skill against the vaunted Luftwaffe.

The record achieved by the Tuskegee Airmen under General Davis's leadership was unique. In 200 escort missions to heavily defended targets, the Tuskegee Airmen never lost a bomber to an enemy fighter. No other fighter unit flying half the missions could claim such success. This triumph is a tribute to the dedication, skill, courage, and discipline of the Tuskegee Airmen, and to the tactical acumen and leadership of General Davis.

It was this unique achievement, along with the other accomplishments of the Tuskegee Airmen—downing 111 enemy aircraft in air-to-air combat; shooting down the second, third, and fourth jet fighters during the war; destroying more than 150 Luftwaffe aircraft on German air bases; destroying many German locomotives and rolling stock; and sinking a German destroyer and numerous river barges—that convinced several Air Corps leaders that segregation was unnecessary and, therefore, an unconscionable waste.

When the Air Force became independent in 1947, its Chief of Personnel studied the disutility of racial segregation. He found no basis for segregation, citing the record of the Tuskegee Airmen and the leadership of General Davis to document the case for integration. Segregation was costly and provoked racial disturbances, and the Air Force (helped by General Davis) proved there was no basis for it in biology or sociology.

General Davis and the Tuskegee Airmen were the reasons the United States Air Force was the first service to integrate, announcing racial integration in April 1948, beginning the process in May 1949, and finishing it two years later in 1951. Before the Army and Marine Corps began the integration process, and well before the Navy passed beyond mere tokenism, the Air Force had integrated.

That reform caused the other services to pay attention. Because it was integrated, the Air Force became overwhelmingly the service of choice for talented blacks. Air Force racial integration, moreover, worked smoothly and improved Air Force operations, setting the example for other services. Soon, with the Air Force in front, the other services, faced with the demands of the Korean War, also integrated.

All the United States Armed Forces completed integration decades before the first black managed a Major League Baseball team or coached a National Basketball Association team. The Armed Forces set the example for American society.

We live in a different America from the one of the 1940s because the armed forces, the school for the nation in the 1940s and 1950s, showed America how to make integration work. General Davis was essential to this transformation.

He did it by exploding myths. He demolished the myth of racial inferiority with his stellar performance at West Point and his outstanding deeds during WWII, the 30 years he spent in the Air Force, and the career he found in the Department of Transportation. He destroyed the myth that blacks lacked courage with his awards for heroism. And so too did the Tuskegee Airmen he led. The Tuskegee Airmen under General Davis's command shattered the myth that black Americans would not follow black leaders.

The last myth he destroyed was the lie that whites would never work for a black supervisor. General Davis at Lockbourne Air Force Base between 1946 and 1949, as commander of the 332nd Fighter Wing and also as base commander, supervised and commanded many civil servants—all of whom were white. Numerous visits by the

Inspector General indicated race relations at Lockbourne were harmonious, and supervision by General Davis was highly effective. Another myth ruined. All of these false notions were barriers to racial integration, and General Davis discredited all of them.

General Davis had a rich career after 1949 at the Air War College, the Pentagon, in Korea, Japan, The Republic of China, Germany, the Pentagon again, Korea, the Philippines, and at United States Strike Command in Florida. He retired from the Air Force in 1970 as a lieutenant general and served as Director of Public Safety in Cleveland and later in the Department of Transportation. He created the Sky Marshall program and drove skyjackings in the United States to zero in short order. He received his fourth star in 1998. He was inspired by flight as a young teenager and used aviation to reform American race relations. He achieved much in his long life, but no accomplishment gave him more satisfaction than leading the effort to integrate the United States Air Force racially.

GENERAL DAVIS'S LIFE

General Davis disliked the term African American and would not speak at events in February (black history month) because he wanted to be acknowledged as an American, not a black American, not an African American. He was not ashamed of being black—there was not a scintilla of self-hate in General Davis—but noted sadly that prominent white Americans were not designated with hyphens, and neither should he. Nobody in the history of America is more responsible for racially integrating American society than Ben Davis, and he did it with his P-40, P-39, P-47, and P-51. He helped integrate America with the skill and discipline of the Tuskegee Airmen who flew 15,000 sorties in WWII and protected American bombers from German fighters flawlessly. He commanded highly successful flying outfits from 1941 into the late 1960s and demonstrated to all except the most bigoted that race did not matter. He fought racism and bigotry at all levels and used aviation to make his point. The title of his autobiography says it all: *Benjamin O. Davis, Jr.: American.*

Benjamin O. Davis, Jr., was inspired by flight. He convinced his frugal father to pay a barnstormer to take him at age 14 on a flight over Washington, DC, and from that moment Benjamin O. Davis, Jr., was captivated by aviation. Many years later when he was about to graduate from the United States Military Academy at West Point, he asserted his desire to fly as a member of the United States Army Air Corps, only to be turned down on the basis of race; the Air Corps had never permitted blacks to join in any capacity.

Unlike many enthralled by aviation whose accomplishments resulted in military achievements, technological advances, or business accomplishments, Davis's fascination with aviation led to a social revolution involving the racial integration of the United

States Armed Forces and, through the influence of the United States military, of American society. After all, it was race bigotry by the United States Army that denied a fully qualified Ben Davis the right to fly.

Ben Davis was born on 18 December 1912 in Washington, DC. He was the son of an Army officer who became the first black brigadier general in the United States Army. His mother died when he was three, and his stepmother became a major influence on his life. She had a graduate degree in English and pushed her stepson and his sisters academically as hard as his father pressed his children to be responsible, honest, and punctual. Davis graduated from the racially integrated Central High School in Cleveland, Ohio, where he excelled academically and athletically and served as an elected class officer.

After high school, Davis attended college at Western Reserve University, Ohio State University, and the University of Chicago before receiving an appointment to the United States Military Academy in 1932. He was the ninth black to enroll at West Point since its founding in 1932 and the fourth to graduate. The Army he joined in July 1932 was a sad reflection of American society. If anything, however, the military academy reflected southern customs, in contrast to northern colleges that were rarely racially segregated. West Point students rejected Davis the way a body rejects a transfusion of the wrong blood type.

From the time he entered until he graduated in June 1936, Davis had no roommate. Upper classmen directed that he be silenced in order to drive him from the academy. Despite the extreme social pressure, Davis graduated 35th in a class of 276. His West Point yearbook showed that some of his classmates admired his persistence; they wrote of him, "The courage, tenacity, and intelligence with which he conquered a problem . . . more difficult than plebe year [the especially arduous freshman year] won for him the sincere admiration of his classmates, and his single-minded determination to continue in his chosen career cannot fail to inspire respect wherever his fortune may lead him."[2] Soon after graduation he married Agatha Scott, his lifelong partner who died only a few months before he passed on.

Because Ben Davis had been smitten by flying from his barnstorming ride when he was 14, he applied for flight training in the Army Air Corps. He even received a positive recommendation from the academy's superintendent, Major General William D. O'Connor, but segregation denied him the opportunity. Headquarters Army Air Corps replied to West Point that since it took no blacks in any capacity, operations or service, there was no place in its organization for a fully qualified West Point graduate near the top of his class who happened not to be white. Davis was assigned to the infantry as commander of a black service company at Fort Benning, Georgia, where he led several hundred black enlisted men engaged in menial activities, and he also coached the unit's athletic teams. Within a

2 *The Howitzer*, United States Military Academy Yearbook, 1936, Davis Entry, Benjamin O. Davis, Jr., Papers, Smithsonian Collection, Silver Hill, MD.

year, Ben Davis was assigned to the Army's Infantry School at Fort Benning. This was earlier than most junior Army officers, but the silencing that had beleaguered him at West Point followed him to Georgia. Silencing by his fellow officers was to continue until he reached the combat theater in 1943 during WWII. There were other indignities suffered by Davis and his wife, such as social snubs and deliberate insults from senior officers. All were borne with dignified silence and the heartfelt goal of altering this abominable prejudice.

Upon graduation from the Infantry School at Fort Benning, Davis became a Reserve Officer Training Corps (ROTC) professor of military training and tactics at Tuskegee Institute, Alabama, embarking on a career pattern similar to his father (then a colonel). In this assignment, he replaced a sergeant! The Army, in other words, had no role for an officer who had excelled at West Point and had graduated from the Infantry School.

But the beginning of WWII in Europe changed the pattern of bigotry. After almost three years at Tuskegee, Davis was assigned as an aide to his father, now a brigadier general, at Fort Riley, Kansas, and soon thereafter to Tuskegee, Alabama, to begin pilot training. Domestic politics played a central role in this move.

In 1940, Franklin D. Roosevelt was campaigning for an unprecedented third presidential term. Because of the economic and moral destruction of the Great Depression, the black vote in 1932 and 1936 had gone to a Democrat for the first time. In 1940, Roosevelt was running against civil rights advocate Wendell Wilkie and was anxious to retain the black constituency. Numerous black leadership organizations, including the largest and most important, the National Association for the Advancement of Colored People (NAACP), and the important black press (most significantly the *Pittsburgh Courier*) campaigned assiduously for blacks to be admitted into military aviation. Roosevelt sensed the pressure and responded by promising, if elected, to initiate a military flying program for blacks. He was reelected, retained a large black majority, and fulfilled his promise.

In December 1940, following the orders of the Commander in Chief, the Army Air Corps submitted a plan to the Secretary of War for creating one, and only one, segregated pursuit squadron that would be composed of 47 officers and 429 enlisted men. The legendary 99th Pursuit Squadron was born on 22 March 1941 at Chanute Army Air Field, Illinois, where the enlisted maintenance men and their small cadre of officers would be trained. Tuskegee Army Air Field was established on 23 July 1941 to train pilots.

Ben Davis, now an Army captain and the only living black United States Military Academy graduate, was chosen to be the commander of the 99th. He had advertised his desire to fly, was physically and mentally qualified, and was a natural leader. He had a racist hurdle to overcome, however. He was at Fort Riley when the War Department order came to him to proceed to Tuskegee, but as a prerequisite he had to pass a flight physical. The local flight surgeon failed him because he asserted, falsely, Davis was an epileptic. Obviously the Army physician had not been told the Army policy refusing all blacks the opportunity to fly had changed.

In August 1941, Davis entered flight training as the commander of the first group of cadets at Tuskegee. There were 13 in the first class, and 5 graduated on 7 March 1942. Several of the first graduating class went on to full military careers. During WWII, Tuskegee Army Air Field graduated 992 black fighter and bomber pilots known forever after as the Tuskegee Airmen.

Racial prejudice had dogged Ben Davis's steps from the time he was born, and graduating near the top of his class at West Point did not end his humiliation. When the 99th had completed its complement of fully trained pilots and enlisted, it could not find a home in the combat zone because successive white commanders refused to accept the 99th. Ben Davis with dignity and patience worked hard to maintain the morale of his troops, stunned by the fact that they were being denied combat and the country was being deprived of their talents because of racial bigotry.

In time, after seemingly endless training missions, the Chief of Staff of the Army General George C. Marshall forced commanders in North Africa to take the unit. When the Tuskegee Airmen departed for North Africa on the USS *Mariposa*, Davis and his men carried with them the sure intelligence that the fate of blacks in military aviation rested on their shoulders. When added to the profound load of combat flying against the Luftwaffe, this unnecessary burden was a great deal to bear.

Yet the Tuskegee Airmen under Ben Davis's leadership did not overtly protest bigotry. There were many enemies of these black pioneers in and out of the military waiting for the opportunity to destroy the entire notion of blacks in military aviation who would have pounced on any dissent. Ben Davis always thought his greatest achievement, from his beginnings at West Point in 1932 to his retirement as Assistant Secretary of Transportation in1975, was keeping the men of the 99th and later the 332nd Fighter Group flying and fighting instead of protesting and diverting themselves from their two-sided mission—to defeat America's enemies in the skies of North Africa, the Mediterranean, and Europe, while overcoming their racist enemies at home. Ben Davis said, "God knows, they had enough to protest about,"[3] but his aim was to prove that his squadron was the equal of white units, and he wanted no energy wasted on protest that might endanger the future of the Tuskegee Airmen.

Ben Davis and the 99th sailed for Africa in April 1943 with a complement of 26 pilots, several support officers, and more than 300 enlisted men. The squadron trained initially in Morocco and was later attached to the 33rd Fighter Group in Tunisia.

The commander of the 33rd Group was hostile to blacks in aviation. From the beginning of their tour in Africa, he insisted on a level of segregation beyond that required by regulation. He ignored the sound tactical guiding principle of providing experienced flight commanders for newly arrived units, asserting War Department policy dictated

3 Interview by Gropman of Benjamin O. Davis, Jr., 1990, the United States Air Force Office of History, Bolling Air Force Base, Washington, DC.

that whites and blacks not be mixed in the same flight formations, which was untrue. It also was patently absurd because in the past all black units had white commanders, and a cadre of white officers supervised black enlisted men. Because of his policy, the first time black pilots flew in combat, they had no experienced pilots flying with them. All were new to the stresses of combat flying.

In the case of the 99th Pursuit Squadron, this was on 2 June 1943, and Ben Davis led the formation. His mission was to strafe enemy targets on Pantelleria Island off the coast of North Africa between Tunisia and the next target of Allied forces, Sicily. On this mission and on subsequent operations over the next week, the men of the 99th did not sight an enemy aircraft; however, on the morning of 9 June, a flight of six P-40 Warhawks from the 99th were attacked by 12 enemy fighters, and a disorganized skirmish occurred in which there were no losses on either side. On 11 June, the enemy forces on Pantelleria surrendered without an invasion, a unique event to that point in any theater in WWII. Ben Davis had encouraged his troops before their first combat flights by telling them to focus on their military mission. "We are here to do a job," he told the pilots and mechanics of the 99th, "and, by God, we're going to do it well, so let's get on with it."[4]

Early in July, Lieutenant Charles B. Hall, flying his eighth combat mission (and sighting his first enemy aircraft), shot down a Focke-Wulf 190, one of the most effective Luftwaffe fighters and one superior to Hall's P-40. Hall's victory was the first for the 99th and one of three victories for Hall during his entire combat tour. Unfortunately, the Tuskegee Airmen did not have another victory until the battle over Anzio, Italy, early in the new year.

The main reason for this lack of aerial combat victories was the mission assigned to the 99th. The Tuskegee Airmen were designated air-to-ground attack missions where they would meet enemy aircraft only by accident. Fighters flying bomber escort operations would much more often encounter enemy fighters. The fact the Tuskegee Airmen rarely saw a Luftwaffe fighter, however, did not stop bigoted Army Air Forces leaders from trumpeting their lack of air-to-air victories as a reason to disband the unit.

In September 1943, barely 90 days into combat, the commander of the 33rd reported the 99th had unsatisfactory discipline in the air and that the unit had crumbled when attacked by the enemy. Ben Davis and his troops also were accused of not being sufficiently aggressive. The 33rd commander alleged (as evidence of the lack of pugnacity) the Tuskegee Airmen would opt to attack secondary targets when their primary targets were supported by anti-aircraft artillery. Ben Davis's troops were accused of a lack of courage and not having the same desire to fight as whites. But they also were condemned in the same report for engaging in dogfights with the well-armed, highly maneuverable and dangerous Me-109s escorting Ju-88 bombers and not attacking the more lightly armed, much less maneuverable bombers. One of the general officers, who endorsed the

4 Ibid.

report that condemned the 99th and recommended it be removed from combat, remarked "the Negro type has not the proper reflexes to make a first-class fighter pilot."[5]

The report that would have ended Ben Davis's dream was approved by every senior officer in the European theater and sent to the Pentagon. There it was sanctioned by the commanding general of the Army Air Forces, H. H. Arnold, and sent to the Army Chief of Staff with the recommendations that the 99th be withdrawn from combat, the 332nd Fighter Group (then in training) be assigned to noncombat roles, and the medium bombardment group (then in preparation) be abandoned. When General Arnold forwarded the report to General Marshall, he attached a letter to be signed by General Marshall to President Roosevelt, stating the plan to remove blacks from combat aviation and explaining the reasons.

Marshall sent the report to the War Department's Advisory Committee on Negro Troop Policies, chaired by John J. McCloy. Ben Davis's father, Brigadier General Benjamin O. Davis, Sr., was a member of the Advisory Committee. McCloy was informed that Ben Davis was back in the United States to take command of the 332nd Fighter Group, then training at Selfridge Army Air Field, Michigan, and he saw to it that Ben Davis was able to brief the Advisory Committee on the Army Air Forces report. The entire future of blacks in combat aviation depended on Ben Davis's ability to convince the men of this committee that criticism of the Tuskegee Airmen made by senior colonels and generals, including the very top of the Air Corps, was inaccurate and unwarranted.

This is an outstanding example of Ben Davis's moral courage. He knew that his testimony would contradict the opinion and recommendation of the most senior commanders in the Army Air Forces. This was dangerous for an officer who intended to make a career in the military. In his statement, he called attention to the handicaps suffered by the 99th that were ignored by its critics. No one in the 99th had any combat experience before it flew against the enemy in flights of four, a practice different from the preparation of white units where in their initial baptisms of fire, experienced flight leaders from other units would lead the troops. There was bound to be a lack of self-confidence initially and also mistakes. Ben Davis acknowledged the Tuskegee Airmen lost formation integrity the first time it was attacked, but argued that first-rate Luftwaffe aircraft attacked his obsolescent P-40s. Despite being outnumbered two to one and being surprised by an attack from above and out of the Sun, none of his men fled the battle. They fought it out until the Germans broke contact, and none of the Tuskegee Airmen were shot down by the more experienced Luftwaffe. This was the only such instance of a lack of formation integrity, but it was given as a reason to eliminate blacks in combat aviation.

5 Alan L. Gropman, "History on Two Fronts," in *Makers of the United States Air Force*, ed. John Frisbee (Washington: Air Force Office of History, 1987), p. 235.

Ben Davis argued that his men were determined to fight, had no lack of a desire for combat, but were exhausted because the 99th was undermanned by comparison to white units, and the men, therefore, were over-employed. His men flew more sorties than did comparable white units, sometimes six sorties per day, another fact left out of the 33rd Fighter Group commander's report. Davis, furthermore, denied that his men ever failed to attack an assigned target because of enemy anti-aircraft artillery.

His deposition prevailed, and the McCloy Committee recommended to Chief of Staff Marshall that the 99th deserved more time to prove itself. They reaffirmed the plans to send the 332nd to the war and to create the 477th Bomber Group. The Chief of Staff agreed that the Army Air Forces leadership had presented insufficient evidence, and he ratified the recommendations of the McCloy Committee. Ben Davis went on to Selfridge Army Air Field to take command of the 332nd Fighter Group and moved it to southern Italy in January 1944.

Meanwhile, the 99th, separated from the commander of the 33rd Group and under the command of the white, tolerant commander of the 79th Fighter Group, flourished. George "Spanky" Roberts, a member of the first graduating class from Tuskegee and an officer carefully prepared by Ben Davis, now had command of the 99th Squadron. Over Anzio on the morning of 27 January, after three weeks of arduous close air support for Allied ground troops, 15 Tuskegee Airmen, still flying underarmed, underpowered P-40s, met a larger number of Luftwaffe flying the superior FW-190 and shot down six and damaged another four. Later that day, another three Germans were shot down, and the fourth was listed as probably destroyed. On 28 January, the men of the 99th shot down four more enemy fighters, and, between 5 and 10 February, the Tuskegee Airmen shot down another four enemy aircraft. The drought was over.

Lieutenant Colonel Benjamin O. Davis, Jr., United States Army Air Forces, in a P-51 cockpit in 1944. (Supplied by the United States Air Force)

Also in January 1944, the 332nd Fighter Group, a three-squadron organization, arrived in Italy flying the export model (flown hard previously by Soviet pilots) of the truly obsolete P-39. This low and slow ground attack fighter had unreliable landing gear, a single cannon located between the legs of the pilot, two machine guns mounted one each wing, and the engine mounted behind the pilot. It was not an easy airplane to fly or land, but the Tuskegee Airmen under Ben Davis's leadership made the most of it.

Several months after arriving in Italy, because of heavy bomber losses, senior leaders asked Ben Davis if he would entertain transitioning to escort duty for the 332nd. Because the glamour mission of Air Corps fighter units was in air-to-air combat escorting (protecting) bombers, Ben Davis seized this opportunity. Soon the unit was joined by the 99th and training to fly the P-47 Thunderbolt, a rugged, long-range fighter that was well armed and armored. Soon thereafter, within weeks, the unit transitioned again, this time to the P-51 Mustang, arguably the best fighter produced by any combatant during WWII.

From that point in the late spring of 1944 to the end of the air war in Europe, Ben Davis's group flew 200 missions escorting B-17s and B-24s to heavily defended targets in Italy, Rumania, the Balkans, and Germany. Ben Davis demonstrated great leadership in several areas. He drove the 332nd to learn to fly, fight, and maintain a completely new airplane (the P-47). He trained them for an entirely novel mission, bomber escort, in a matter of days. Then again in a matter of days they learned to fly another thoroughly different airplane, the P-51. Throughout WWII, the 99th and 332nd maintained its airplanes as well as white fighter units and had the same percentage of aircraft ready to fly and fight as any other unit in the theater. As importantly, the pilots thoroughly mastered the escort mission. For evidence of the mastery of the Tuskegee Airmen, we will examine some of their missions.

During the two years in the combat zone, the Tuskegee Airmen flew more than 1,500 missions, about 15,000 sorties. In its last 10 months, the 332nd, joined by the 99th, flew 200 escort missions (as well as ground attack missions), roughly 10,000 sorties, and its record was unique. The Tuskegee Airmen never lost a friendly bomber to an enemy fighter despite the defenses Ben Davis's men had to overcome. On 9 June, for example, 39 P-47 Tuskegee Airmen "Redtails" (called that because the entire tail assembly was painted bright red, and each fighter unit had a distinctive paint scheme) led by Ben Davis took off from Ramitelli Air Base on the east coast of central Italy to escort B-24 Liberators bombing targets near Munich, Germany, a well-defended German city. In one part of the mission, Ben Davis led a formation of 8 P-47s attacking 18 Me-109s. Over the entire mission, the Tuskegee Airmen shot down five German fighters, losing one P-47. Ben Davis was awarded the Distinguished Flying Cross for his leadership and bravery on this mission.

Ben Davis also had to overcome the oft-repeated statement that he was the only Tuskegee Airman with command presence and management ability to lead. All the other black officers according to this racial slur were not intelligent, brave, or capable enough to lead men in combat. This example of prejudice is easy to refute because Ben Davis's lieutenants led most of the 200 escort missions, and their records as leaders were outstanding too. On 25 June 1944, for example, the pilots of the 332nd led by Captain Joseph Elsberry sank a German destroyer in the Gulf of Venezia, another unique achievement. This successful attack was the only sinking of a major combatant solely by machine gun fire during the entire war, despite numerous attempts to do so.

Similarly, on 18 July, 66 P-51s led by Captain Lee Rayford flew from central Italy to southern Germany on a bomber escort mission. The Tuskegee Airmen reached the rendezvous point on time, but the bombers were late. The 322nd stayed in the area, burning precious fuel, because the men understood the costs to the bombers if they were not escorted. Fighting outnumbered, the Tuskegee Airmen destroyed nine Me-109s and two FW-190s without the loss of a single American fighter. In another mission later that month also led by Captain Rayford, the Tuskegee Airmen, outnumbered by attacking German fighters again, destroyed eight Luftwaffe fighters.

These successes, and there were many more, needed to be emphasized because a senior commander under whom the 332nd served testified before a group of senior generals that all the success of the 332nd should be credited to Ben Davis solely because, this senior general asserted, unless Davis led a mission, there was doubt that it would be completed. He argued further that the other officers in the 332nd lacked leadership, initiative, aggressiveness, and dependability. This was really no compliment for Ben Davis. What kind of a leader would he have been if he could not mentor junior leaders? In any case, this statement is an example of racial bigotry of the worst kind. The general officer who made the comment always called black officers "niggers" in private conversation and went to his grave arguing that Air Force racial integration "ruined the Air Force."[6] I cite this to indicate what bigotry and racism the Tuskegee Airmen had to overcome despite their achievements. This particular general went on to achieve four-star rank, retired more than a decade after WWII, and never lost his biased attitude of blacks, living well into the 1970s.

On 31 March 1945, the Tuskegee Airmen shot down 13 Luftwaffe fighters in air-to-air combat, and a week earlier Ben Davis led his men on a 1,600-mile roundtrip escort mission to Berlin, the most heavily defended target in the Third Reich. For this 24 March mission, the 332nd assembled 54 P-51s to defend B-17s on a bomber raid of the Daimler-Benz tank assembly factory. The Tuskegee Airmen were directed to escort the bombers on the run into the target up to the "initial point" from which the bombers would fly straight to the target in tight formation for mutual protection.

The 332nd fought its way to the target against the best the German air force could marshal against them, including at least 30 of the new jet fighter aircraft, the Me-262, and the newest rocket fighters. Ben Davis and his men were supposed to be relieved just short of the target by another escorting unit, but that group of fighters failed to arrive. Despite the fatigue that came from flying a Mustang for 800 miles at the end of which was a furi-

6 Alan L. Gropman, *The Air Force Integrates, 1945–1964* (Washington, DC: Air Force Office of History, 1979), pp. 91, 92.

ous air battle, in spite of the fact that his men had barely enough fuel to fly back to Italy, Davis ordered the Tuskegee Airmen to remain with the bombers, and, of course, despite the peril, they did. Ben Davis's men shot down three of the Me-262 jets, the third, fourth, and fifth jets to be destroyed in the history of air warfare. The 332nd was awarded the Distinguished Unit Citation for this mission. The pilots and their legion of support personnel were cited for their "enthusiasm" and "esprit de corps," and also for "conspicuous gallantry, professional skill, and devotion to duty."[7]

Three weeks later, on 15 April 1945, Ben Davis led 36 P-51s through low clouds and mountainous terrain to attack a railroad line near Munich, Germany, and also Salzburg, Austria. The Tuskegee Airmen struck well-defended targets and made repeated strikes destroying or damaging six locomotives and only left the target area when there were no more targets to destroy. For persistently striking these key logistics targets in marginal weather and against solid anti-aircraft artillery fire, Ben Davis was awarded the Silver Star, the third most prestigious combat decoration.

On 26 April, the Tuskegee Airmen had the honor of killing the last four enemy aircraft destroyed in the Mediterranean theater of operations during WWII, and 11 days later Germany surrendered. The Tuskegee Airmen had destroyed more aircraft then they had lost—111 enemy aircraft killed to 66 Tuskegee Airmen losses from the earliest days in North Africa, through the campaign in Sicily, to the ground attack P-39 missions, to escort and ground attack Mustang sorties. Of those 66 casualties, only 6 or 7 had been shot down, for an outstanding kill ratio of about 15 to 1. Significantly, no Luftwaffe fighter ever shot down a single American bomber during 200 Tuskegee Airmen escort missions. This unique record is evidence to more than the skill, dedication, and discipline of the Tuskegee Airmen; it is proof of Ben Davis's leadership.

His bomber-escort tactic, moreover, differed from almost all other fighter commanders. Ben Davis insisted the Tuskegee Airmen remain with the bombers they were escorting and not hunt for enemy fighters in nearly empty skies. The men of the 332nd remained in sight of the potential German attackers and would even break off several of the fighters to escort wounded bombers back to a friendly base. The Luftwaffe eagerly pounced on crippled bombers that fell behind the formation because of a lost engine or for other mechanical reasons. Ben Davis's approach to bomber escort was certainly a reason for the Tuskegee Airmen's unique record of never losing a bomber to an enemy fighter. Because of its success, bomber units began to request the 332nd for escort duty, and Ben Davis changed the name of his fighter from *Agitha Jo* (after his wife Agatha) to *By Request.*

With the war in Europe winding down, in May Ben Davis was rushed back to the United States to take command of the 477th Medium Bombardment Group to prepare

7 Ulysses Lee, *The Employment of Negro Troops* (Washington, DC: Office of the Chief of Military History, 1966), p. 521.

it for combat in the Pacific. This unfortunate organization of Tuskegee Airmen bomber crews had mutinied in April 1945 because of the bigotry and discrimination practiced by its group commander and all other white officers who served him. Ben Davis quickly overcame the severe morale problems that had provoked the mutiny and brought the unit to full combat readiness by passing all inspections clearing the organization to ship to the Pacific when the atomic bombings of Hiroshima and Nagasaki ended WWII.

Ben Davis remained in command of the 477th and moved it Lockbourne Army Air Field near Columbus, Ohio. The unit soon converted to P-47 fighters and was renamed the 332nd Fighter Group and later the 332nd Fighter Wing. In September 1947, the Army Air Forces were split off from the United States Army, becoming the United States Air Force. Ben Davis and the 332nd remained at Lockbourne Air Force Base until the late spring of 1949 when the United States Air Force racially integrated.

Racial integration came to the Air Force before the other services mainly because of the record of Ben Davis and the Tuskegee Airmen. Very few senior officers, uniformed or civilian, saw the merits of racial integration during WWII and virtually none before the war. In 1945, the Secretary of the Navy, for example, believed integration would improve the Navy and directed that all career fields become open to people irrespective of race; but his command failed to move the Navy, and only token blacks were permitted to serve in integrated fashion. As soon as the Air Force became independent, however, the Deputy Chief of Staff for Personnel, Lieutenant General Idwal Edwards, ordered one of his sharpest staff officers, Lieutenant Colonel Jack Marr, to investigate segregation. Marr proved to himself and to Edwards that blacks with the same aptitude as whites and given the same training as whites performed as well as whites, and the focus of his intense study was the Tuskegee Airmen. He recognized Ben Davis's leadership, the capability of his subordinate commanders who maintained his tactics when he was not in the air with them, and who, more significantly, sustained the escorting success. The enlisted men, furthermore, had the same record as white mechanics and armorers. Marr examined other black service units and found that blacks with the same aptitude as whites who were given the same training got the same results as whites. He recommended to General Edwards that racial integration become the personnel policy because the foundation of segregation (racial inequality) was false. General Edwards was eager to integrate racially because segregation—maintenance of two sets of facilities—was costly, and also because it provoked discrimination, which in turn fomented disorder, with which General Edwards was intimately familiar.

General Edwards convinced the Air Force Chief of Staff, General Carl Spaatz, the Vice Chief of Staff, General Hoyt Vandenberg (who soon succeeded Spaatz), and the Air Force Secretary, Stuart Symington (a race-relations pioneer as a businessman and a future senator from Missouri), that Air Force effectiveness would be enhanced were it to integrate racially. With the senior-most leaders on board, the Air Force announced in April

1948 that the Air Force would soon racially integrate its people. Lieutenant Colonel Marr and General Edwards relied on Ben Davis to assist them through the practical steps of racially integrating. The first black unit to be dissolved was the 332nd, and its men were sent to all corners of the world performing the same missions in formerly white units that they had performed in the 332nd. Four of the men, for example, went to Luke Air Force Base in Arizona to teach new pilots how to fly fighter aircraft, almost all of whom were white.

On the eve of the disbanding of the 332nd, Ben Davis called his unit together and told them of the new racial policy and explained that it meant equality of opportunity not just for them, but also for all blacks enlisting in the Air Force. No longer would quotas or any other form of discrimination on an Air Force base be allowed to hamper the advancement of blacks. He told his troops that he expected the other services would follow in time. He wanted his troops to be proud of their contribution to this reform. Their achievement had accomplished this dream for all blacks. He told the unit they had nothing to fear from integration; they were not just "colored troops," but fully qualified men and women of the United States Air Force, as ready and mission-proficient as any whites. The Tuskegee Airmen had proved themselves with their valor in the skies over North Africa, the Mediterranean, and Europe during WWII. He told the troops, "I have faith in you. I have the fullest confidence in you. I know what you can achieve, and I know that you will succeed."[8] Racial integration in the Air Force became reality 5 years before *Brown v. the Board of Education of Topeka*, 15 years before the public accommodations Civil Rights Act of 1964, and 16 years before the revolutionary Voting Rights Act of 1965. Ben Davis, inspired by flight as a teen, had used aviation to open fully one of the first mass organizations in the world to black people.

GENERAL DAVIS'S CAREER AFTER AIR FORCE RACIAL INTEGRATION

Ben Davis soon left Lockbourne Air Force Base for the Air War College in Montgomery, Alabama, where he won the praise of his classmates and his instructors, and from War College to the Pentagon. All of his classmates at Air War College and all of his subordinates in the Pentagon were white. He moved from Washington to Korea in 1953 via jet fighter and gunnery schools. In Korea he was the commander of the 51st Fighter Interceptor Wing at Suwon Air Base. Virtually every other man in his wing was white.

8 Interview of George "Spanky" Roberts by Gropman, February 1974, Washington, DC.

Indeed, everywhere Ben Davis traveled in the 1940s, 1950s, and 1960s, he was a pioneer. After his assignment in Korea, he moved to Japan and Far East Air Force Headquarters where he was promoted to Brigadier General, moving from there to the Philippines very briefly, and then on to Taiwan where he was the commander of Air Force Task Force 13, an organization he built from scratch to defend Taiwan from an attack by the People's Republic of China. He grew his organization in short order from 5 officers and enlisted men to more than 430 people.

From Taiwan, Ben Davis moved to Europe where he held several positions, the highest of which was the very demanding Deputy Chief of Staff for Operations. It was at Wiesbaden Air Base in Germany where Ben Davis pinned on his second star as a major general. He moved from Germany in the spring of 1965 back to the Pentagon, where he was the Director of Manpower and Organization. In 1967, he was promoted to lieutenant general and became the vice commander of United States Forces, Korea. From that crucial billet, he was moved to become the commander of 13th Air Force, supervising and directing more than 55,000 people all over Southeast Asia. Many of the fighter wings engaged in combat during the Vietnam War came under Ben Davis's command. He moved from that key position in the late 1960s to become the deputy commander of United States Strike Command, and in 1970 he retired from the Air Force while holding that central position. He had advanced further than any other black man in military history. Interestingly, his highly distinguished career had begun with a $5 barnstorming ride in a biplane that took off from Bolling Air Field and flew over Washington, DC.

Assistant Secretary of Transportation Benjamin O. Davis, Jr., shaking hands with President Gerald Ford in 1974. (Supplied by the United States Air Force)

After leaving the Air Force, Ben Davis continued to serve his country, first as Director of Public Safety in Cleveland, Ohio, then Director of Civil Aviation Security in the Department of Transportation (where he ended hijackings of aircraft by training air marshals and introducing metal detectors at airports, among other reforms), and later as Assistant Secretary of Transportation for Environment, Safety, and Consumer Affairs. President Richard M. Nixon appointed him to that position. Ben Davis retired as

Assistant Secretary in 1975, receiving a warm letter from President Gerald Ford: "Under your direction . . . security of the American airways has been restored with an absolute minimum of disruption to the freedom of movement of travelers. You can take pride in this accomplishment and a large measure of personal satisfaction in the physical and emotional well being the program brought to millions of American travelers." The letter was signed, simply, "Jerry Ford."[9]

Ben Davis's contribution to air travel safety was significant, but it pales in comparison to his role in the transformation of the Air Force and the other armed forces. The achievement of racial integration is of immense (truly incalculable) benefit to his people and country. He set out in 1941 to demolish the nucleus of racial segregation, the false notion that blacks were inherently inferior to whites, and he succeeded. The integrated armed services today are infinitely more capable than the segregated ones of 1941, when Davis soloed at Tuskegee. When Davis moved from Fort Riley to Alabama, the Army Air Corps and United States Marine Corps were keeping out 10 percent of the population, more than 10 million people, simply because of race. In addition to depriving the military services of capable black fighting men and women, equally as important, racial segregation provoked race riots. There were race riots in every state and all combat theaters during WWII. The Air Force suffered a black officer mutiny in 1945 at Freeman Army Air Field, Indiana. Racial discrimination also provoked race riots at MacDill Army Air Field and Fort Worth Army Air Field. These ended in 1949 with the integration of the Air Force.[10] When Ben Davis went to Tuskegee, he was the only black officer in the Air Corps; today there are more than 3,500 black officers filling every role in the Air Force and all other services. Numerous black officers have achieved four-star general rank, and one such four-star general served most ably as Chairman of the Joint Chiefs of Staff. General Colin Powell never fails to acknowledge that he climbed to that position on the backs of the Tuskegee Airmen.

Ben Davis continued to serve his country on numerous commissions. Because of his achievements in the military and his contribution to the integration of the armed forces, and because military integration led to integration in the broader society, President Bill Clinton, wearing a red blazer presented to him by the members of the Tuskegee Airmen in late 1998, promoted Ben Davis to a four-star general. Nobody would deny he deserved that promotion.

9 Letter from President Ford to Davis, Benjamin O. Davis, Jr., Papers, Smithsonian Collection, Silver Hill, MD.

10 See Gropman, *The Air Force Integrates*, pp. 17–31, 62–64, 64–71, for information on the Freeman Field mutiny, racial disorders in the military during and after WWII, and the MacDill and Fort Worth race riots. See also Lee, *Employment of Negro Troops*, pp. 348–379.

7

Curtis E. LeMay and the Ascent of American Strategic Airpower

TAMI BIDDLE

AN OHIO NATIVE OF FRENCH-CANADIAN DESCENT, CURTIS EMERSON LEMAY ROSE FROM HUMBLE ORIGINS TO BECOME THE COMMANDER OF THE MOST POWERFUL MILITARY ORGANIZATION THE WORLD has ever seen. Aloof, focused, clipped in speech, and often blunt in manner, LeMay came to symbolize American aerial prowess; he became, over time, a kind of one-man embodiment of the daunting, unflinching military might the United States had achieved by the mid-20th century. The aim of this essay is to assess LeMay's role in the development of American aviation. It is, as well, to peel away some of the layers of myth that now blanket the man—or the memory of the man—and to portray him as he really was.

LeMay was, without question, one of the most proficient, influential, and successful field commanders in modern history. During the Second World War his ideas and innovations had immediate and lasting effects on the course of American strategic bombing.

After the war, he oversaw the Berlin airlift before becoming the legendary head of the Strategic Air Command (SAC), the frontline force in America's Cold War with the Soviet Union. Because of the professionalism and proficiency he displayed as a young officer, LeMay was called on repeatedly to perform challenging and high-profile tasks for his young service. The inheritance he left behind for the U.S. Air Force was one of exacting standards, personal bravery, aggressiveness, technical prowess, and commitment to duty.

General LeMay, while Commander in Chief of the Strategic Air Command, is shown in unfamiliar surroundings aboard the aircraft carrier USS Midway. *In 1956, the general joined 50 high-ranking NATO officials for a one-day indoctrination cruise with a U.S. Sixth Fleet carrier taskforce. (Photo courtesy of the United States Air Force Museum)*

Because LeMay became so firmly fixed in the American mind as a living symbol of strategic air power, and because he seemed to fit the role and play the role so well, he was easily caricatured. His Manichaean worldview only increased the tendency of observers to draw him as a kind of cardboard cutout—grim-faced and steely-eyed, clenching a cigar between his teeth—a cigar that was similar in shape to the fuselage of the immense B-36 bombers he commanded in the early years after WWII. While those depictions reflect key elements of his persona, and are in line with the image he was happy to cultivate in the public mind, they do not capture the full texture of the man or reveal the tensions that sometimes lurked just beneath the surface of the scowling exterior.

YOUTH AND AIR CORPS

The year of his birth, 1906, meant that Curtis LeMay would, quite literally, grow up with airplanes. It meant, too, that the progress of his life would track in parallel with the rise of American power through the early years of the 20th century. He was, from the first, drawn to aviation. He clearly remembered seeing his first airplane, in the sky over west Columbus, Ohio, on a winter's day; he was still a small child. Instinctively, he ran after it, trying to chase it down and catch it as one might try to catch a butterfly. The disappointment he felt in his failure was consoled only by the thought that he might have another chance on another day. Reflecting back on that crystalline memory, he would later say: "I wanted not only the substance of the mysterious object, not only that part I could have touched with my own hands. I wished also in a vague yet unforgettable fashion for the drive and speed and energy of the creature. Also I needed to understand and possess the reason and purpose for this instrument—the *why* of it as well as the *what*."[1] His need to understand the "why"—including the science and technology behind it, and its larger role and purpose—would distinguish LeMay from many of his contemporaries who were drawn to flying mainly for its inherent thrill. LeMay was intrigued by the science; no detail of the machine escaped his attention, none was too insignificant to rouse his curiosity. He wanted to know and master the technology of the machines that so fascinated him.

Since money was always tight in his family, LeMay worked his way through school and then college. If he did not quite have a sense of destiny about him, he surely had determination. He worked for the *Columbus Citizen*, receiving newly printed newspapers and dividing them among the boys riding the afternoon paper routes. He also delivered telegrams for both Western Union and Postal Telegraph, transported candy from a local shop to drugstores and confectionery shops, and hauled packages to department stores. Even at a young age he was enterprising and entrepreneurial. He found some time for the Boys Scouts of America, and he bought himself a .22-caliber rifle that filled the few spare hours of leisure time he had. The rifle competed, though, with his developing interest in radios. He passed up, sometimes regretfully, the typical adolescent socializing in favor of work opportunities. But he found pleasure in the opportunity he had, now and then, to reach far-off stations on his homemade radio set. (Much later, during his tenure as Chief of Staff of the U.S. Air Force, he would build himself a color television set from scratch.) Taking responsibility seemed to come naturally to him. The oldest child in a family that expanded every few years, he often went fishing in order to provide his siblings with the evening meal that his father seemed complacent about providing regularly. He learned, early on, to be purposeful and deliberate in the use of his time.

In 1924, LeMay enrolled in the Ohio State University in his hometown of Columbus; he hoped for a degree in civil engineering. Six days a week he went to work in a nearby foundry. His shift, which ran from 5 p.m. to 2 or 3 a.m., left him so tired that he regularly slept through his 9 a.m. class. He also enrolled in the Reserve Officer Training Corps, not only because he believed that young men ought to have some military training, but also because he hoped it might be a route to pilot training. He later would write, "With me it was flying first and being in the service second."[2] His enthusiasm was fueled

1 Curtis E. LeMay and MacKinlay Kantor, *Mission with LeMay* (Garden City, NY: Doubleday, 1965), p. 14.
2 Ibid., p. 39.

by the exploits of those like Charles Lindbergh and Amelia Earhart, who filled the newspapers and magazines with the adventure and romance of newly emergent aviation. He watched in awe when six Army de Havilland airplanes flew into town one day. He recalled, "Every time I closed my eyes for a long while afterward all I could see were leather flying helmets and goggles; all I could hear was the sound of those Liberty engines."[3]

Through a combination of ingenuity, good luck, and sheer determination, LeMay won himself a slot as an Army flying cadet in 1928. That autumn he took the train from Ohio to March Field, California, where he would try his hand at flight school. He was accompanied on the trip by his schoolmate Francis Hopkinson Griswold, who would later serve as the Chief of Staff of the Eighth Fighter Command during WWII, and then as LeMay's deputy commander at SAC. As his train rolled westward, he had no idea what lay in store for him. He knew only that each mile brought him a little closer to his dream of flying.

On 12 October 1929, not long before the stock market crash would deliver a fearful blow to the once buoyant and booming American economy, LeMay graduated as a commissioned officer in the Air Corps. His life in the young and still subordinate service was affected by those events shaping the institution as a whole, including the trials and travails of flying the mail in 1934 and, later, the thrill of receiving the first B-17 bombers. Throughout the 1930s, the Air Corps was still seeking its own identity. If the Army thought its air arm ought to serve the needs of ground troops first and foremost, the flyers had other ideas. LeMay developed an immense respect for Lieutenant Colonel Robert Olds, who came to command the Second Bomb Group shortly after LeMay had been posted there. He inspired in LeMay a forward-thinking sense of mission (to fight and win wars with air power) and a proactive attitude toward preparing to fulfill that mission, including close attention to detail. Olds soon discovered that LeMay never had to be told anything twice. Once Olds had pointed out the necessity of knowing the day's weather, for instance, LeMay began each of his mornings in the weather room, poring over data and charts.[4]

In the 1930s, the isolationist mood of the nation, the concern with domestic affairs, and the defensive national security posture meant that military budgets and national preparedness were far down on the list of President Franklin D. Roosevelt's immediate priorities. Like many of his colleagues in the Air Corps, LeMay grew to resent the entrenched service rivalries that were only intensified by scarcity. And he was deeply aggrieved by what he felt were the Army's efforts to constrain the role and the potential of its own air arm. But if progress in military aviation slowed, it surely did not cease. And the aviators found ways to end-run official proclamations about supporting ground

3 Ibid., p. 41.
4 Ibid., pp. 132–133.

troops. Indeed, as the 1930s progressed, the Air Corps developed an increasingly sophisticated and independent vision of its role in future wars.[5]

More than many of his contemporaries, LeMay understood instinctively that the future of bombardment aviation would depend on mastering the science and technology of both flying and aerial bombing. A tireless advocate of navigation (in a service that often failed to value it adequately), LeMay sought out opportunities to learn and then teach dead reckoning and celestial navigation. He also trained in instrument flying and worked to hone all the skills required of those who would fly the new four-engined bombers. He took it upon himself to understand the details of advanced aviation and to teach them to others.

His skills as a navigator prompted Olds to assign him as lead navigator for the August 1937 exercise pitting B-17s against a flotilla of ships, including the battleship *Utah*. He served as lead navigator once again, during a special flight of six B-17s to South America, in February 1938. His professionalism and competence on these missions earned him the respect of his peers. They also earned him the job of chief navigator for a high-profile, high-publicity exercise that Air Corps leaders felt would bring attention to the long-range capabilities of the U.S. bomber forces—the aerial interception of the inbound Italian liner, *Rex*. Daunted by the task, LeMay nonetheless understood what it might mean for the future of the Air Corps; never before had so much ridden on the skills he had carefully developed since making his first few novice flights at March Field some 10 years earlier. LeMay would later write of his reaction to hearing the news that he had been handed this exalted task, "Responsibilities and puzzles inherent in these glad tidings were enough to cool me all the way down to my toes."[6]

Dreadful weather—"low, storm-driven, opaque" clouds—greeted LeMays's crew on the morning of 12 May 1938, adding to the disquiet he felt about the *Rex*'s failure to radio her position the previous evening.[7] The position report came in at the last possible moment—literally as the plane was getting ready to taxi to the runway. The *Rex* was progressing westward more slowly than had been forecast, and LeMay's B-17 had to fly hard and fast through the terrible turbulence stirred up by dozens of thunderstorms; he did his navigational calculations in between giant downdrafts that would pull the aircraft toward the ocean, sending his pencils, charts, and data upward (momentarily weightless) in midair. Since his plane also was fighting a strong headwind, LeMay knew he had no margin for error; the NBC radio team on board was due to broadcast at 12:25 p.m. If the *Rex* was nowhere to be found, the Air Corps would suffer an embarrassment they felt they could ill-afford.

---

5 For an overview, see Tami Davis Biddle, *Rhetoric and Reality in Air Warfare: The Evolution of British and American Ideas about Strategic Bombing, 1914–1945* (Princeton: Princeton University Press, 2002), pp. 128–175.

6 LeMay and Kantor, *Mission with LeMay*, p. 184.

7 Ibid., pp. 186–187

Under the most dramatic of circumstances, LeMay came through, finding the *Rex* right on time for the NBC broadcast. The newspapers and magazines took note, and the *New York Times* printed the words that the Air Corps longed to hear: "valuable lessons about the aerial defense of the United States" will be drawn from the interception of the *Rex*, "which already has furnished . . . a striking example of the mobility and range of modern aviation."[8] The chapter about the *Rex* in LeMay's autobiography reads like a thriller; even those who know the outcome find themselves turning the pages quickly and compulsively. The chapter is also a marvelous case study in determination, courage, and leadership for young servicemen and servicewomen.

WORLD WAR II

Curtis LeMay was still a young man at the start of the Second World War, but his 12 years of experience in the Air Corps (later, as of 20 June 1941, called the "United States Army Air Forces," or USAAF) ensured that he would be handed key responsibilities quickly. Even as the United States remained neutral, he was called upon to ferry passengers across the Atlantic, from Canada to Prestwick, Scotland, and back. He made his first crossing in a new B-24, having flown the aircraft exactly one time before heading out into the heavy overcast, stormy weather, and wing icing so characteristic of the North Atlantic. There was nothing by way of air traffic control at that point—no fixed altitudes for planes headed east or west; aircraft in the increasingly crowded skies avoided each other through pilot skill and sheer luck. LeMay would later remark, "We old ferry types look back on those days now and wonder why the surface of the North Atlantic was not fairly crawling with dismembered airplanes, crews, and passengers."[9] They flew low through the notoriously turbulent skies, since there was no oxygen provided for the passengers.

In his memoir, LeMay describes the turmoil, hurry, and shoestring nature of air operations at that point in time. By the late 1930s, President Roosevelt had decided, at last, to invest heavily in military aviation; the Air Corps/USAAF was thrown into a frenzy of expansion as it tried to meet its own needs as well as those of its future ally, the Royal Air Force. Expansion was often haphazard and frustrating for all involved. Still, there was little alternative. Pearl Harbor, despite its tragic toll in human life, came as something of a relief to LeMay; now, after so many months of tense uncertainty, the path was clear—the route was charted.[10] It would, however, be a hard road indeed.

8 *New York Times*, quoted in LeMay and Kantor, *Mission with LeMay*, p. 192.
9 LeMay and Kantor, *Mission with LeMay*, p. 204.
10 Ibid., p. 208.

In May 1942, LeMay was ordered to take command of the newly formed 305th Bomb Group. His novice, rather ragtag unit was ordered from Salt Lake City to Spokane, Washington so that it might intercept any Japanese forces daring to approach the West Coast. LeMay could think only about the shortages and shortcomings: he did not have enough airplanes; he did not have properly trained men; and he did not have adequate supplies or spare parts. When the group came east to Syracuse, New York, in preparation for heading overseas, they found themselves facing unseasonably cold autumn weather with not a shred of warm clothing. Illness was rampant, and LeMay himself was stricken with a virus that would leave the right side of his face partly paralyzed. With no time to seek out cures or feel self-pity, LeMay drove himself and his men ever harder. He did everything he could think of to make his unit war ready as quickly as possible, handing his men a rigorous training schedule that earned him the nickname "Iron Ass."[11] The pride he felt as he led his men across the Atlantic and into war was tempered by his acute knowledge of their inadequacies—novice pilots (barely out of basic trainers), novice navigators (many with no experience over water), and novice bombardiers. All he could do was continue to train and train hard.

Arriving in England, LeMay was troubled by reports of bombers engaged in evasive maneuvering right up to the moment that they dropped bombs on their targets. The whole point of aerial bombing, he thought, was to "drop bombs where they would do most harm to the enemy." Trying to evade enemy defenses directly over the target could only ensure poor aiming and bomb dropping. Even though he had, at that point, no experience of flying over targets in war, LeMay knew instinctively that a long straight shot was necessary if bombardiers were to have even half a chance at accuracy. Whenever he identified problems, he immediately sought ways to work them out. And working them out usually meant applying logic and math. His was a mind of endless calculations, assessments, and comparisons; he believed that if he could get the data and the right equation, he could work out a solution—at least the best possible solution. He knew that this particular problem needed working out, and he wrestled with it.

One autumn night, prior to the 305th's first run at a target, he got up out of bed and fished out the old artillery manual he had packed in his foot locker. He would do the math for a precision fire problem: a German flak gun versus a B-17 at 25,000 feet. How many rounds were necessary to achieve a single hit on a plane while it was within the anti-aircraft artillery's field of fire? His answer, based on the speed of the guns and the numbers of guns available (according to available intelligence), was "372 rounds in order to hit a B-17 flying dead level, straight in." Though he had not calculated the exact gun

11 Captain B. W. Crandell, "Iron Ass Was the Name," *Air Force* magazine (August 1943); typescript in Papers of Gen. Curtis E. LeMay (hereinafter LP), Box 39 (Misc. Biographical Data), Library of Congress Manuscript Room, Washington, DC; and LeMay and Kantor, *Mission with LeMay*, p. 217.

the Germans used, he thought his number was close enough—close enough to counter the widespread but wrongheaded notion that flying level for more than 10 seconds would guarantee being shot down.[12]

When he took the 305th over its first real target, at St. Nazaire on 23 November, he flew the lead bomber. He took his planes in, long and straight over the target. Not a single one fell to flak, and they put more than twice the number of bombs on target as any other group that went to St. Nazaire that day. "Now I knew just how we were going to bomb," LeMay remarked, laconically, in his memoir.[13]

His mind was never at rest. Shortly after his arrival in the European theater, LeMay also had worked out a bombing formation that would become standard for crews flying out of England. It was effective, but simple enough for green pilots to master. In addition, he worked out a scheme that helped lead crews become highly familiar with specific targets through the use of reconnaissance photographs. The crews that had studied a given target would lead the way to it when it had been designated as the primary target of the day and announced as such at the mission briefing. This idea evolved into a Lead Crew School.[14] All of these innovations brought with them important improvements in the long-range bombing performance of the U.S. Army Air Forces in Europe.

In 1943, LeMay was promoted again and, in the reshuffling and renaming so characteristic of the burgeoning AAF at that time, emerged as the commander of the Third Air Division. He moved to Elvenden, in East Anglia. At the time, he felt wholly underqualified for the job, since, as he believed, he had barely begun to absorb the lessons taught to him as a group commander. But he threw himself into the task with characteristic determination and energy. Not long after taking up his new post, LeMay was called upon to play a leading role in the joint assault on the cities of Schweinfurt and Regensburg, both deep inside Germany. The First Air Division of the Eighth Air Force would head to Schweinfurt, home to most of Germany's antifriction bearings production, and the Third Air Division would aim for Regensburg with its Messerschmitt factory. The theory was to use the Third Division to draw defenders away from the First Division attack so that the latter could go to Schweinfurt less hindered by defenders than would otherwise be the case. Instead of turning around and heading back to England, the Third Division would continue south to Allied-held North Africa, where the planes would land, rest, refuel, and reload for another attack on Germany to be waged on the way home.

LeMay had some questions and concerns about the mission—particularly regarding the facilities he would find in Africa—but he went ahead with his planning and trained

12 LeMay and Kantor, *Mission with LeMay*, pp. 237–238.

13 Ibid., p. 245.

14 Ibid., pp. 235, 256–257.

his crews hard for the bad weather takeoff he rightly assessed they would be called upon to make on the day of the raid. Despite pea-soup fog on the morning of 17 August, the Third Division took off, formed up, and headed for Regensburg. But the larger First Division, due to come in about 10 minutes behind LeMay, did not get off the ground at its appointed time. Indeed it was held up so long that the timing of the mission was thrown off, and LeMay ended up flying to Regensburg without fighter cover.[15] He lost 24 of the 127 planes attacking the target. He managed, however, to hit the target with an impressive degree of accuracy, using highly explosive and incendiary bombs to maximize damage to all the buildings in the industrial complex. The base facilities in North Africa were, as LeMay had feared, inadequate, and his Third Division was held up for days awaiting adequate supplies and attention. Needless to say, this experiment did not sell either LeMay or the Eighth Air Force on the "shuttle base" idea, but LeMay's masterful command of the mission, under less than ideal circumstances, won him further acclaim among his colleagues.[16]

Nothing LeMay had done since arriving in the European theater went unnoticed by his superiors. They had high regard for the good record of his units, his personal bravery, and his leadership skills; they knew that he had won the full confidence of the men under him. Most importantly, perhaps, they respected LeMay's prowess as an innovator and problem-solver. He had consistently sought new and better ways to reach and attack targets from the air, thus helping to enhance the performance of the entire Eighth Air Force. He wrestled with problems constantly; nothing, it seemed, escaped his attention. If there was a job to be done, he would find a way to do it.[17]

Indeed, it was LeMay's "can-do" attitude and analytical approach to problem-solving that led the USAAF's Commander in Chief General Henry "Hap" Arnold to conclude that he might be the right man for the difficult and challenging task of running the 20th Bomber Command in the Pacific theater. The problems inherent in the job had quickly overwhelmed the 20th's first commander, General Kenneth Wolfe, who was fired by Arnold less than a month after launching his first combat mission in the theater. But the problems were daunting indeed. Fuel, ordnance, and supplies had to be brought in "over

---

15 LeMay and Kantor, *Mission with LeMay*, p. 293.

16 Interview of Col. LeMay by Bruce Hopper, Eighth Air Force Historical Section (7 September 1943), III 3 (d), Biographical Facts on Maj. Gen. Curtis E. LeMay [1944], Box 39, LP; and LeMay and Kantor, *Mission with LeMay*, pp. 288–298.

17 The public relations biography prepared on Maj. Gen. LeMay in 1947 stated: "General LeMay probably did more to improve bombing accuracy and to develop the most effective techniques for bombing operations than any other group commander in the Eighth Air Force. In his final tour of duty with the Eighth Air Force, as commanding general of the Third Bombardment Division, General LeMay was credited with heading up the best of the heavy bomber divisions in that air force." See Public Relations Division, Headquarters, Army Air Forces, Biography: Major General Curtis E. LeMay (21 March 1947), Box 39 (Misc. Biographical Data), LP.

the hump" of the Himalayas; to deliver 8 tons of fuel and cargo, a converted bomber required 28 tons of gas for the round trip. Distances were long, and crews had trouble reaching and finding their targets, and then getting home again. The majority of the 20th's bombers had been lost due to causes other than enemy action.[18]

General Arnold felt that the USAAF needed an outstanding performance in the Pacific in order to guarantee postwar autonomy from the Army. Much was riding, especially, on the performance of the AAF's big, new, high-profile bomber, the B-29. Once again the young LeMay—at 38, one of the youngest major generals in the Army—was being asked to step up to the plate for his organization. The press release read: "HEADQUARTERS, XX BOMBER COMMAND, SOMEWHERE IN INDIA—Major General Curtis Emerson LeMay, who for two years played a direct and important part in the strategic bombing of Hitler's Fortress Europe, has arrived at this headquarters to direct the XX Bomber Command in similar operations against the Japanese Empire."[19]

LeMay immediately went to work trying to whip his organization into shape. He had no compunction about removing people who were not up to the task, and he was equally prepared to put in special requests for those officers he knew and trusted. He was well aware that Arnold was looking over his shoulder and breathing down his back. In September, Arnold pressed him to have the B-29s carry maximum bombloads, stating "there has always been a tendency to plan missions so that the weakest pilot will be assured of a safe gasoline reserve on return to base. With the B-29, this tendency must be eliminated completely. Pilots that are weak must either be replaced or trained to a point where they can obtain the maximum from their airplane."[20]

Whenever Arnold posed a problem or raised a question, LeMay sent back a detailed answer indicating that he had not only thought about the issue, but had discerned aspects of it that others had overlooked. His ability to stay a step ahead of everyone, and to eschew complaints and excuses, was a quality that Arnold deeply appreciated.[21] The USAAF chief understood that the logistics of what was called the "China-Burma-India" theater posed monumental difficulties for any commander. And he also appreciated that the B-29, rushed into production straight from blueprints, was full of kinks and bugs that needed to be worked out. The engines, for instance, had a nasty tendency to catch fire. LeMay was frank about the challenges, but wrestled with them energetically and uncomplainingly. He urged Arnold, however, to find a better basing arrangement—one that would ease the logistics problems and bring the bombers a bit closer to the Japanese islands.

---

18 Kenneth Werrell, *Blankets of Fire* (Washington, DC: Smithsonian Institution Press, 1996), pp. 98, 112.

19 Public Relations Office, Hdq. XX Bomber Command [1944], Box 39 (Misc. Biographical Data), LP. "XX Bomber Command" was a designation sometimes used during WWII for the 20th Bomber Command.

20 Letter from Arnold to LeMay (22 September 1944), Box B11, LP.

21 Letter from LeMay to Arnold (19 October 1944), Box B11, LP.

In November, LeMay was able to report that the morale of his crews, despite the B-29's engine problems, was very high. He stated, "We are finally beginning to collect dividends on our investment."[22] Arnold responded promptly, noting, "I think the number of airplanes that you were able to keep in operating condition was impressive. I appreciate the difficulties that you must have encountered and know full well this problem must have been solved largely by the determination of you and your maintenance people." He added, "Your record of operations has also been most gratifying."[23] Afterward, Arnold would use "Dear Curt" instead of "Dear LeMay" for the opening address in his correspondence with LeMay.[24]

In the meantime, the 21st Bomber Command had been established under General Haywood Hansell, an early architect of the American theory of precision daylight bombing against key industrial targets. Hansell was to operate his B-29s out of bases in the Marianas in order to place heavy pressure on the Japanese and their emperor. In mid-November, Arnold told LeMay, "The fine work your people have been doing is providing a standard for the other B-29 units. We are passing to Hansell everything of interest from the XX Bomber Command, and, he, in a recent letter here, stated that he would have to push his people pretty hard to stay in the same league with your command."[25]

But at the end of the day, Arnold did not believe that Hansell, as a field commander, had stayed in the same league as LeMay. The poor weather and the jet stream winds over Japan hindered Hansell's operations and made formation flying and daylight precision bombing virtually impossible propositions. But Arnold was in no mood to hear excuses. By January he had decided to sack Hansell and put LeMay in command of the B-29s based in the Marianas. LeMay was under no illusions about what was, once again, being asked of him. In December, Arnold had written, "As I told you before you went out to India, the B-29 project is important to me because I am convinced that it is vital to the future of the Army Air Forces. I think progress has been made, and you have contributed materially to this .... The report you sent me proves that you have the right attitude in this matter." He added, "I think we can do better bombing with the B-29 than has been done by any aircraft up to this time, and I expect you to be the one to prove this."[26]

But bringing LeMay down to the Marianas did not change the weather over Japan, nor still the winds of the jet stream. LeMay had to find a way to cope with these problems

22 Letter from LeMay to Arnold (6 November 1944), Box B11, LP.

23 Letter from Arnold to LeMay (13 November 1944), Box B11, LP.

24 Letter from Arnold to LeMay (17 November 1944), Box B11, LP.

25 Ibid.

26 Letter from Arnold to LeMay (9 December 1944), Box B11, LP. Arnold wrote, "Of all the successes of the AAF during the past year, none have been more gratifying to me than those of XX Bomber Command." Letter from Arnold to LeMay (1 January 1945), Box B11, LP.

and, somehow, get good results from his bomber crews. He went to work again, shuffling personnel, setting up rigorous training programs, and, above all, thinking his way through the obstacles. The previous December, while still in command of the 20th, LeMay had led an incendiary attack on Hankow, a Chinese city being used as an operating base by the Japanese. The raid was considered a success by the Americans.[27]

Planning for the use of incendiary bombs in the Pacific theater had been underway for some time; the Americans were well aware that the wood and paper construction of Japanese cities would be highly susceptible to fire. General Hansell had undertaken an

Pinpointing ground zero on a scale model of Bikini Atoll, Maj. Gen. Curtis E. LeMay reviews U.S. Air Force participation in Pacific atom bomb tests. Looking on are Brig. Gen. William F. McKee and Maj. Gen. Earle E. Partridge. (Photo courtesy of the United States Air Force Museum)

27 Conrad C. Crane, *Bombs, Cities, and Civilians* (Lawrence, KS: University Press of Kansas, 1993), pp. 127–128.

incendiary raid on Nagoya in December, but he did so only reluctantly since he believed that daylight precision bombing had not yet had an adequate trial.[28] A year earlier, improvements to the M-69 incendiary bomb had made it the most successful incendiary employed against a mock Japanese village at the Dugway Proving Ground in Utah.[29]

In March, LeMay decided to try a radical experiment. He stripped the B-29s of their defensive armament, filled their bomb bays with incendiaries, and flew them over Japanese cities at low levels—some 5 miles lower than their usual altitude. He knew just how risky these tactics were, and he could not rule out the possibility that his losses would be heavy. "We might lose over 300 aircraft and some 3,000 veteran personnel in this attack," he later explained, adding, "It might go down in history as LeMay's Last Brainstorm."[30] But LeMay never had a brainstorm that was not analyzed right down to the last decimal point. He ran the numbers, backward and forward, before setting out on any new course of action. He was always ready to innovate, but he never took random or impulsive risks. Arnold was deeply pleased by the result in Tokyo; on 21 March, he wrote, "I want you and your people to understand fully my admiration for your fine work. . . . Your recent incendiary missions were brilliantly planned and executed." At the same time, though, he kept the pressure on his field commander, stating, "I want you to put the maximum weight of effective bombs on Japanese targets, consistent with sound operating practice. A study of the effect of the Tokyo attack of 10 March and knowledge of the fact that by 1 July you will have nearly a thousand B-29s under your control leads one to conclusions which are impressive even to old hands at bombardment operations. Under reasonably favorable conditions, you should then have the ability to destroy whole industrial cities should that be required."[31]

LeMay still tried to do precision bombing when he could, but weather continued to thwart him. Knowing that Washington wanted results, he proceeded with his nighttime incendiary raids, burning out some 67 Japanese cities in a campaign of vast scope and fury. LeMay oversaw, as well, the use of the two atomic bombs dropped on Japan, on 6 and 9 August 1945. Arnold had made it clear that civilian casualties were not a barrier to the prosecution of these raids. Indeed, there was very little discussion of the issue of civilian casualties at the time, and virtually no protest was made by the American people as they read graphic descriptions of the air attacks in their daily newspapers. Against the backdrop of the terrible, protracted battle for Iwo Jima, General LeMay said of the 9–10 March Tokyo raid, "If the war is shortened by a single day, the attack will have served its

---

28 Michael Sherry, *The Rise of American Air Power* (New Haven: Yale University Press, 1987), pp. 257–258.
29 Werrell, *Blankets of Fire*, pp. 48–49.
30 LeMay, *Mission with LeMay*, p. 10.
31 Letter from Arnold to LeMay (21 March 1945), Box B11, LP.

purpose."[32] The ethical questions raised by the incendiary raids and by the atomic bombs would arise later. In 1945, General Arnold and President Roosevelt wanted the war over with as quickly as possible. LeMay knew this.

Indeed, Arnold's final war dispatch to American citizens was not only unapologetic, but celebratory in tone. It included a map of Japan, showing each of the cities that had been firebombed. To aid readers in understanding the meaning of the map, each city had next to it the name of an American city of approximately the same size. He was proud of the bombing campaign in the Pacific and confident that it would help forward the cause of air force autonomy in the United States. At the end of the war, his victory telegram to LeMay read, "The part you played in developing and commanding the 21st Bomber Command represents one of the outstanding personal achievements of this war. You and the men under your command have indeed made clear to the world the full meaning of strategic bombardment. Your imagination, resourcefulness, and initiative have reflected credit on the entire Army Air Forces. We are intensely proud of what you have done."[33] By the end of the war, LeMay had earned, in addition to a Distinguished Service Cross, a Distinguished Service Medal and Cluster, Silver Star, Air Medal and three Clusters, British Distinguished Flying Cross, Russian Order of the Patriotic War First Class, Presidential Unit Citation of Third Bombardment Division, Legion of Honor (France), and Belgian Croix de Guerre with Palm.[34]

EARLY POSTWAR SERVICE

A few weeks after the war ended, General LeMay was ordered back to the United States. He flew a B-29 nonstop from Hokkaido to Chicago, then on to Washington. LeMay's deep concerns about rapid postwar mobilization and the seriousness of the Soviet threat caused him to consider a dip into politics. President Harry S. Truman had appointed Ohio Senator Harold H. Burton to the Supreme Court, leaving a vacancy in Congress. It was the responsibility of Ohio Governor Frank Lausche to appoint a successor to serve out the remainder of Burton's term. Lausche thought LeMay was a natural candidate for the position, and the general was interested, especially since it would only be a brief appointment (Burton's term was due to expire in 1947). LeMay figured he

32 Warren B. Moscow, "Not a Building Is Left in Fifteen Square Miles," *New York Times* (11 March 1945): 1, 13.

33 Telegram from Arnold to LeMay ([15] August 1945), Box B11, LP.

34 Public Relations Division, Headquarters, Army Air Forces, Biographical Data, Maj. Gen. Curtis E. LeMay, Box 39 (Misc. Biographical Data), LP.

would take a year of leave from military service and then return to the Air Force. But Truman's new Secretary of War, Robert Patterson, told LeMay that it would be inappropriate to take a post in the Senate unless he resigned his commission. That was too high a price, and LeMay chose to remain a serving officer.[35]

LeMay's first postwar position was Deputy Chief of Air Staff for Research and Development. General Carl Spaatz, who would take over for Arnold as the senior Air Force officer, recommended the establishment of the position in an October report he headed on the impact of atomic weapons on Air Force strategy and organization; he specifically requested LeMay for the position. Spaatz knew that research and development were near and dear to General Arnold; he believed that the modern, autonomous U.S. Air Force would have to be equipped with the finest aircraft, rockets, and missiles that scientists and engineers could produce.[36] In his new position, LeMay would have a direct say in shaping the Air Force of the future. He would work to coordinate the efforts of scientists, industry, and the military. He would participate, as well, in Operation Paperclip, a program designed to scoop up talented German scientists before the Soviet Union could get to them. Wernher von Braun and Walter Dornberger, along with many others, were brought to the United States through this means; they would have a direct role in shaping and advancing rocket and missile programs in the United States in the postwar years.[37]

Though LeMay considered himself a field commander, he was later grateful for the experience he gained in Washington during 1946–1947. Though his job sometimes frustrated him, he believed it was of enormous help to him later in his career when he needed to understand the politics and bureaucracy of the Pentagon. As it turned out, LeMay "flew a desk" in Washington only briefly before being tapped to take command of the United States Air Forces in Europe. On 1 October 1947, LeMay earned his third star, making him the youngest lieutenant general in the Air Force. By this time, Air Force leaders had won the coveted autonomy they had sought ever since the interwar years; no longer subordinate to the will of the Army, they could chart their own course for the future.

LeMay's most important role during his tenure in Europe was to head the Berlin airlift, supplying the beleaguered city when rising Cold War tensions and economic reform in the western sectors of Germany caused the Russians to cordon off road access. Under LeMay's careful guidance, the Berlin airlift became one of the most successful and admired aerial operations in American history. Once again, LeMay's skills and attention

---

35 Thomas Coffey, *Iron Eagle: The Turbulent Life of General Curtis LeMay* (New York: Crown, 1986), pp. 249–250.

36 Arnold did not place LeMay in the job immediately, as he initially had another post in mind for LeMay. But Spaatz soon had his way. See Coffey, *Iron Eagle*, pp. 250–251.

37 LeMay, *Mission with LeMay*, pp. 397–398.

to detail paid handsome dividends. During the time it operated, the airlift carried 2,223,000 tons of fuel and food and other supplies into Berlin. Under LeMay's direction, planes were soon landing every 5 minutes. That rate slowed a bit in bad weather, but not by much since the commanding general was constantly at work arranging for better navigation aids and traffic procedures. He also saw to it that several new airfields were built to help supply the city.[38]

STRATEGIC AIR COMMAND

In the autumn of 1948, LeMay was recalled to the United States to take over leadership of the Strategic Air Command (SAC), the first line of defense in the growing Cold War conflict with the Soviet Union. He assumed this crucial task in a period of low ebb. Postwar demobilization and sharply curtailed budgets had left all the services scrambling for men and materiel, even as American commitments abroad continued to expand. In its first two years, SAC had floundered. By 1948, when it became increasingly obvious to American decision-makers that the Soviet Union was likely to be a formidable and long-term foe, the strength, security, and reliability of U.S. strategic forces became a high priority for the nation. Once again, LeMay was called upon to step up and deliver.

Despite the hurdles he faced, LeMay had several things working in his favor. In 1948, fiscal year 1950 budget ceilings had forced reliance on a strategic air offensive, since more costly conventional alternatives were out of reach. In the same year, the Berlin crisis forced President Truman to take seriously the possibility of having to deliver an atomic attack against the Soviet Union. Finally, technological developments in nuclear weapons tests undertaken in the spring of 1948 indicated that the "doctrine of scarcity" that had governed nuclear weapons planning up to that point might not apply for much longer.[39]

Upon taking up his new post, LeMay was appalled at the state of SAC. A simulated radar bombing mission against Dayton, Ohio, revealed just how unprepared for war his new command was. The results, at least, made it easier for LeMay to win widespread cooperation within SAC for the major overhauls he believed were necessary. He would later say of that trial, "Not one airplane finished that mission as briefed. Not one. . . . This really shook up our people, when they realized what sad shape they were in."[40] In his usual fashion, he set right to work identifying problems and devising solutions. He devel-

38 Ibid., pp. 416–417.

39 David A. Rosenberg, "The Origins of Overkill: Nuclear Weapons and American Strategy, 1945–1960," in *Strategy and Nuclear Deterrence*, ed. Steven Miller (Princeton: Princeton University Press, 1984), p. 129.

40 Quoted in Conrad C. Crane, *American Airpower Strategy in Korea, 1950–1953* (Lawrence, KS: University Press of Kansas, 2000), p. 19; and LeMay, *Mission with LeMay*, pp. 432–433.

oped training programs and kept track of every statistic on every item he deemed important. He established safety standards and reduced accidents. In addition, he again followed his usual pattern of cleaning house and handpicking subordinates he knew and trusted from past experience. These included Major General Thomas S. Power, who became the new SAC deputy commander, Brigadier General John B. Montgomery, who became SAC's operations chief, and Major General Emmett "Rosie" O'Donnell, who took command of SAC's 15th Air Force. He also brought in the best enlisted men he could find. In an effort to improve force morale and retention rates, LeMay took a detailed and active interest in base facilities and family housing.

At a key conference held at the Air University in Alabama in December 1948, the Air Force high command supported LeMay's position that the service's highest priority had to be the delivery of the SAC atomic offensive "in one fell swoop telescoping mass and time." From then on, as historian David Rosenberg writes, "LeMay's budgetary and programming requirements for building SAC into a 'cocked weapon' were given top priority."[41] LeMay's uplift and improvement program was grueling: crews trained around the clock in all weather conditions. *Time* magazine would later write, "LeMay picked the outfit up by the neck, shook it in a way none of the old-timers will soon forget, and flung it across the United States to get ready for its mission." He worked tirelessly and demanded no more of others than he demanded of himself. He insisted that if he called his men out to war, he would lead them in the first plane.[42]

While keeping an eye on visual bombing, LeMay insisted that radar bombing rates improve. Concerned about readiness and security, he demanded fast and obvious improvements in both realms. The results were almost immediate. Six months after taking control, LeMay had transformed SAC from a hollow threat to a formidable and disciplined striking force. In April of 1949, *Newsweek* placed LeMay on the cover and announced, in a triumphal tone, "The man who made SAC into this pulverizing force is scowling, black-haired, olive-eyed Lt. Gen. Curtis Emerson LeMay, the 42-year-old genius who supervised the bombing of Japan, including the atom-bomb raids on Nagasaki and Hiroshima, and who, after the war, built up the amazing Berlin airlift as the answer to the Russian blockade. Now he has done it again."[43]

SAC's nuclear capable aircraft increased from 60 in December 1948 to 250 by June 1950. By the autumn of 1951, SAC had 28 bomber and 7 fighter wings, with 15 counted as combat ready.[44] New, immense B-36 bombers entered the SAC arsenal, and old B-29s, reclassified

41 Rosenberg, "The Origins of Overkill," p. 129.
42 "Man in the First Plane," *Time* (4 September 1950): 18.
43 "Bombers at the Ready," *Newsweek* (18 April 1949): 25.
44 Rosenberg, "The Origins of Overkill," p. 129.

as "medium bombers," were replaced with B-50s. The bombers of the future, jet B-47s and B-52s, were in developmental stages, a few years away from service. The ascendance of SAC rankled the Navy, long used to being the first line of defense in the United States, and tension broke out into something approximating all-out service warfare in 1949 when Secretary of Defense Louis Johnson canceled the Navy's super-carrier, the *United States*, in favor of air power. But LeMay was adamant that the strategic bomber force deserved the attention it received. In an article called "Why the Air Force Wants the B-36," *U.S. News & World Report* stated, "Defense of the U.S. in the period ahead is to center around the big B-36 bomber. This is the weapon, determined upon recently by the Air Force, that will carry the initial offensive to the enemy in the event this country is attacked."[45]

By mid-1950, SAC's 60,000 personnel made up 15 percent of USAF strength, and SAC held the biggest piece of Air Force modernization funding. Even as defense spending cut group strength from 60 to 48, SAC's strength rose from 18 to 19, with 6 bomb groups approved to maintain a wartime level of 45 instead of 30 aircraft. These benefits were due to a consensus, within the Joint Chiefs of Staff, that the nation's strategic bombing force took priority in the defense budget.[46] In testimony during the summer of 1949, Chairman of the Joint Chiefs, General Omar Bradley, had testified that, "At the instant of aggression, the United States must fling the full force of its strategic air offensive against the enemy's heartland."[47]

In assuming command of SAC, LeMay became the nation's premier warrior in the increasingly entrenched battle with the Soviet Union. Already legendary due to his role in the Second World War, LeMay seemed, in many ways, an ideal cold warrior—focused, determined, and unflinching. He fit the role, and he quickly came to define it in the eyes of most Americans. But when a real war came, LeMay found himself rather more constrained than he would have liked. As soon as North Koreans began flowing over their border and into the south, LeMay commenced his preparations. He called General Rosie O'Donnell of the 15th Air Force to SAC headquarters and poured over available intelligence on North Korean targets. As the Far Eastern Air Force (FEAF) was activated, LeMay readied crews to prepare for relocation to Yokoto and Kadena air bases in Japan and Okinawa. General Douglas MacArthur, in command of the United Nations effort, hoped that the announcement of SAC's participation would be enough to deter Chinese intervention.[48]

---

45 "Why the Air Force Wants the B-36," *U.S. News and World Report* (17 June 1949): 18.

46 Crane, *American Airpower Strategy*, p. 21.

47 "How the U.S. Will Fight the Next War: Air Attack Only a Beginning," *U.S. News & World Report* (26 August 1949): 11.

48 Crane, *American Airpower Strategy*, p. 27.

In sending SAC units out to the Far East, LeMay made it known that this diminution of SAC's overall strength could be justified if the necessity of striking targets north of the 38th parallel was recognized. And, instead of using demolition bombing, LeMay and O'Donnell wanted to use the urban incendiary techniques that had been used against Japan five years earlier.[49] Arriving in the theater, O'Donnell briefed his plan to MacArthur, who replied that he was not prepared to go that far yet. When the Chinese entered into the war in the autumn, however, FEAF was given permission to proceed with incendiary raids. Historian Conrad Crane has pointed out that, in a 1972 interview, LeMay complained that his plan, if it had been implemented immediately, might have convinced the communists of U.S. seriousness and ended the war. Instead, LeMay argued, the war dragged on, and we destroyed "every town in North Korea and every town in South Korea" anyway. LeMay believed that a decision for war ought to allow for immediate escalation to "overwhelming military force" in order to save resources and lives on both sides.[50]

LeMay and his colleagues were frustrated by the politics of the limited war, which prevented attacks on sources of supply outside of North Korea. After the war, the head of SAC would complain that, "We never did hit a strategic target" in Korea.[51] With the war behind them, he and his colleagues were happy to focus elsewhere. FEAF's 1954 final report on the Korean War asserted that the war contained so many unusual factors that it was a poor model for planning. Instead, the Air Force was anxious to reassert its priority, preparing for strategic air war against the USSR. With increased funding flowing to all the services as a result of the Korean conflict, SAC had grown in size and strength. Between 1954 and 1962, the United States' total nuclear arsenal grew from 1,750 to 26,500 weapons. SAC, which owned most of them, planned to deliver them in a "massive pre-emptive bomber assault."[52]

By early 1951, LeMay had won increasing control over the process of targeting and planning nuclear war. He sold the argument that the USAF should concentrate on industry located in urban areas so that bombs missing their targets could derive a "bonus" effect. In this way, the United States could conserve its stockpile of nuclear weapons while ensuring that SAC would cause maximum damage to the Soviet war-making capacity. LeMay insisted on the priority of operational considerations as well. His forceful and detailed arguments meant that, in the future, target lists would be sent to SAC for comment before being forwarded to the Joint Chiefs of Staff for approval.[53]

---

49 Ibid., pp. 31–32.

50 Ibid., p. 32.

51 Thomas Hone, "Strategic Bombardment Constrained: Korea and Vietnam," in *Case Studies in Strategic Bombardment*, ed. R. Cargill Hall (Washington, DC: Air Force History and Museums Program, 1998), p. 517.

52 Biddle, *Rhetoric and Reality*, p. 296; and Conrad C. Crane, "Raiding the Beggar's Pantry: The Search for Airpower Strategy in the Korean War," *Journal of Military History* 63, no. 4 (October 1999): 920.

53 Rosenberg, "The Origins of Overkill," p. 128.

LeMay, who earned his fourth star in 1951, served as the commander of SAC until 1957. During his unprecedented 10-year command, SAC became a highly professional, well-honed instrument. He would say later that if you could look inside SAC as you might look inside a machine, its personnel "would be as the intricate electronic physiology of an airplane today: each functioning, each trained, each knowing his special part and job—knowing what he must do in his groove and place to keep the body alive, the blood circulating. Every man a coupling or a tube; every organization a rampart of transistors, battery of condensers. All rubbed up, no corrosion. Alert."[54]

Once LeMay had gotten SAC into shape, he did not ease up on his demands or requirements. He knew that if the force was to remain an effective deterrent—or war-fighting force if that became necessary—it had to be constantly modernized and maintained in a high state of readiness. As Soviet nuclear capabilities expanded, LeMay sought to keep his own side ahead. He was concerned, especially, to disperse the strike force so as to preserve it as an effective deterrent; in order for the United States to successfully deter war, it had to maintain a credible "second strike" force that would be able to survive a Soviet "first strike" and respond with a level of retaliation that would be unacceptable to the enemy.

In an interview for *U.S. News and World Report*, LeMay explained in December 1955 that "we must build our forces so that they are dispersed, not vulnerable, are trained and on the alert, so that we minimize in every way that we can the possibility of being completely surprised." In response to a question about improvements in the Air Force since World War II, he stated:

> There have been major technical advances, of course. Our equipment is very much better than what we had in the last war, especially in the electronics field, radar bombsights, and things of that sort. Our training is very much better, and above all we have experience in our units now. The units that we have at the present time are trained—under as near war conditions as we can simulate the missions they would actually perform in wartime. . . . If we do go to war, the flight with the bomb on board is just the same flight that they have made hundreds of times before, except flying over different geography.[55]

Ever insistent on training and preparation, LeMay was determined that none of his SAC crews would have to fly wartime missions when they were still green and lacking flight hours—as so many of his WWII crews had been. He had felt immense stress and anxiety about sending those young crews into battle, and he was determined never to relive that experience.

In addition to improved equipment and better training, SAC continued to gain access to better and better intelligence about the Soviet Union. This allowed it to specify Soviet

54 LeMay, *Mission with LeMay*, p. 496.

55 "We Must Avoid the First Blow," *U.S. News & World Report* (9 December 1955): 40, 42.

targets with greater certainty and develop more effective countermeasures to Soviet defenses. In 1955, the revolutionary U-2 spy plane made its debut; it began mapping missions over the USSR a year later. It was not unusual in this period for SAC reconnaissance aircraft to fly routine patrols along the Soviet border and, sometimes, to penetrate into Soviet airspace to check response times and gain information on Soviet defenses. At this time, the United States was beginning to develop, as well, its space satellite program.[56]

In light of improved intelligence, SAC targeting was, by the mid-1950s, increasingly oriented to "counterforce" targets (specific military and industrial installations) as opposed to "countervalue" targets (enemy cities). But it was not until the end of the decade that SAC had all the technology and intelligence required to make a true counterforce strike feasible. Until then, SAC targeting combined a concentration on counterforce targets with important industrial and governmental centers, which, when destroyed, would undermine the enemy's capability and will to fight. LeMay sought to find a war-fighting doctrine that would allow him to do the maximum amount of damage to the Soviet military-industrial complex in the shortest possible time.[57]

Some of LeMay's critics have suggested that he was so determined to keep SAC from being caught on the ground in war, and so distrustful of civilian authorities, that he was prepared—in a crisis—to take into his own hands the decision to launch his force. Some even have asserted that he was looking for an excuse to go to war and was deliberately provocative in SAC flights over Soviet territory.[58] These criticisms stem from concerns that were not wholly unfounded. LeMay did worry a great deal about the ability of his force to react instantly to a Soviet launch, and he did have antagonistic relationships with some civilian authorities. And U.S. flights over Soviet territory were a source of profound irritation and anxiety to the Soviet Union.

By 1955, when the U.S. nuclear stockpile was 2,280 bombs, an American atomic attack was expected to cause the loss of 118 out of 134 Soviet cities and to kill 60 million people. In the late summer of 1960, Eisenhower's last secretary of defense created the Joint Strategic Target Planning Staff to consolidate the military services' nuclear war planning efforts. The staff's Single Integrated Operational Plan (SIOP) was completed in December to enter into force at the start of the new fiscal year. It called for a massive strike using some 2,244 bombers and missiles, carrying 3,267 weapons yielding more than 7,800 megatons. Execution of the plan would have killed approximately 285 million people in the Soviet Union and the People's Republic of China.[59] As historian David Rosenberg has pointed out:

56 Steven L. Rearden, "U.S. Strategic Bombardment Doctrine since 1945," in *Case Studies in Strategic Bombardment*, p. 405.

57 Ibid., pp. 408–409.

58 Richard Rhodes, "The General and World War II," *New Yorker* (19 June 1995): 47–59.

59 David A. Rosenberg, "Nuclear War Planning," in *The Laws of War: Constraints on Warfare in the Western World*, ed. Michael Howard (New Haven: Yale University Press, 1994), p. 169.

> SIOP 62 represented a technical triumph in the history of war planning. In less than 15 years the United States had mastered a variety of complex technologies and acquired the ability to destroy most of an enemy's military capability and much of the human habitation of a continent in a single day. SIOP 62 incorporated operational choices that aimed to reduce the friction of war, coordinate and protect bomber forces, and integrate bomber and missile forces at the cusp of two eras in warfare.

But, Rosenberg added, "it was also inflexible and with little basis in political and military realities."[60] SIOP 62 was a manifestation and culmination of the principles that LeMay had inculcated into SAC, and which had come to guide U.S. nuclear planning. It was designed to ensure effective deterrence and to provide for a prompt and massive assault on the Soviet Union in the event that a war did come. LeMay had been asked to prepare to fight and win a nuclear war if necessary, and this was how he planned to do it. He was determined that his crews would not be caught on the ground and that they would successfully find their way to as many Soviet targets as possible. But if the plan looked daunting to the Soviets, it very likely would have proven unwieldy and "self-deterring" to American policymakers in a moment of crisis. The dynamics that drove the SAC planning process took nuclear war planning in a direction that gave statesmen few options other than a massive strike—a strike that would have had grave consequences for all the world's peoples.

From 1945 to 1974, the United States had no detailed national nuclear weapons employment policy. NSC 30, the 1948 U.S. guidance on atomic warfare, concluded that, "in the event of hostilities, the national military establishment must be ready to utilize promptly and effectively all appropriate means available, including atomic weapons, in the interest of national security and must plan accordingly." It pointed out that the "decision as to the employment of atomic weapons in the event of war is to be made by the Chief Executive when he considers such decision to be required."[61] This very general guidance gave the military the authority to plan for nuclear war—planning that was embodied in a special annex to what became the annual Joint Strategic Capabilities Plan. Under President Dwight D. Eisenhower, guidance remained general and vague; indeed, the Eisenhower administration discouraged the development of constraints on nuclear war planning.[62] General LeMay was happy to take advantage of the leeway that his Commander in Chief was willing to give him.

---

60 Ibid., pp. 174–175.
61 Quoted in Rosenberg, "Nuclear War Planning," p. 169.
62 Ibid., pp. 169–171.

Embracing a national security policy that placed heavy reliance on nuclear weapons, the Eisenhower administration also expanded the number of nuclear weapons available for planning purposes to NATO commanders; between 1952 and 1958, that number grew from 80 to more than 3,500. In 1956, Secretary of Defense Charles Wilson had directed that the Joint Chiefs should plan that "in a general war, regardless of the manner of initiation, atomic weapons will be used from the outset." A year later, President Eisenhower had begun the process of providing for "predelegation"—preauthorization for theater commanders to employ nuclear weapons under a range of emergency conditions. Eisenhower believed that deterrence would be bolstered by maintaining the threat of unconstrained nuclear war and leaving to the imagination of Soviet policymakers the kinds of situations that might prompt such a response from the Americans.[63]

As plans for nuclear war grew more complex and more dependent on the timing of attacks and bomber routes to targets, any assumption that presidential control could be fully maintained grew less and less robust. There is no doubt that, even under the best circumstances, launch procedures would have been liable to confusion brought on by the need to make profoundly consequential decisions under exceptionally tight time constraints. It is surely not inconceivable to imagine that events might quickly have slipped from the control of politicians and into the hands of the military—just as they had during the First World War under the demands of mobilization schedules. The operational details of the war plan would have affected its execution. This circumstance, which did indeed allow LeMay to concentrate authority in his own hands, was unsettling in that it pulled planning for war further and further from the political realities that shape the way states actually enter into wars. SAC's desire to prevail led to war plans that were large and unwieldy and contained little flexibility for limited strikes.

Since the operational details of the U.S. war plans are still tightly held, it is impossible to know precisely how much power the SAC commander had at any given time or the degree to which various Cold War presidents could have maintained control in the event of nuclear exchange.[64] But to argue that LeMay took advantage of the general guidance and the loose rein that Eisenhower gave him is not to argue that he was a warmonger bent on provoking a conflict that would eliminate the Soviet Union. Surely LeMay and his SAC successors were cold warriors who deeply opposed the ideology of the Soviet Union, but arguments that LeMay was actively seeking a war, or that he and subsequent

63 Ibid., pp. 171–173.
64 Ibid., p. 161.

SAC commanders were operating well outside political authority, stretch the available evidence. LeMay had lived through a long and brutal war—and had written hundreds of letters to distraught parents of pilots and crewmen. If he was unrepentant for the attacks on Japan, he was not looking for an opportunity to start a conflict that would be vastly more devastating in its consequences. LeMay was under no illusions about what a nuclear war would have meant, for the United States and for the world. He worried a great deal about the possibility that an accident or incorrect intelligence would lead to war, and he worked to ensure against that possibility.

Finally, to assume that LeMay was looking for opportunities to usurp civilian authority underestimates the degree to which the principle of civilian control is inculcated in American military officers. It underestimates, too, the degree to which LeMay sought peace through effective deterrence. He was very proud of the fact that SAC served, through all the years of his command, as an effective deterrent force. He believed that SAC personnel were motivated by that mission (heralded by their slogan: "Peace Is Our Profession") and that they maintained the highest professional standards because they were carrying out that mission. In 1955, he told an interviewer, "I know I can only get the work out of them [SAC personnel] because they believe in what they are doing and are sustained by the success they have had with it so far." Asked by the same interviewer, "So, is it not correct to assume that all a military man wants is a chance to fight?" LeMay answered tersely, "Not this one."[65]

FINAL YEARS OF SERVICE

In 1957, LeMay was named Vice Chief of the Air Force, but his legacy at SAC continued on. By 1959, the United States deployed its first Atlas D intercontinental ballistic missiles. A year later, SAC had dispersed its aircraft to some 66 bases around the continental United States and Canada and had brought one-third of its bombers and tankers to 15-minute alert. Four years later, LeMay was sworn in as Chief of Staff of the U.S. Air Force under President John F. Kennedy. The *New York Times* article announcing LeMay's ascendance to the top post was titled simply, "Air Chief of Jet Age, Curtis Emerson LeMay."[66] LeMay had, sometimes, a testy relationship with the young President Kennedy and with his Secretary of Defense Robert S. McNamara. A cold warrior of the old school, LeMay often found himself out of step with Kennedy's more liberal worldview. During the Cuban Missile Crisis, for instance, LeMay was the most vigorous advocate of a swift air strike on the Caribbean island—an option that the administration wisely rejected in

65 "We Must Avoid the First Blow," *U.S. News and World Report* (9 December 1955): 38.
66 "Air Chief of Jet Age: Curtis Emerson LeMay," *New York Times* (23 May 1961): 4.

General LeMay reached the summit of his Air Force career on 30 June 1961, when he was sworn in as Chief of Staff by Secretary of the Air Force Eugene M. Zuckert. Observing the ceremony in the rose garden of the White House are the late President John F. Kennedy and then Vice President Lyndon B. Johnson. (Photo courtesy of the United States Air Force Museum)

favor of naval quarantine. And LeMay deeply resented what he believed was McNamara's know-it-all attitude about things military.[67]

Some of LeMay's critics argued at the time that he was opposed to the integration of ballistic missiles into the emerging strategic triad (of bombers, missiles, and submarines). He repeatedly denied this. He agreed, however, that he was a military conservative, stating, "I believe we shouldn't discard a proven, reliable weapon system or concept unless we have something that is able to replace it and do a better job. In short, I believe in protection along with progress."[68]

67 LeMay, *Mission with LeMay*, p. 8.
68 "Protection with Progress," *Time* (29 September 1961): 15.

Early in 1963, LeMay gave another of his many appearances before Congress. This testimony was particularly close to his heart, since it argued in favor of maintaining the manned bomber even as the USAF became increasingly reliant on ballistic missiles. He was asking for some $300 million to be restored to a procurement law so that the USAF might accomplish the early stages of its program for a new reconnaissance strike aircraft, the RS-70 (formerly the B-70). LeMay had been at loggerheads with Secretary McNamara on this issue; opposing the administration view before Congress enabled him to use the power inherent in the constitutional system of checks and balances to appeal to the body in charge of the purse strings. In the end, LeMay won the day; Congress voted in favor of funding the RS-70. Though LeMay knew that the money might not, in the end, be spent as authorized, he felt good in his successful effort to put forward the case of the USAF and Joint Chiefs of Staff. He remarked, "Consciences of the Joint Chiefs of Staff, in majority, are clear. And so will be the consciences of the Committee on Armed Services."[69]

The debate over the RS-70 was only one of many issues that kept LeMay's tenure as Chief of Air Staff a turbulent one. Though he learned the ways of Washington, he was no natural politician. The skills that had made him an excellent field commander—being able to fix on a problem and locate the single best solution—were the same skills that complicated his life at higher levels, where problems were rarely amenable to a single best solution. But LeMay served as his conscience dictated. He gave the advice he believed to be the best military advice—not the advice he felt the politicians would want to hear. He carried out his duties within the bounds established by the civil-military relationship defined in the Constitution—a relationship that LeMay respected and believed in.[70] LeMay retired from service in February 1965. In his brief final speech he told the members of the Air Staff, "Make sure you are right before you move and then stick to your guns and keep fighting for what you want. It takes a long time here to get things done; however, water wears away the stone. Right prevails in the end in our form of government."[71]

CONCLUSION

Curtis LeMay lived for nearly 30 more years. American policy in Vietnam grieved him deeply, and, in 1968, he made an injudicious and unsuccessful entry into politics as the vice presidential candidate for Governor George Wallace's third-party run for the U.S. presidency. Toward the end of his life he admitted that choice had been a mistake, but,

---

69 LeMay, *Mission with LeMay*, p. 12.
70 Ibid., pp. 553–554.
71 Coffey, *Iron Eagle*, pp. 437–438.

he added, he considered it his "last chore for the public."[72] The decision would dog him and would deeply influence the public's memory of him.

Not long before his death, LeMay made a final public speech at the National Air and Space Museum in Washington, DC. After indicating that he would speak only briefly, he went on to give a long, articulate, and impassioned speech on behalf of air power. He was a staunch advocate to the end.[73]

As much as any other individual, Curtis LeMay shaped the evolution of American air power in the 20th century. And he shaped, as well, the image of the United States during the height of the Cold War. He came to represent—through his position as head of SAC and through his persona as the determined and unflinching proponent of deterrence through intimidation—the power and authority of the United States in the era following the Second World War. He was a consummate field general, perhaps the best one fielded by any nation between 1939 and 1945. He led the Berlin airlift and then led SAC through its crucial and formative years during the Cold War. He was less successful in Washington, where deft interpersonal skills and strong political instincts are required.

The U.S. Air Force called on Curtis LeMay repeatedly to fulfill its most demanding tasks. They relied upon his professionalism, technical skill, ingenuity, fearlessness, determination, and willingness to accept heavy responsibility. LeMay was later criticized for the fire attacks he led against Japan, for his part in building the nuclear arsenal into a vast and unwieldy entity, and for war plans that would have been deeply destabilizing, highly stressful, and perhaps tragic in a crisis. But LeMay faced his critics unflinchingly. He believed that the campaign over Japan helped to save American and Japanese lives and that the powerful strategic force he built was the rockbed of successful deterrence throughout the Cold War. Indeed, he believed these things so completely that he was more than prepared to endure the personal criticism that came along with his professional record. He felt that standing fast for what he believed in was simply part of the profession of arms and part of his responsibility to the nation he served.

Whether one ultimately sees LeMay as a hero or as a threat to world peace depends on one's interpretation of the dynamics of deterrence during the Cold War and one's interpretation of the role of a massive nuclear bombing force in keeping the peace between the United States and its Soviet adversary. If one believes that hard-nosed and robust deterrence kept the peace, then one surely can see LeMay as a "force multiplier"—a military term for an individual (or weapon, or condition) who disproportionately enhances the prospects and performance of a combat organization. There is little doubt that the

72 Ibid., pp. 446–447.
73 Tami Davis Biddle, personal notes from LeMay speech, National Air and Space Museum (June 1990).

Soviets took him very seriously. Indeed, one might argue that if Curtis LeMay had not existed, we might have had to invent him. Perhaps the very size, inflexibility, and danger of his war plans ensured that they would never have to be tested. We will never know. But, whatever it was that kept the Americans and the Soviets from coming to blows during the Cold War, we must all be grateful for it. LeMay will live on as a major figure in the history of the 20th century, and Americans will view him with profoundly mixed feelings. The American people were proud of the unprecedented power they achieved, but they also were ambivalent about owning it and deeply concerned about what it might have wrought if deterrence had failed.

8

Willy Ley

Chronicler of the Early Space Age

TOM D. CROUCH

WILLY LEY DIED OF A SUDDEN HEART ATTACK AT HIS HOME ON 77TH STREET IN THE JACKSON HEIGHTS SECTION OF QUEENS, NEW YORK CITY, ON 24 JUNE 1969. It didn't seem fair. Only 62 years old, Ley had been in the best of spirits, preparing to fly to Houston, where he would be NASA's honored guest when human beings first walked on the Moon. One of a handful of enthusiasts who had given birth to this dream 40 years before, he had made it his life's work to infect others with his enthusiasm for spaceflight. "For Willy Ley," the editor of *Popular Mechanics* noted, "man's greatest triumph came a month too late." It would be fitting and proper, he continued, for the astronauts of Apollo 12 to scatter Willy's ashes on the Moon.[1]

Willy Ley was a native Berliner, born on 2 October 1906.[2] His father, Julius Otto Ley, was a wine merchant from Königsberg, East Prussia. His mother, Frida May, was the

---

1 Robert Crowley, "Send Willy Ley's Ashes to the Moon!" *Popular Mechanics* (September 1969): 10.

2 Elements of this essay were drawn from Tom D. Crouch, "Willy Ley," in *American National Biography* (New York: Oxford University Press, 1999). The principal source was the Willy Ley Papers, National Air and Space

continued on next page

daughter of a Lutheran Church sexton. Leaving his wife and son with her parents, Julius Ley traveled to New York City in 1910, then to London, where he established a German delicatessen in 1913. Frida left her son with her three sisters and joined her husband in England. She returned to Germany at the outbreak of war in 1914, when the British interned Julius as an enemy alien on the Isle of Man.

Unable to find employment in Berlin, Frida deposited her eight-month-old daughter Hildegarde with the sisters who were already caring for Willy and found work as a milliner outside the city. For the next four years, the children would see their mother only once or twice a month. Released from internment at the end of the war, Julius Ley returned to Germany, reunited his family, and reestablished himself as a wine merchant in Königsberg.

The Ley youngsters thrived in spite of their family upheavals. Berlin public schools provided Willy with an extraordinarily sound education. He earned consistently high marks and excelled at soccer and tennis. He had a talent for languages—studying Latin, French, and English—and was fascinated by science. He also took full advantage of the rich opportunities that Berlin offered for informal education. "I grew up," he recalled many years later, "in the shadow of the Museum of Natural History in Berlin" and spent "most of my youthful Sundays there."[3] He discovered science fiction, as well, devouring stories by Jules Verne, Edgar Allen Poe, and the German master of the genre, Kurd Lasswitz.

Willy's desire to attend university came as a surprise to his parents. Times were hard, and it was an unusual course for the son of a tradesman. He would pay for much of his own education, working 8 hours a day as a bank clerk and attending evening classes. He entered the University of Berlin in 1924, where he studied paleontology, zoology, botany, anatomy, mathematics, physics, and astronomy. When family fortunes took a turn for the worse, he shifted to the University of Königsberg, where his father was working. With both the family and national economies in collapse, Willy was finally forced to leave the university for good in 1927.

In fact, the course of his career had been established in 1925 when he discovered a copy of the second edition of *Die Rakete zu den Planetenraumen* (*The Rocket Into Plane-*

continued from previous page

Museum (NASM), Smithsonian Institution Archives. The Special Collections Department of the library at the University of Alabama, Huntsville, preserved the 5,000 books and journals that made up Ley's private library. Both institutions maintain extensive biographical and bibliographic files on Ley. Additional files of Ley correspondence are in the papers of G. Edward Pendray at Princeton University and of Wernher von Braun at the University of Alabama, Huntsville. Biographical material appears in "Willy Ley," *Die Rakete* (15 August 1928): 128; Steve Bland, "Sky Guy," *Philadelphia Inquirer Magazine* (9 September 1951): 12; P. E. Cleator, "A Tribute to Willy Ley," *Spaceflight* 11 (November 1969): 408–409; and Lester del Rey, "The First Citizen of the Moon," *Galaxy Magazine* (September 1969): 151–157. The standard obituary can be found in *New York Times* (25 June 1969).

3 Willy Ley, *Exotic Zoology* (New York: The Viking Press, 1959), p. xi.

tary Space) in a Berlin bookstore. In 92 short pages, densely packed with mathematical proofs, Rumanian physicist Hermann Julius Oberth demonstrated that the rocket was capable of traveling into space. Originally published in 1923, the book sold well and ignited a flurry of interest in spaceflight. Max Valier, a journalist and veteran of the Austro-Hungarian Air Force, published a popular account of Oberth's ideas in 1924 entitled *Der Vorstoss in den Weltenraum* (*The Dash Into Space*).

Dissatisfied with Valier's effort, Willy Ley wrote his own account of spaceflight for laymen in 1926 entitled *Fahrt ins Weltall* (*A Trip Into Space*). The 20-year-old university student received a hefty advance for his first book, a 64-page paperback issued by a Leipzig publisher. The volume sold well and established its author as one of the most knowledgeable men in the field.

Not satisfied with publicizing their dream, the young space enthusiasts of Weimar Germany took practical steps toward its realization. In the spring of 1927, Max Valier enlisted Ley in the effort to establish a rocket society. The Verein fur Raumshiffahrt (VfR) (Society for Space Travel) was founded in the Golden Scepter, a Breslau beer hall, on 5 June 1927. Within a year, the organization boasted a membership of 500.

Ley, who was elected vice president of the organization in 1929, explained to an American correspondent that the members of the VfR were determined "to spread the thought that the planets were within reach of humanity, if humanity was only willing to struggle a bit for that goal." In order to support and publicize the new organization, he convinced the biggest names in the spaceflight movement to prepare a series of essays, which he edited and published in 1929 as *Die Möglichkeit der Weltraumfahrt* (*The Possibility of Space Travel*).[4]

The wave of popular enthusiasm for spaceflight peaked on the evening of 15 October 1929, when Fritz Lang, Germany's best-known motion picture director, premiered his latest film, *Frau im Mond* (*The Woman in the Moon*). A complex melodrama, and the last major silent film produced in Germany, it told the story of the first trip to the Moon. Determined to make his film believable, Lang hired Oberth as a technical consultant. Willy Ley was hired to write 12 articles explaining the scientific principles of the production. Lang also provided financial support for a rocket that Oberth was to launch as part of the premiere activities.[5]

Oberth's rocket was not a success, but Ley and his fellow enthusiasts were determined to continue experimenting with liquid-propellant reaction motors. They acquired permission to transform an abandoned military storage depot in the Berlin suburb of

4 Willy Ley, *Die Möglichkeit der Weltraumfahrt: Allgemeinverständliche Beiträge zum Raumschiffahrtsproblem* (Berlin: Hachmeister and Thal, 1928).

5 Willy Ley, "Review in Retrospect," in the Willy Ley Papers, NASM Archives.

Reinickendorf into a raketenflugplatz—a rocket-testing field. Between March 1931 and April 1932, the members of the VfR completed 270 static liquid-propellant rocket engine tests; 87 flights; 23 demonstrations for other organizations; and 9 presentations for the press. Their rockets reached altitudes of up to 4,922 feet.

Ley, the single-most visible member of the VfR, communicated news of the organization's research program to other rocket enthusiasts around the world. He wrote articles, lectured, corresponded widely, and hosted young rocketeers from other nations, including both G. Edward Pendray of the American Rocket Society, who visited the VfR in the spring of 1931, and Englishman Phil Cleator, who arrived in 1934, not long after the organization of the British Interplanetary Society.[6]

By 1933, a series of problems brought an end to the golden age of VfR rocketry. The deaths of several rocket experimenters, including Valier and Reinhard Tilling, underscored the dangers inherent in liquid-propellant rockets. Moreover, Rudolph Nebel, the man in charge of VfR rocket experiments, was creating problems for the organization. In the spring of 1933, Ley and VfR president Major Hans-Wolf von Dickhuth-Harrach discovered that Nebel had signed a contract with the city fathers of Magdeburg, promising to launch a man-carrying rocket to high altitude. Fearing that the VfR might be charged with fraud, Ley and Dickhuth-Harrach attempted to force Nebel out of the organization. Failing that, the two men announced their own resignations and attempted, unsuccessfully, to establish a new society.

Split by internal dissension, the VfR finally succumbed to government pressure. German army interest in rocket weapons had resulted in the creation of a small military rocket research team headed by the young Wernher von Braun, whom Ley had drawn into VfR membership. A curtain of military secrecy was drawn across all rocket experiments. Private individuals were forbidden to build or launch rockets, or to write articles on the subject. By the end of 1934, Ley was barred from lecturing or publishing on his favorite subject. The future looked dim for science journalists in the Third Reich, and there were other things to consider. "How I look like, you know," the dark, wavy-haired Ley had remarked to Pendray the previous May, "[I] could be blonder for the time being (don't mention the last!)."[7]

---

6 Cleator, "A Tribute to Willy Ley," p. 408.

7 Willy Ley to G. E. Pendray (15 May 1934), G. Edward Pendray Papers, Princeton University. The comment raises a question regarding Ley's heritage. On his mother's side, the family had a long tradition of service to the Lutheran Church, and Willy was certainly raised as a Lutheran. When he announced that he wanted to attend university, the first member of his family to do so, it caused such a furor that the Leys presented the matter to a Lutheran clergyman, presumably the family priest, for a decision. He laughed and advised the family to do what they could to support their ambitious son. At the same time, it is possible, even likely, that there was a Jewish connection on Julius Ley's side of the family. Willy might have been alluding to this possibility in his comment to Pendray, or he might simply have been calling attention to the fact that, with his dark, wavy hair and broad features, he did not look particularly "Aryan."

Willy Ley, a prolific and visionary author captivated by the dream of spaceflight, published more than 30 popular books and articles on rocketry between the end of WWII and his death in 1969, shortly before the first human beings walked on the Moon. (Photo from the NASA Headquarters Historical Reference Collection)

The time had come to leave Germany. Ley made use of his broad contacts in the international astronautical community, traveling first to England in January 1935, where he stayed at the Liverpool home of Phillip Cleator, a member of the British Interplanetary Society, while waiting for passage to the United States. He arrived in America on 21 February and lived for a time with the Pendrays, whose letters of support had convinced U.S. officials to provide Ley, who was almost blind in one eye, with a tourist visa. The couple found him an entertaining house guest, given to singing Wagnerian arias while accompanying himself on the piano. With his wide range of interests, Ley was an engaging conversationalist, if sometimes inclined to pontification. "If you asked him a question," one friend recalled, "you got a lecture."[8]

With the assistance of Pendray and other American friends, Ley made the acquaintance of a number of important engineers interested in rocket propulsion, including Alexander Klemin of New York University. As a result of these contacts, Ley was hired to serve as flight operations supervisor for an experimental winged rocket designed to carry small packets of mail across frozen Greenwood Lake in upstate New York. Two of the rockets were launched on 23 February 1936. The first rocket climbed to an altitude of 1,000 feet, then spun to the ground when the combustion chamber burned through. The wings of the second rocket ripped off only 15 seconds into the flight.[9]

Forbidden by immigration regulations from accepting full-time employment, Ley made his living as a freelance writer and lecturer. He spent the years 1936 to 1940, as he later explained to a *New York Times* reporter, "writing day and night, turning out articles for scores of publications both here and in Europe."[10] A friend estimated that he contributed at least 90 articles to science-fiction magazines alone between 1935 and 1950. Most of these

8 Walter Sullivan, "Willie Ley, Prolific Science Writer, Is Dead at 62," *New York Times* (25 June 1969): 47.

9 Wernher von Braun and Frederick I. Ordway, *History of Rocketry and Space Flight* (New York: Thomas Y. Crowell Co., 1975), p. 75.

10 Sullivan, "Willie Ley."

treated aspects of science, although he did write a few science fiction stories under the pseudonym Robert Wiley.

Ley joined the staff of the liberal tabloid newspaper *PM* as science editor in 1940. The following year he married ballet dancer Olga Feldman, a Russian immigrant who wrote a physical fitness column for *PM*. The pair would have two daughters, Sandra and Xenia. The Ley apartment, one visitor reported, was "just as out of the ordinary as one might have expected for two such outstanding personalities—full of a prodigious number of books on every subject under the Sun and tanks containing small fish and reptiles, which testified to the owner's other interests."[11]

The year 1941 also marked Ley's emergence as an author of popular books on science. His first English language book, *The Lungfish & the Unicorn*, a volume of essays on natural history, was followed by *Bombs and Bombing*, *Shells and Shooting*, and *The Days of Creation*. He was launched on a career as one of the most prolific and successful writers of the era on aspects of science.[12]

In 1944, Ley became a naturalized U.S. citizen and published the first edition of his best-known and most influential book, *Rockets*.[13] Based on the author's 20-year search for material on the subject, his own experience in Germany, and his correspondence with virtually all of the pioneering figures in the field, the book traced the history of rocketry from the black powder era through the 1930s and explained the basic physical principles that would govern spaceflight. Over the next 28 years, Ley would produce three major new editions of the book: *Rockets and Space Travel* (1947); *Rockets, Missiles and Space Travel* (1951); and *Rockets, Missiles and Men in Space* (1968).[14] In all, the book went through 20 printings during Ley's lifetime.

For all of his expertise, the advent of the space age took Ley by surprise. A. V. Cleaver, a British weapons expert visiting the United States in the fall of 1944, remembered that Ley refused to believe reports that long-range German rockets were falling in London. His old colleagues, Ley argued, "were most unlikely to have developed such a weapon, which would be inaccurate and uneconomical, and probably impossible to achieve at that date, in any case."[15]

---

11 A. V. Cleaver, "Tribute to Willy Ley," *Spaceflight* 11 (November 1969): 408.

12 Willy Ley, *The Lungfish & the Unicorn: An Excursion into Romantic Zoology* (NY: Modern Age Books, 1941); *Shells and Shooting* (NY: Modern Age Books, 1942); and *The Days of Creation* (NY: Modern Age Books, 1941).

13 Willy Ley, *Rockets: The Future of Travel Beyond the Stratosphere* (NY: Viking Press, 1944).

14 Willy Ley, *Rockets and Space Travel* (NY: Viking Press, 1947); *Rockets, Missiles, and Space Travel: The Future of Flight Beyond the Stratosphere* (NY: Viking Press, 1951); *Rockets, Missiles and Men in Space* (NY: Viking Press, 1968).

15 Cleaver, "Tribute to Willy Ley," p. 408.

Ley had underestimated the German rocketeers. During the nine years since his departure from Germany, the Nazi government had established a great research center at Peenemünde on the Baltic Coast. There the rocket team headed by Wernher von Braun, whom Ley had known as one of the brightest and youngest members of the VfR, had succeeded in developing the A-4, or V-2, the world's first large ballistic missile capable of carrying one metric ton of high explosives for a distance of 330 kilometers. At the peak of its ballistic path, the rocket was coasting along the roof of the atmosphere, 60 miles high.

The wartime record of the A-4, and well-publicized postwar rocket tests at White Sands, New Mexico, fueled public interest in spaceflight. Ley remained a leading commentator on the subject for the rest of his life. He held a variety of positions during the early postwar years, serving for a time as a research engineer with the Washington Institute of Technology in College Park, Maryland; a lecturer on scientific topics at Farleigh Dickinson University; an information specialist with the Office of Technical Services, U.S. Department of Commerce; a technical consultant to the producers of the pioneering science-fiction television series *Tom Corbett, Space Cadet*; and, from 1950 to the end of his life, as science editor of the science-fiction magazine *Galaxy*.

In 1951, Ley and Hayden Planetarium Director Robert Coles organized the First Annual Symposium on Space Travel. Held in New York on 12 October 1951, the symposium featured papers on spaceflight by leading American scientists and engineers. Intrigued by the gathering, Cornelius Ryan, a writer for *Collier's* magazine, began work on what would become a series of eight feature articles on spaceflight. Ryan drew on the expertise of a large number of leaders in the field, but Ley and von Braun were the central figures in the project. With illustrations by artists Chesley Bonestell, Fred Freeman, and Rolf Klep, the articles, which appeared between March 1952 and April 1954, were an enormous success.

Viking Press, which had published Ley's *The Conquest of Space* (1949), transformed the *Collier's* articles into three bestselling books—*Across the Space Frontier*, *The Conquest of the Moon*, and *The Exploration of Mars*.[16] Ley was also an important contributor to three Walt Disney television programs on spaceflight that were inspired by the *Collier's* article. "Man in Space," "Man and the Moon," and "Mars and Beyond" aired on the *Disneyland* television program beginning in September 1955.

Ley continued to produce popular books on science and aspects of spaceflight, including *Dragons in Amber*, *Lands Beyond*, *Salamanders and Other Wonders*, *Exotic Zoology*, *Harnessing Space*, *Beyond the Solar System*, *Watchers of the Skies*, *Ranger to the Moon*,

16 Cornelius Ryan, ed., *Across the Space Frontier* (New York: Viking Press, 1952); Wernher von Braun, Fred Whipple, and Willy Ley, *The Conquest of the Moon* (New York: Viking Press, 1953); and Willy Ley and Wernher von Braun, *The Exploration of Mars* (New York: Viking Press, 1956).

and *Mariner IV to Mars*.[17] At the time of his death, Ley had a total of six new books under contract. He served as an adviser to the National Aeronautics and Space Administration and was preparing to leave for the launch of Apollo 11 at Cape Kennedy, Florida, when he died of a heart attack at his home in Jackson Heights, Queens, New York.

Reporting his death, the *New York Times* remarked that Ley had "helped usher in the age of rocketry and then became perhaps its chief popularizer."[18] Captivated as a youth by the dream of spaceflight, he communicated that dream to others in the more than 30 books and countless articles that he produced during his 40-year career as a writer. He was the first important historian of the space age and one of its most eloquent advocates. Through his books, articles, and television appearances, he was one of the most familiar spokesmen for spaceflight. The crew of Apollo 12 did not carry Willy's ashes to the Moon, as the editor of *Popular Mechanics* had suggested. The year after his death, however, the International Astronomical Union named a lunar crater in his honor. With a detailed map of the Moon and a good telescope, you can find it at 42.2° north latitude and 154.9° east longitude. That is the sort of scientific immortality that would surely have pleased him.

17 Willy Ley, *Dragons in Amber: Further Adventures of a Romantic Naturalist* (New York: Viking Press, 1951); Willy Ley and L. Sprague De Camp, *Lands Beyond* (New York: Rinehart and Company, 1952); Willy Ley, *Salamanders and Other Wonders: Still More Adventures of a Romantic Naturalist* (New York: Viking Press, 1955); Willy Ley, *Exotic Zoology* (New York: Viking Press, 1959); Willy Ley, ed., *Harnessing Space* (New York: Macmillan, 1963); Willy Ley, *Beyond the Solar System* (New York: Viking Press, 1964); Willy Ley, *Watchers of the Skies: An Informal History of Astronomy from Babylon to the Space Age* (New York: Viking Press, 1963); Willy Ley, *Ranger to the Moon* (New York: New American Library, 1965); and Willy Ley, *Mariner IV to Mars* (New York: New American Library, 1966).

18 Sullivan, "Willy Ley."

9

Who Was Hugh Dryden and Why Should We Care?

MICHAEL GORN

DURING HIS LIFETIME, HUGH LATIMER DRYDEN (1898–1965) EARNED AN ENVIABLE REPUTATION AS ONE OF THE WORLD'S PIVOTAL FIGURES BOTH IN AERONAUTICS AND SPACEFLIGHT. Today, almost two generations after his death, his achievements are not well known and even forgotten. In part, Dryden contributed unwittingly to his own disappearance. Reserved and modest, he worked quietly to resolve scientific and bureaucratic problems, enabling others to benefit from his ideas. Moreover, although he made a number of crucial scientific discoveries, he never produced the one conceptual breakthrough that might have elevated him to enthronement among the visible saints of science.

Nevertheless, Hugh Dryden deserves to be remembered for a number of reasons. One of the rare individuals whose career transcended the panorama of 20th-century flight—from the earliest days of aeronautics to the initial human and robotic forays into space—he wedded an unparalleled knowledge of the requisite science and engineering with keen management and bureaucratic instincts. He knew virtually all of the prominent persons in his field, and not just in the United States; his foreign contacts covered nearly every part of the globe. Perhaps more than any other figure, he prepared America for the space age. During his 18 years at the top of the government's leading civilian air and space institutions (as Director of the National Advisory Committee for Aeronautics and as Deputy Administrator of the National Aeronautics and Space Administration), he became a controlling force in federally sponsored research on flight. His own scholarship made impressive and long-lasting contributions to the field. Finally, Hugh Dryden's personal story is a compelling one about an obscure child prodigy who rose ultimately to international prominence.

COMMON ORIGINS

The forbears of Hugh Dryden showed no sign of fame. His ancestors originated in Scotland, in the vicinity of Edinburgh. William and Agnes Dryden (or Dredden, as the family then called itself) immigrated to the colony of Maryland in 1682 seeking religious toleration in the face of Charles II's imposition of Anglican bishops on the Presbyterian Church. The Drydens settled in Somerset and Worcester Counties at the southern tip of the state. Townsmen in the old country, they rapidly accustomed themselves to farming in the New World, although many continued to pursue skilled trades like tailor, shipwright, shoemaker, and cooper. For the most part, they settled near the quiet Pocomoke River, a tributary of the Chesapeake Bay. During the 19th century, many of the Drydens embraced Methodism. For 200 years, then, this family—people of average wealth and station—worked at their occupations, paid taxes, and produced offspring.

Then, during the late 19th century, one Dryden broke ranks with family tradition. Samuel Dryden, the son of Isaac F. and Hester Ann of Pocomoke, Maryland, showed uncommon mathematical ability, and his instructors encouraged him to pursue it as a career. Accordingly, when he earned his high school diploma he also received a teaching certificate. After settling in Pocomoke City, he began a career in the classroom and soon became a vice principal. Samuel seemed destined for a life of at least local distinction. But after seven promising years in the profession, he resigned abruptly, removed his family from Pocomoke City, and relocated to a rural crossroads known as West Postoffice, Maryland. Perhaps the redirection occurred because of his reputed hot temper; perhaps it reflected his desire to better provide for his family. In any event, Samuel and his brother opened a general store in West Postoffice in 1900. Unfortunately, it failed in the panic of 1907, prompting Samuel, his wife Zenovia, and their two sons to leave southern Maryland for good.

Some years before this family crisis, Zenovia Dryden gave birth to her first son. She and Samuel named him for a Methodist clergyman popular in the region, Hugh Latimer Elderdice. Hugh Latimer Dryden entered the world on 2 July 1898 with some distinct advantages, as well as some equal disadvantages. On the positive side, he lived among a sea of relatives—gregarious and often inclined toward music—and witnessed the birth of no fewer than 28 paternal first cousins as he matured. Moreover, his mother, known by the nickname Nova, inspired Hugh with some of her best qualities, especially patience, modesty, and discernment. If young Dryden possessed his mother's temperament, he inherited the mental agility of his father. Precocious, he could read by the age of four, and by eight he had already begun to master the fifth-grade curriculum. But as a boy and young man, Hugh Dryden also lived with some less-than-satisfactory conditions. Try as he might, Samuel Dryden failed to provide his family with more than the bare essentials. Regarded as brilliant by his neighbors and well known for charity and generosity, he nonetheless lacked the toughness for commerce. In addition, the humble one-room schoolhouse in West Postoffice provided poor preparation for a student of Hugh's promise.

When the Drydens left West Postoffice in 1907, they settled in Baltimore, then a thriving port and home to many recent immigrants. The big city represented opportunity for Hugh but the end of ambition for his father. Samuel Dryden took a job with the Baltimore United Railways and Electric Company as a streetcar conductor and remained one the rest of his working life. While steady, his income just met the household's needs. Hugh, on the other hand, found a new world. He became a *Baltimore Sun* paperboy and took various other part-time jobs (working on an assembly line at the United Biscuit Company, sorting packages for an express delivery service, and toiling in a canning factory).

But struggle as Hugh might to augment the family fortunes, education remained the object of his desire, and he continued to be accelerated beyond his years. He enrolled first in Public School Number 85, where he completed the remainder of the fifth and then the sixth grade in just over one year. In School Number 52, where he studied with boys dressed in coats and neckties and girls trimmed in hats and dresses, he required only 18 months to finish grades seven and eight. About three years younger than the rest of his classmates, he started intermediate school at age 9 and graduated at 12. Yet, he continued to excel.[1]

1 Michael H. Gorn, *Hugh L. Dryden's Career in Aviation and Space* (NASA Monograph in Aerospace History No. 5) (Washington, DC: NASA History Office, 1996), pp. 1–2; Dryden's remarks on paperboys quoted in Shirley Thomas, *Men of Space*, vol. 2 (Philadelphia: Chilton, 1961), p. 66; interview with Mary Ruth (Dryden) Van Tuyl, Silver Spring, MD (3 August 1994); interview with Nancy (Dryden) Baker, Rockville, MD (9 August 1994); interview with Nancy Baker by telephone (25 July 1994); and autobiographical sketch by Hugh L. Dryden (27 September 1965), Hugh L. Dryden (HLD) Papers, Ms. 147, Series 2.2, Milton S. Eisenhower Library, Johns Hopkins University (referred to hereafter as JHU). All interviews cited are in Michael Gorn's possession.

Most of his assignments involved brief essays on history, civics, geography, and the natural sciences. Although the writings of a child, they still show clear exposition and a logical sequence of ideas. Despite his age, he generally earned high marks. In an essay written in April 1908 entitled "The Fairyland of California," he revealed a charming sense of wonder about a far-off place.

> There are parts of California where it is summer all the year round. The flowers always bloom, and the trees are always green. In Los Angeles they sometimes have a festival of roses to celebrate the New Year. On a Christmas morning there you could go to the seashore, take a bath, and come back and set your Xmas [sic] dinner under the orange trees. Then you could go up on the mountains and see some of the finest Xmas [sic] trees in the world.[2]

The young pupil also took many spelling tests and demonstrated high proficiency. He impressed Mrs. Mary Kennedy, his arithmetic teacher, with his capacity to convert various types of weights and measures and to calculate weekly wages and deductions. Finally, in recognition of excellent attendance, he received a prize—a book of religious proverbs entitled *Many Thoughts of Many Minds: A Treasury of Quotations*. He wrote in it the words "a treasure worth keeping" and checked a few entries that especially appealed to him. One, quoted from the English clergyman and author Dr. Jeremy Taylor, sheds light on his developing mind: "Hope is like the wing of an angel, soaring up to heaven and bearing our prayers to the throne of God."[3]

While Hugh Dryden's education proceeded, Wilbur and Orville Wright's epochal flight in December 1903 became known, but not much heralded. Yet, once the Wrights demonstrated the capabilities of their aircraft to the U.S. Army and to European throngs, a wave of curiosity swept America. Just as Hugh Dryden entered puberty, Baltimore's residents beheld one of the aerial spectacles gripping the nation. The *Baltimore Sun*

---

2 Dryden, "The Fairyland of California," school essay (15 April 1908), HLD Papers, Ms. 147, Series 1.1, JHU. See also the following school essays by Dryden in the same archive: "Indian Corn and the Corn Belt" (3 February 1908); "Robert Fulton and the Steamboat" (3 March 1908); "Life in the Timber Regions" (4 March 1908); "The Wonders and Treasures of the Rocky Mountain Region" (11 March 1908); "A Visit to a Gold Mine" (18 March 1908); "Eli Whitney" (20 March 1908); "San Francisco and the Chinese" (1 May 1908); "Our National Capital" (15 May 1908); "Baltimore Oriole" (21 May 1908); "A Visit to the President and to the Halls of Congress" (22 May 1908); "The Departments of the Government" (28 May 1908); "Baltimore and Our Oyster Beds" (5 June 1908); "In Philadelphia: A Visit to the Mint" (11 June 1908); "John C. Fremont and Kit Carson" (17 June 1908); "Review of Hawthorne Works" (29 October 1909); and "Science" (22 February 1910).

3 Dryden's collected spelling tests (4 May 1908–19 June 1908), HLD Papers, Ms. 147, Series 1.1, JHU; collected arithmetic tests (5 March 1908–10 June 1908), HLD Papers, Ms. 147, Series 1.1, JHU; and Louis Klopsch, comp., *Many Thoughts of Many Minds: A Treasury of Quotations from the Literature of Every Land and Every Age* (New York: The Christian Herald, 1896), pp. 3, 135, in the possession of Mary Ruth Van Tuyl.

offered a prize of $5,000 to the first flier to navigate the skies over the city. Sportsman and aviator Hubert Latham accepted the challenge. On 7 November 1910, he strapped himself into his French-made Antoinette monoplane and took off. Light at 1,300 pounds, the little vehicle performed well, attaining a high speed of 40 miles per hour as Latham looped again and again around the *Sun*'s downtown offices during his "great flight." Packed in with the rest of the population, Hugh Dryden also witnessed this modern wonder. Nova Dryden noticed that her son reacted to the demonstration not merely with youthful enthusiasm, but with a deeper appreciation. Partly, he must have caught the sense of awe present in the crowd. But more importantly, the event seemed to capture him intellectually. He wanted to know what forces sustained Latham in the air, to understand the mechanical workings of his plane, and to grasp the principles of flight itself. The sight formed a lasting picture in his mind, one that he recalled later in life.

Four days later, the young student wrote an essay based on Latham's flight. "The Advantages of an Airship Over an Aeroplane" revealed an independent mind at odds with the fervor of the day and with his teacher's preconceptions. He received an F and the comment "Illogical," because rather than write an homage to the airplane, he concentrated on the very real deficiencies of the new invention. Dryden grasped that these awkward machines were prone to failure, "the least break in [which] will hurl the aviator to the ground." He also understood that the frail Antoinette (like the other aircraft of the time) had very narrow capacities, limiting travel to short distances and greatly inhibiting the transport of people and products. Furthermore, they offered no competition to the speed or capacity of the existing railway structure. As a consequence, Dryden argued for the superiority of airships over airplanes "for commerce and exploration," due to their simplicity and potential to haul cargo. In short, the youngster understood that astounding though the Wright's invention may have been, its success depended on years of improvement and refinement. Of course, Hugh could not foresee that his own life would become the embodiment of this process. Nonetheless, during his boyhood and adolescence, no technology generated such exhilaration and excitement as aeronautics. This enthusiasm, coupled with Dryden's innate curiosity about the technical mysteries of flight, soon persuaded him to enter this daring new field.[4]

4 Autobiographical sketch by Dryden (27 September 1965), HLD Papers, Ms. 147, Series 2.2, JHU; Richard K. Smith, *Hugh L. Dryden Papers, 1898–1965: A Preliminary Catalogue* (Baltimore: Johns Hopkins University, 1974), pp. 19–20; Thomas, *Men of Space*, p. 66; interview with Nancy Baker, Rockville, MD (9 August 1994); Dryden, "The Advantages of the Airship Over the Aeroplane" (11 November 1910), HLD Papers, Ms. 147, Series 1.1, JHU; and Eugene M. Emme, "Astronautical Biography: Hugh Latimer Dryden, 1898–1965," *Journal of the Astronautical Sciences* 25 (April–June 1977): 152–153.

FIRST IN HIS CLASS

The next step in Dryden's education led him to one of the city's finest high schools, known as Baltimore City College. He began at age 12 and experienced some setbacks, including Fs on two different English compositions. But by the end of 1911, he had improved enough to earn an "Excellent" for a dual biographical sketch of two 18th-century figures, essayist Samuel Johnson and actor David Garrick. From that point on, his troubles with writing diminished (although he still faced such humbling remarks as "Have you heard of the article "the"? Use it occasionally—you are not writing a guide-book."). Finally, by the end of his tenure at Baltimore City College, his handwriting—and indeed his narrative style—began to assume the trademark compactness and simplicity associated with his later years. He also developed a fondness for books, formed his own little library, and for his 13th birthday received a book as a gift from a classmate.[5]

During his high school years it became clear that his greatest intellectual strength lay in mathematics. Like his father, who showed so much early promise in the subject, he excelled beyond all expectations. An envious fellow student proclaimed in Dryden's school yearbook, *The Green Bag*, "behold the future professor of mathematics in general. This kid is some shark when it comes to handling the mystic 3-x, y, z—and what's more, he knows it. Hugh, sweet child, always studies his lessons. Then if he doesn't know them, the prof takes what he says, anyhow." The dimensions of his achievement became apparent at the June 1913 commencement exercises. He not only matriculated in three years rather than the usual four, but also became the youngest ever to win a diploma at the school. Because of his age and small size, and because he still dressed in knickerbockers, he looked out of place on the graduation stage; yet he ranked first among 172, the largest class in the history

---

5 Autobiographical sketch by Dryden (27 September 1965), HLD Papers, Ms. 147, Series 2.2, JHU; autobiographical sketch by Dryden (15 September 1962), HLD Papers, Ms. 147, Series 2.2, JHU; interview with Mary Ruth Van Tuyl, Silver Spring, MD (3 August 1994); interview with Nancy Baker by telephone (25 July 1994); and Dryden, "An Evening Spent Alone in an Old House" (30 September 1910), HLD Papers, Ms. 147, Series 1.1, JHU. See also the following essays by Dryden in the same archive: "My Hero in Fiction" (14 October 1910); "A True Halloween Party" (28 October 1910); "How To Set Up A Tent" (9 December 1910); "A Winter Scene" (9 January 1911); "Resolved, That a College Should Be Located in a Small Town, Not in a Large City" (20 January 1911); "Evil Is Wrought by the Want of Thought as Well as by Want of Heart" (19 May 1911); "Resolved, That Moving-Picture Parlors of Instruction Should Displace the (common, everyday) Moving-Picture Parlor" (2 June 1911); "How the Debt Was Paid" (29 September 1911); "German Exercise" (11 October 1911); "The Veal Butcher's Stall" (27 October 1911); "Samuel Johnson and David Garrick" (8 December 1911); "How To Do Your Christmas Shopping" (22 December 1911); "A High School Paper" (1 March 1912); "A Railroad Accident" (15 March 1912); "Time Flies Over Us, But Leaves Its Shadow Behind—Hawthorne" (26 April 1912); "Brutus Was Justified in Killing Caesar" (10 May 1912); "The Greatest Obstacles to World's Peace" (24 May 1912); and "Botany" (27 March 1913). Dryden, "The Stage in Shakespeare's Time" (12 April 1912), and letter from Glen Owens to Dryden (22 August 1911), both in the possession of Leona "Peggy" Dryden, Hyattsville, MD. Rev. Charles Kingsley, *Hereward: The Last of the English* (New York: A. L. Burt Publishers), title page in the possession of Mary Ruth Van Tuyl, Silver Spring, MD.

Shown around 1919, Dryden earned his doctorate in applied physics under one of the eminences in the field, Joseph Ames, later President of Hopkins. Just 20 years old upon graduation, Dryden became the youngest person to earn a doctorate at the university. Ames—one of the original members of the National Advisory Committee for Aeronautics—found a position for Dryden at the National Bureau of Standards (NBS). Dryden remained at the NBS for 29 years, assuming positions of increasing responsibility. (Collection of Nancy (Dryden) Baker, Rockville, MD. Reproduced with her permission.)

of the college. During the ceremonies, he also received both the Peabody Prize for mathematics and the special notice of Baltimore's Mayor Preston. "I am so glad to hear of your fine record," wrote a classmate less jealous than the one who marveled at Hugh's mathematical talents. "You have gained an honor that could not be bought. Congratulate your mother and father for me. I know they must be very proud of you."[6]

During mid-1913, Samuel and Nova's son turned his attention to higher education. He had narrowed his choices to Johns Hopkins, a relatively new university founded 37 years earlier. By Dryden's freshman year, it had already laid claim to distinction.

Hugh Dryden had good reason to set his sights on the campus just northwest of his own neighborhood. For one, it had achieved national recognition in several subjects, including mathematics, the course of study he wanted to pursue. Continuing to live in the same city as his parents also enabled the teenager to maintain his regular patterns of family life and at the same time save money otherwise required for room and board. His sparkling academic record left no doubt about his admission to Hopkins. But clearly, without extensive financial assistance to compensate for the lack of family income, the young scholar would find the university outside his grasp.[7]

---

6 The envious remarks by one of Dryden's fellow students quoted in Thomas, *Men of Space*, pp. 65, 67; the remarks by a sympathetic classmate quoted in a letter from Glenn Owens to Dryden (7 June 1913), in the possession of Leona Dryden; Gorn, *Hugh L. Dryden's Career*, pp. 1–2; and autobiographical sketch by Dryden (27 September 1965), HLD Papers, Ms. 147, Series 2.2, JHU.

7 Interview with Nancy Baker, Rockville, MD (9 August 1994).

During the summer of 1913, he searched for the needed backing. He received invaluable assistance from two of his Baltimore City College math teachers, S. F. Norris and Richard Uhrbrock. They persuaded their protégé to apply for a Maryland scholarship, which paid all expenses for four years. Both men coached him for the examination. Meanwhile, Uhrbrock wrote a splendid reference to the Maryland Scholarship Board on Dryden's behalf. Even accounting for the inflated claims inherent in such letters, Uhrbrock made a powerful appeal for his pupil. "In my teaching experience of 25 years," he wrote, "I have never had a student superior to Hugh Latimer Dryden." Chronologically still a boy, Dryden nonetheless manifested some of the extraordinary personal qualities and intellect that foreshadowed the mature man.

> [Dryden] was in my classes in mathematics for nearly three years. At all times his conduct has been that of a gentleman, and his scholarship has been of the very highest order. His presence in the class has been a source of inspiration. He is a quiet and unobtrusive leader. By his work and conduct he has aroused and kept alive interest and enthusiasm in his classmates, without exciting the least jealousy or envy.[8]

Thanks to the strategies of his mentors, as well as his own talents, he won the scholarship and entered Hopkins in the fall of 1913 with full support for his entire undergraduate career. Not only that, during the summer he completed and passed two other tests—one oral, one written—under the direction of professor L. S. Hurlburt of the Johns Hopkins Mathematics Department. These exams, taken in lieu of coursework, freed him from taking analytic geometry, so that when he walked for the first time onto the university grounds at Charles and 34th Streets, the 15-year-old entered with advanced standing.[9]

OUT OF THE SHADOWS

Early in his career at Hopkins, Dryden encountered one of the great synthesizers in American physics, professor Joseph Sweetman Ames, chair of the department and later president of the university. In Ames, Dryden chose a formidable teacher. Hard driving and gruff—in part because of a lifelong stammer—Ames understandably lived by action, rather than conversation. He exercised a profound, almost parental influence over

8 Quoted in a letter from Richard H. Uhrbrock to the board awarding Maryland scholarships (3 June 1913), and S. F. Norris to Dryden (17 June 1913), both in the possession of Leona Dryden.

9 Letter from L. S. Hurlburt to Dryden (15 June 1913), in the possession of Leona Dryden; autobiographical sketch by Dryden (27 September 1965), HLD Papers, Ms. 147, Series 2.2, JHU; handwritten biographical statement by Dryden, HLD Papers, Ms. 147, Series 2.2, JHU; and personal history statement of Dryden (28 May 1947), HLD Papers, Ms. 147, Series 2.2, JHU.

In his early 20s, Dryden married a lively young woman whom he met at the Appold Methodist Church in Baltimore. Hugh and Libby (Travers) gave birth to Hugh, Jr., shown here. By this time (about 1923) Dryden became Director of the Bureau of Standards' Aerodynamic Section and, as such, managed its modern wind tunnel. (Collection of Nancy (Dryden) Baker, Rockville, MD. Reproduced with her permission.)

Dryden, considering him "the brightest young man [I] ever had, without exception." Ames mentored him through his bachelor's in mathematics (in three years) and guided him toward his master's in applied physics. Ultimately, Hugh Dryden earned a doctorate in physics in 1919 at age 20, the youngest Hopkins graduate to earn a Ph.D. His dissertation concerned the scale effects of air flowing around columns perpendicular to the wind, a subject of exceptional importance in the rising field of aerodynamics. The conditions under which he earned his degree affected the remainder of his career.[10]

Actually, Dryden earned his degree while he worked. Near the end of the First World War, Ames won a position for him at the National Bureau of Standards (NBS), situated in the Maryland suburbs, testing munitions gauges. After pursuing this project for a short period, Dryden's career took root. He quickly transferred to the bureau's newly opened Aerodynamics Section, equipped with one of the country's most advanced wind tunnels. Because Ames had recently been appointed to a seat on the fledgling National Advisory Committee for Aeronautics (NACA), which met in Washington, DC, he found it convenient to offer courses to Dryden and some of his other students living near the nation's capital. Meantime, in 1920—at the age of 22—Dryden became the first Chief of the Aerodynamics Section, in charge of wind tunnel research (a program subsidized in part by the NACA, starting with a grant of $40,000 in 1921 and rising to more than $100,000 annually during World War II). Perhaps even more impor-

10 Gorn, *Hugh L. Dryden's Career*, pp. 1–2. For an essay about the life of Joseph Ames, see N. Ernest Dorsey, "Joseph Sweetman Ames: The Man," *American Journal of Physics* 12 (1944): 135–148.

tant than his livelihood, during the same year the reserved young scientist married a vivacious, dark-haired woman named Mary "Libby" Travers whom he had met years before in Sunday school at the Appold Methodist Church in Baltimore. The couple relocated from Baltimore to a modest row house inside Washington, DC's verdant Rock Creek Park and began to raise a family.

Hugh Dryden eventually became Chief Physicist and, finally, Associate Director of the National Bureau of Standards. But on the retirement of George W. Lewis (Director of the NACA since 1919), Dryden became his successor. During his 10 years on the job, he transformed it from a purely aeronautical institution to one equally involved with spaceflight. He is shown here in September 1947, at the age of 49, just as he assumed Lewis's mantle. (Collection of Nancy (Dryden) Baker, Rockville, MD. Reproduced with her permission.)

Dryden achieved distinction soon after starting at the bureau. During the first six years he embarked on a research collaboration with Dr. Lyman J. Briggs, a mentor and friend who eventually became Director of the agency. Dryden decided to concentrate on transonic flight, a subject far ahead of its time and a theoretical curiosity in an age when the winner of the 1925 Schneider Cup Race flew at 233 miles per hour. Together with Briggs, Dryden focused on the problem of compressibility. During the early 1920s, the phenomenon manifested itself in the behavior of propeller blade tips. Attached to aircraft engines of increasing power, the blades rotated at transonic and supersonic speeds, resulting in unexplained boundary layer separation and buffeting. Dryden and Briggs predicted the effects in a seminal NACA Report (Number 207) published in 1925, entitled *Aerodynamic Characteristics of Airfoils at High Speeds.* The absence of research tools available to Dryden and Briggs suggested the vision inherent in the project; no wind tunnel yet existed to replicate the required wind velocity. They found a substitute in Lynn, Massachusetts, where the General Electric Company made available its large centrifugal compressors for the experiments. Here Dryden and Briggs made some of the earliest experimental observations about aerodynamic drag approaching the speed of sound, the effects of compressibility on aerodynamic lift and drag, and airfoil design modifications for propeller manufacturers. The results presented aerodynamicists with some of the earliest credible findings about travel approaching Mach 1, a generation before Captain Chuck Yeager's celebrated 1947 flight that exceeded the speed of sound in the X-1 aircraft. Dryden and Briggs reported five physical phenomena as airfoils neared supersonic speed:

1. The lift coefficient for fixed angle of attack decreased quickly as the speed increased.

2. The drag coefficient increased rapidly.

3. The center of pressure moved rearward, toward the trailing edge of the wing.

4. The speed at which such changes occurred declined by increasing the angle of attack and by increasing the camber ratio.

5. "The angle of zero lift shifts to high negative angles up to the "critical speed" and then moves rapidly toward 0 degrees."[11]

Several decisive research papers followed, the results of which prompted the aeronautics community to lay plans for wind tunnels capable of transonic airflows and above.

Hugh Dryden established his reputation with the release of this and several other NACA Reports about the aerodynamics of high-speed flight. His next project placed him in proximity to one of the pivotal figures in global aeronautics. In his study of compressibility, Dryden became conversant with one of the persistent flaws of existing wind tunnels—incidental turbulence generated as air caromed off the tunnel walls, thus compromising the accuracy of the aerodynamic measurements. He and Arnold Kuethe, his assistant at the bureau, found the answer in a simple but ingenious hot-wire anemometer, capable of detecting even slight variations in the speed of airflow. Since even minute fluctuations might cause wide experimental discrepancies—the data on one airship model tested in the bureau's tunnel differed 100 percent from the results obtained in the Washington, DC Navy Yard tunnel—Dryden adapted the instrument to detect these disturbances and to reduce their unwelcome effects. Using this device, he developed criteria by which to diminish turbulence in existing tunnels and to further reduce it in the construction of new ones. Dryden then tested these techniques in the bureau's own machine. Benefiting from the improvements, he succeeded in isolating the workings of the boundary layer (in particular the transition from laminar to turbulent flow), thus enabling him to verify experimentally Ludwig Prandtl's landmark boundary layer theory of 1907, one of the benchmark discoveries in the field of aerodynamics.

11 Gorn, *Hugh L. Dryden's Career*, pp. 3–5; Alex Roland, *Model Research: The National Advisory Committee for Aeronautics, 1915–1958*, vol. 2 (Washington, DC: National Aeronautics and Space Administration, 1985), p. 478; John T. Greenwood, ed., *Milestones of Aviation: Smithsonian Institution National Air and Space Museum* (New York: Hugh Lauter Levin Associates, 1989), p. 98; Lyman Briggs, G. F. Hull, and Hugh L. Dryden, *Aerodynamic Characteristics of Airfoils at High Speeds: National Advisory Committee for Aeronautics Report Number 207* (Washington, DC: Government Printing Office, 1925), pp. 3, 16.

The publication of Dryden's findings, which had profound research implications, caught the eye of one of the world's greatest and best-known aerodynamicists, the Hungarian-born Theodore von Kármán, Director of the Guggenheim Aeronautical Laboratory at the California Institute of Technology. The two men met during the early 1930s and became close friends. Their relationship gave new credence to the proverb that opposites attract. The laconic and understated Dryden and the expressive and flamboyant Kármán embarked on a 35-year personal and professional collaboration. Although a union of equals, Dryden did owe a debt in at least one important respect. Kármán, 17 years his senior, opened the wide realm of international science to him, introducing him to countless aeronautics practitioners in many nations. Once known, Dryden quickly developed his own following in these circles, due as much to his personal qualities as to his unusual scientific capacities.[12]

Indeed, Hugh Dryden began to assume an honored role in the profession just after meeting Theodore von Kármán, but not because of him. The American Association for the Advancement of Science elected him a member, as did the Washington Philosophical Society. He won promotion at the NBS, becoming Chief of the Mechanics and Sound Division, which included his own Aerodynamics Section. He also found his judgments sought after by some of the giants of aeronautics. Yet, Dryden did not seek recognition. Diligent, reserved, and self-effacing, he worked quietly on his own projects at the NBS, guided those of his subordinates, and earned a reputation for effectiveness. His lifelong and intense devotion to the Methodist faith may have contributed to his distaste for self-promotion.

Dryden's stature became evident at the Fifth International Congress of Applied Mechanics in 1934. Here—in his first trip outside the United States—he presented an important paper called "Boundary Layer Flow Near Flat Plates," one that added significantly to the understanding of the mechanics of laminar flow. He subsequently became a founding member of the newly formed Institute of the Aeronautical Sciences in New York, helped inaugurate its scholarly journal, the *Quarterly of Applied Mechanics*, and served on its board of editors. Dryden won further recognition in 1938 when the institute invited him to deliver the annual Wilbur Wright Lecture, the first American so honored. The NBS rewarded his rising eminence by naming him Chief Physicist.[13]

---

12 For more about Theodore von Kármán, see Theodore von Kármán with Lee Edson, *The Wind and Beyond: Theodore von Kármán, Pioneer in Aviation and Pathfinder in Space* (Boston, Toronto: Little Brown, 1967), and also Michael Gorn, *The Universal Man: Theodore von Kármán's Life in Aeronautics* (Washington, DC and London: Smithsonian Institution Press, 1992). Hugh Dryden and A. M. Kuethe, *The Measurement of Fluctuations of Air Speed By the Hot-Wire Anemometer: National Advisory Committee for Aeronautics Report Number 320* (Washington, DC: Government Printing Office, 1929), pp. 359–361; Gorn, *Hugh L. Dryden's Career*, pp. 3–4, 6; and interview with Nancy Baker (17 June 1994).

13 Gorn, *Hugh L. Dryden's Career*, pp. 3–4.

A UNIQUE PERSONALITY

Although an introvert, Hugh Dryden's temperament did not hinder his rise. At social functions, for instance, he learned to mingle and relax even though he probably regarded such events with indifference at best. But he did adapt and developed some habits that helped. One involved alcoholic drinks. When he ordered, he always asked bartenders for the same simple mix—water and ice in a highball glass. Dryden then sipped and circulated, never telling anyone about his teetotal preferences. He also dressed well. He wore nicely tailored business suits, well-polished shoes, and in general made a refined appearance. His voice—clear, distinct, and pleasing—also contributed to his persona. Finally, although not a charismatic personality, he did project an amalgam of keen intellect and surprising personal warmth.

Dryden's religious faith played a profound role in his life. As a teenager, he felt drawn toward the ministry. Had one Methodist seminary accepted applications from 15-year-olds, Johns Hopkins, Joseph Ames, and the world of aeronautics might never have encountered Hugh Dryden. Despite his choice of the secular life, Dryden nonetheless pursued his spiritual calling as a lay Methodist minister and a persuasive preacher, witnessed by his many surviving sermons that expressed a fervent devotion (and submission) to God.

> You want sharpness and keenness to come into your brain; you want courage and strength to make decisions and carry them through. This is the secret. Yield yourself to God. "The Kingdom of God is within you." You do not need to hunt it from the outside, just release it, it is within you. You have been defeated in some situation. You have been educated, you have ability, you work hard, but you are defeated by worry, anxiety, and frustration. Establish the contact through faith.

Moreover, Hugh Dryden bore little resemblance to the solemn figure often depicted in his photographs. He had a fine sense of humor that often showed itself in wickedly accurate imitations of his friends and coworkers. He also possessed due admiration for the opposite sex. By present standards, Dryden's enjoyment of female company seems almost quaint. Apparently, he often assumed the task of grocery shopping to relieve his wife of one of her burdens, and his younger daughter Nancy often accompanied him. When they got inside the supermarket, he asked her to take part of the shopping list and fill it. Meantime, he headed for the meat counter where, as virtually the only man in sight, he got all of the attention of the women making their selections. Nancy Dryden called this harmless practice "being single with one eye."

Among Hugh Dryden's formative influences, his father Samuel played a decisive part. The elder Dryden possessed keen intelligence, but he also suffered a number of self-inflicted reversals early in his career. The effects of such behavior must have been perceived by young Dryden, who learned from his father's failings the qualities necessary to fashion his own career—patience, tolerance, and the capacity to collaborate with others. Dryden not only practiced these arts with rare skill, but also recognized and appreciated them in others.

Finally, Dryden also developed a profound commitment to international scientific cooperation, in part the result of his close association with Theodore von Kármán, but more importantly due to his uncanny instinct for merging complementary talents and personalities. Both men felt that alliances with foreign colleagues served as instruments of peace and of technical advancement. Dryden began during the 1930s and in time not only became the equal of his mentor, but even surpassed him. During the late 1950s, President Eisenhower appointed him a representative to the United Nations Ad Hoc Committee on the Peaceful Uses of Outer Space. This assignment led to his role as chief negotiator with Soviet academician Anatoly Blagonravov, a dialogue that resulted during the early 1960s in limited but unprecedented superpower cooperation on meteorology, communications, and magnetic field research. "I am persuaded," wrote Dryden with characteristic simplicity, "that there are very great values to the United States in this cooperation."[14]

AN INTERNATIONAL PROJECT

Not surprisingly, one of Dryden's most admired achievements occurred in the international arena. It originated with a long and fruitful collaboration between Theodore von Kármán and Army Air Forces General Henry H. Arnold. The two men had become acquainted during the mid-1930s when Arnold commanded March Field, California, in close proximity to Caltech. They established a friendship based on Arnold's desire to stay abreast of the latest aeronautical developments. As the Second World War approached, the general prevailed upon Kármán to devote an increasing proportion of his time to advising the Army Air Forces about the future of flight. On one occasion, the scientist flew to Dayton, Ohio on the invitation of General Frank Carroll, who asked him to calculate the likely success of flying at the speed of sound. Kármán's highly technical but affirmative answer set the wheels in motion for the development of the X-1 aircraft.

14 The overall assessment of Hugh Dryden's personality is based on research for a forthcoming book by Gorn entitled (tentatively) *To Ride the Air: NASA's Hugh L. Dryden*. In addition, see Gorn, *Hugh L. Dryden's Career*, pp. 4, 14. For a statement of Dryden's religious views, see "The Importance of Religion in American Life," letter written in March 1950, reprinted in *Hugh L. Dryden's Career*, pp. 123–130.

During the summer of 1944, when it seemed clear that the allies finally held the key to victory in Europe, Arnold contacted Kármán for the biggest assignment yet. The general asked for a complete, global review of the technologies underpinning the wartime advances in aeronautics and missilery. To achieve this end, he asked Kármán to travel to the far corners of the world to obtain data firsthand. More daunting, Arnold also wanted a comprehensive blueprint for future development based on the most recent discoveries. Realizing that this project required many hands, Kármán turned first to Hugh Dryden to be his partner and deputy. Dryden's own work during World War II raised his stature as a science advisor. Officials at the Office of Scientific Research and Development engaged him to supervise the development of the Bat Missile, a highly advanced naval air-to-surface weapon. A self-correcting, launch-and-leave gravity bomb guided by radar and steered by its own control surfaces, it posed not only formidable technical challenges, but also involved complicated management responsibilities for Dryden. To achieve success, he harnessed the research talents of scientists at the Bureau

This photo hints at Hugh Dryden's temperament. On the right is professor Theodore von Kármán, one of the most renowned aeronautical scientists of the 20th century. The flamboyant and highly animated Hungarian offered a sharp contrast to his close friend Dryden. Seen here around 1960, the two struck characteristic poses: Kármán entertaining his dinner companions, Dryden sitting quietly with a look of mild amusement and perhaps some discomfort. Much as he admired Kármán, Dryden preferred the self-effacing and undemonstrative approach. (Collection of Nancy (Dryden) Baker, Rockville, MD. Reproduced with her permission.)

of Standards, the Bureau of Naval Ordnance, and academia. Moreover, his assignment entailed not just the realization of a workable concept; rather, the Office of Scientific Research and Development instructed him to supervise the Bat's design, fabrication, and test, and then to field it as an operational weapon capable of destroying enemy shipping. Placed in the U.S. arsenal during the final months of the war, it succeeded beyond all expectation, actually sinking several Japanese ships at the Battle of Okinawa.

Fresh from the experience of managing a large federal research program, Dryden joined his friend in December 1944 and for the better part of a year collaborated with him in fulfilling Arnold's enormous task. They met in the recently completed Pentagon building and compiled the names of some of the most eminent American scientists to enlist for the project. After some discussion, they selected six researchers to organize the team, including Kármán's brilliant student Hsue-shen Tsien for rocketry and Dr. Louis Alvarez for radar. The more difficult job involved picking subject-area experts for the technical panels, a small cadre who converted the hard data gleaned in the U.K., France, Germany, Italy, the Soviet Union, Japan, and other nations into comprehensive reports and forecasts. They chose 21 individuals, most of whom from academia, and especially from Caltech.[15]

The group departed for Europe in April 1945 prepared to assess the current state of international aeronautics. In London, Kármán and Dryden donned uniforms bearing the simulated ranks of major general and colonel, respectively. They arrived in Paris (the embarkation point of the mission), traveled to and crossed the German border, and stopped at a huge aeronautics laboratory near Braunschweig, in the village of Volkenrode. Concealed from public view in a setting of forests and farms, it just had been discovered by U.S. forces. Dryden and Kármán interviewed some of the leading German officials about their guided missile and jet propulsion research and amassed some 1,500 tons of materials relating to swept-wing aerodynamics and high-speed human physiology, a treasure shipped back to the United States where it profoundly influenced postwar aeronautics. Dryden and Kármán then divided forces. Dryden and part of the group went to Munich, where he conducted intense interrogations about the V-1 and V-2 missiles with Dr. Wernher von Braun, his boss General Walter Dornberger, and some 400 scientists who had been relocated from the Peenemünde rocket facility. Dryden then returned home after retracing his steps through Germany, France, and England.

Rather than participate in the next round of globetrotting, Hugh Dryden decided to remain in Washington, DC and serve as the general editor of the technology forecast promised to General Arnold. In order to save time, Dryden assembled the report as

15 Michael H. Gorn, *Harnessing the Genie: Science and Technology Forecasting for the Air Force, 1944–1986* (Washington, DC: Office of Air Force History, 1988), pp. 11–23; and Gorn, *Hugh L. Dryden's Career*, pp. 5–7.

events unfolded, that is, as the sections arrived by cable from overseas. Meanwhile, Kármán and his team went abroad again, this time aboard Arnold's own C-54 transport. They embarked on a whirlwind world tour that lasted much of the fall of 1945, visiting scientists and laboratories in the U.K., France, Holland, Switzerland, Sweden, and Italy. But age and exhaustion at last caught up with Kármán. He broke from the group and encamped in the luxurious Prince of Wales Hotel in Paris while the rest of his party journeyed to Australia, India, China, and Japan. Kármán took the opportunity to write the first chapter of the report, entitled "Science, the Key to Air Supremacy," a prescient account of airpower as it existed and as it might evolve.

At the same time, in Washington, DC, Hugh Dryden assumed the heavy responsibility of shaping Kármán's and the many other essays into an accurate and coherent summary of worldwide airpower developments. In essence, he assumed control of the project. In the end, the sprawling forecast arrived on General Arnold's desk on 15 December 1945. Called *Toward New Horizons*, this seminal collection of 33 essays by 25 scientists served not merely as a blueprint for American survival in an age of potentially devastating attack from the air, but as a guide for civilian aeronautics—and indeed space research—for decades to come.[16] It looked to the future under broad headings such as aerodynamics and aircraft design, power plants, fuels and propellants, radar, weather and flight, and aviation medicine. *Toward New Horizons* also made recommendations to the Army Air Forces, urging its leaders to become more scientifically oriented and to reorganize in order to take advantage of the recent and future breakthroughs in aeronautics. But perhaps more importantly, Hugh Dryden's work offered military leaders and industrialists (as well as younger scientists and engineers just leaving military service) a clear agenda for postwar research, one that offered direction to generations of aerospace practitioners.[17]

SEEDS PLANTED ON EARTH

Meantime, at the NACA, its venerable Director of Research George W. Lewis—worn and ill after 28 years of service, as well as the rigors of intense wartime research—announced his retirement in the summer of 1947. To no one's great surprise, the unassuming yet eminently qualified Hugh Dryden, recently named Associate Director of the Bureau of Standards, succeeded him. His appointment initiated a remarkable transformation of the NACA.

16 Theodore Von Kármán, *Toward New Horizons* (Washington, DC: Army Air Forces Scientific Advisory Group, 1945).

17 Gorn, *Harnessing the Genie*, pp. 23–42; and Gorn, *Hugh L. Dryden's Career*, pp. 6–7.

Dryden began almost immediately to reorient the Agency toward high-speed flight. He assumed Lewis's job on 2 September and, by the end of the month, crossed the country to visit the Muroc Flight Test Unit, located in the isolated high desert of Southern California. Dryden made this trip for a good reason. Here, Walter Williams and his small NACA team prepared to challenge the sonic barrier, and, as a leading theorist on the subject since his youth, Dryden wanted to encourage and recognize their important work. More than mere recognition, however, his presence symbolized the new direction of his leadership—toward supersonic and eventually toward hypersonic flight. Until this point, the NACA's desert outpost existed as a temporary station of the Langley Memorial Aeronautical Laboratory, whose personnel first assembled in California solely to fly the X-1 and D-558 experimental aircraft. To affirm the importance of Muroc's role, Dryden directed that it join Langley and the other NACA facilities—the Ames and Lewis Laboratories in Northern California and Cleveland, respectively, and the Wallops Island, Virginia station—as a permanent NACA complex.

Dryden also took this measure because of his more recent involvement with high-speed flight. He served prominently on the joint NACA-Army-Navy Research Airplane Committee that, in mid-1944, launched the X-1 and D-558 projects in the first place. Once he became Director of the NACA, he naturally lent his support to these planned supersonic flights. A few years later, his support once again proved decisive in initiating and sustaining an even more advanced aircraft, the North American X-15. After a two-year gestation period at Langley, in 1954 Dryden approved the concept of a Mach-7 hypersonic airplane capable of achieving altitudes of 300,000 feet. He reconvened the Research Airplane Committee in October 1954, consisting of Air Force Brigadier General Benjamin Kelsey and Rear Admirals Lloyd Harrison and Robert Hatcher, the leaders of their services' research and development programs. Experienced at dealing with the armed forces, Dryden not only persuaded these officers to underwrite this expensive and risky venture (that offered no clear military value), but reserved the role of chairman for himself. Once in command, Hugh Dryden guided the fortunes of the first aircraft to fly hypersonically, to enter the realm of space, and to serve as a test bed for later, more daring activities outside of the atmosphere. He did so by controlling the X-15 committee, retaining for himself the power to call committee meetings, to organize X-15 technical conferences, and to assume direct control of the NACA's role in this project.[18]

The X-15 strained the NACA's resources like few other endeavors. It required intensive flight planning and pilot preparation, as well as unprecedented technical demands such as a computer-assisted control system, simulators more complicated and realistic than any ever attempted, a powerful new rocket engine, and state-of-the-art equipment necessary to conduct high-speed wind tunnel experiments. Dryden worked energetically to equip his Agency with the resources necessary for the X-15 to succeed.

18 Hansen, *Engineer in Charge*, p. 385; and Arnold S. Levine, *Managing NASA in the Apollo Era* (Washington, DC: National Aeronautics and Space Administration, 1982), p. 11.

The implementation of the National Unitary Wind Tunnel Plan illustrates just one of his travails and successes. Originating conceptually with *Toward New Horizons*, the proposition of massive government research centers for aeronautics gained credence among the NACA, the Air Force, and the aircraft manufacturers. All of them pressed for federally sponsored, high-speed wind tunnel facilities, but the parties soon found themselves at odds over the location and control of these installations. Hugh Dryden played a decisive part in brokering these disputes. Had he objected to Air Force claims for a huge hypersonic wind tunnel complex, the entire initiative might have collapsed; instead, he offered no resistance to the establishment of the USAF Arnold Engineering Development Center in Tullahoma, Tennessee. In return, the NACA agreed to abandon its dream of a National Supersonic Research Center and instead accepted $136 million from Congress to build three supersonic and hypersonic tunnels at its existing laboratories. But the National Unitary Wind Tunnel Plan Act of 1949 forced Dryden to make a further concession; it reserved these NACA machines mainly for industry use. In the end, however, his compromise proved more than satisfactory. The aircraft industries never accounted for more than a fraction of the new tunnels' work schedules. Indeed, from the beginning, the new tunnels teemed mainly with NACA experiments. In fact, these machines opened just in time for some essential research. By the mid-1950s, X-15 models underwent critical aerodynamics tests at the Mach 2 tunnel at Lewis, the Mach 3.5 at Ames, and the Mach 5 at Langley.[19]

Other advances blazed a path for hypervelocity and, ultimately, for spaceflight. Ames Aeronautical Laboratory researcher H. Julian Allen announced in 1950 a surprising discovery. As they fell through the atmosphere, objects with blunt shapes generated lower temperatures than those with pointed noses or protuberances. Apparently, pressure drag caused rounded bodies in flight to dissipate heat into the atmosphere. Frictional drag, on the other hand, induced bodies with sharp angles to absorb the heat into themselves during reentry. Allen's colleague Alfred J. Eggers concluded that a conical design best embodied Allen's theory. These concepts—demonstrated in the Ames wind tunnels—also manifested themselves in actual flight conditions at the NACA's Wallops Island, Virginia Pilotless Aircraft Research Station, an outpost of the Langley Aeronautical Laboratory. Here, missiles and rockets attained speeds up to Mach 12, and the engineering staff experimented with such space-related developments as the Ames rounded nose cone, heat-resistant materials (like Inconel, the alloy used to fabricate the X-15), reaction controls,

19 Roland, *Model Research*, vol. 1, pp. 211–221; and notes from a telephone conversation between Michael Gorn and Dr. James O. Young, Air Force Flight Test Center historian (12 October 1998), relative to Dryden's role in the passage of the National Unitary Wind Tunnel Plan Act of 1949. Dr. Young conducted an interview with General Laurence C. Craigie, one of the chief Air Force figures involved in the Unitary Wind Tunnel Plan, in 1991 at the Air Force Village of the West in Riverside, CA. Craigie related to Young that in his capacity as NACA Director, Dryden could have derailed the negotiations at any time, but instead supported conciliation, as explained in the narrative above.

and dynamic stability. From 1945 to 1957, roughly 3,000 firings occurred at Wallops (an average of 250 launches per year during Dryden's tenure) with an approximate success rate of 80 percent. The power plants and fuels propelling these vehicles often originated in the test stands at the Lewis Flight Propulsion Laboratory (now the Glenn Research Center) in Cleveland, Ohio. By 1957, roughly one-fourth of its engineers and scientists devoted themselves to rocketry. The 10-foot-by-10-foot Unitary tunnel, opened a year earlier at Lewis, produced realistic data unobtainable by any other means. It complemented a new 20,000-pound thrust test stand complex, which allowed the firing of rocket engines double the size of the ones they replaced. The latter facility also made possible safe runs using high-energy fuels like hydrogen-fluorine and nuclear material.[20]

All of these space-related activities scattered across the NACA landscape required Hugh Dryden to find the human and material wherewithal to accomplish them. Unfortunately, Congress failed to appropriate more than token additional resources. Between 1951 and 1957, the NACA's budget rose only modestly, from $63 million to $77 million, leaving little for new ventures after accounting for inflation and large capital investments. Therefore, Dryden turned to the only means at hand: reallocating the assets under his control. During 1954, roughly 10 percent of the NACA's research related to space. Two years later, by scaling back aeronautics work, Dryden managed to dedicate one quarter of the Agency's total projects to investigations involving space travel. The Director's last budget estimate, submitted six months before NASA came into being, requested that half of all NACA activities concentrate on flight beyond the atmosphere. Yet, a simple budgetary profile fails to give a complete picture of Dryden's transformation of the NACA's mission. It ignores programs like the X-15, an endeavor funded mainly by the military services, guided by the NACA, and rich in technical achievements fundamental to the later space program. Also, the raw numbers alone do not illuminate the inherent obstacles faced by Dryden as he realigned the NACA. For one, he carried out his quiet revolution without the slightest change in the NACA charter. Until its last moments, it officially remained an institution devoted to aeronautical, not to space, research. Even more telling, Dryden began and then accelerated this momentous shift long before October 1957, the subsequent Sputnik panic, or any national consensus had emerged about the value of spaceflight. Essentially, he embarked on this program on his own initiative.[21]

20 Gorn, *Expanding the Envelope,* pp. 256–257; Anon., "Wallops Gathers Hypersonic Flight Data," *Aviation Week and Space Technology* (3 June 1957): 14–15; and Robert Cushman, "Lewis Pushes Work on Rocket Engines," *Aviation Week and Space Technology* (3 June 1957): 6–9.

21 Levine, *Managing NASA in the Apollo Era*, p. 11; and Roland, *Model Research*, vol. 2, p. 475.

Yet, his reallocation efforts failed to win laurels. Once the American public realized the importance of Sputnik, many in Congress blamed the NACA and Hugh Dryden for failing to prepare the nation for this eventuality, even though most members of the House and Senate showed little or no interest in the subject until Soviet satellites flew over American soil. Indeed, in an atmosphere of denial and recrimination, both the NACA and its Director made a poor showing. Hugh Dryden's undemonstrative personality and his unobtrusive, incremental progress toward spaceflight not only epitomized his own style of operation, but that of the NACA as well. Throughout its history, the Agency's engineers worked almost anonymously, removed from public scrutiny as they documented and disseminated knowledge related to speed, altitude, stability and control, efficiency, and flight safety. But in a time of crisis, steadiness and self-effacement seemed out of season. As a consequence, on 1 October 1958, the NACA found itself supplanted by NASA. Moreover, T. Keith Glennan, President of the Case Institute of Technology, assumed the position of the Agency's first Administrator, but only on condition that Hugh Dryden remain as his deputy. In his 40th year of government service, a disappointed Dryden agreed to stay.[22]

Still, Dryden's main initiative as Director of the NACA—to carve out a role in spaceflight—paid handsome dividends during the formative years of NASA. Because of past preparation, Project Mercury, for instance, exemplified NASA's capacity to launch the nation into the space race with speed and sureness. During the summer of 1958 (still some months before the birth of NASA), Dryden persuaded Robert Gilruth, Langley's former Chief of Stability and Control and subsequently the director of the laboratory's Pilotless Aircraft Research Division, to transfer to NACA Headquarters and lead Mercury. Here, Gilruth worked with a small task group comprised of scientists and engineers from Langley and Lewis to create the essentials of the initial U.S. man-in-space program. Less than three months later, all of the main elements of Project Mercury became known: an orbiting capsule populated by a single human being; ballistic missiles pressed into service as launch vehicles; a reentry into the atmosphere made possible by the capsule's blunt body shape and heat shield; and a parachute landing at sea whereupon the U.S. Navy retrieved the ship and crew. Indeed, as Arnold Levine points out in *Managing NASA in the Apollo Program*, the "NACA was well on its way to becoming a space agency even before Sputnik."[23] Hugh Dryden, then, deserves much of the credit for laying the foundation of the early U.S. space program before NASA existed, and perhaps even more recognition for guiding it after the Agency came into being.

---

22 Roger D. Launius, *NASA: A History of the U.S. Civil Space Program* (Malabar, FL: Krieger, 1994), pp. 30–31, 38; and Gorn, *Hugh L. Dryden's Career*, pp. 12–13.

23 Hansen, *Engineer in Charge*, p. 385; and Levine, *Managing NASA in the Apollo Era*, p. 11.

CONCLUSION

Hugh L. Dryden represented an indispensable ingredient in the metamorphosis of American flight, from its beginnings as a purely aeronautical pursuit to its gradual evolution as a field that routinely transported human beings outside the atmosphere. A man born in obscurity and almost lost to history due to his mild manner and to his abhorrence of self-promotion, Dryden commands a place among the most prominent figures in American aeronautics and spaceflight. His own research influenced the basic theories of flight. His leadership at the Bureau of Standards and later as Director of the NACA affected the research agenda of countless engineers and scientists, as well as that of the nation as a whole. During his decade at the NACA, he transformed it slowly but surely into an institution capable of sophisticated space research, the bedrock—not merely the forerunner—of NASA. His role in international scientific cooperation resulted in landmark contributions like *Toward New Horizons* and led to space agreements among the United States, the NATO nations, and the USSR during the Cold War. Finally, during the last portion of his career, Hugh Dryden played a decisive role in shaping the Mercury, Gemini, and Apollo programs, not only in structuring their managerial and technical priorities, but also as a pivotal technical advisor to the American Presidents on whose political authority the U.S. space program ultimately rested.

10

Wernher von Braun

A Visionary as Engineer and Manager

ANDREW J. DUNAR

DURING THE 1960S, AS NASA'S APOLLO PROGRAM PREPARED TO PLACE MEN ON THE MOON AND RETURN THEM SAFELY TO EARTH, WERNHER VON BRAUN WAS UNDOUBTEDLY the most well-known nonastronaut in the American space program. An immensely talented man, he had a rare combination of the vision to project the potential for human spaceflight in the 20th century, the engineering skills to develop the technology needed to make such dreams reality, and the managerial ability to direct accomplished scientists and engineers by motivating them, earning their loyalty,and organizing their energies into a cohesive enterprise that pressed the limits of new technology. In Germany during World War II, he developed the notorious V-2 rocket, which also became the first rocket to lift an object constructed by humans into space. After the war, von Braun helped stimulate interest in space travel in the United States and the West in the 1950s through articles in popular magazines, speeches, and appearances on television. He directed the development of the rocket that launched the first American satellite, Explorer I, into space in January 1958 and the first American, Alan Shepard, into space on 5 May 1961. After the establishment of the National Aeronautics and Space Administration, he became the Director of one of its two largest Field Centers, the Marshall Space Flight Center in Huntsville, Alabama. There he directed development of the powerful Saturn rocket series that served as the launch vehicles for the Apollo program, the American program of lunar exploration.

Yet von Braun was a complex man whose critics never let him forget that his earliest notoriety came from his work on behalf of Hitler's Nazi regime, developing the V-2 missiles that fell on London during WWII; that he had been a member of the Nazi Party, and as was later discovered, of the SS; and, as the story developed in the late 1960s and early 1970s, that slave labor built the V-2 rockets. The German background was always present, for even in the United States von Braun built his team on a foundation of German engineering talent that had worked with him developing the V-2 in Peenemünde during WWII and accompanied him to the United States after the war ended. One of the reasons for his success was his ability to blend Germans and Americans into a successful organization, incorporating the best of contrasting approaches in engineering methodology, testing, and development.

SOCIETY FOR SPACE TRAVEL

Von Braun was born on 23 March 1912 in Wirsitz in Posen, a territory east of the Oder River; the town Wirsitz became part of Poland after WWI. His father was a government administrator, the equivalent of a county commissioner in Wirsitz, and later held positions in the German government in Berlin. Young Wernher was confirmed into the Lutheran church at the age of 13. His mother stimulated his first interest in space when she gave her son a telescope.[1] Wernher recalled an early experiment with rockets, in which he fastened skyrockets to a wagon—an unmanned vehicle, he remarked—and watched in fascination as it careened wildly about. "The police, who arrived late for the beginning of my experiment, but in time for the grand finale, were unappreciative," he recalled.[2]

During von Braun's adolescence, general interest in rocketry in Germany developed into a national fascination. Hermann Oberth, an ethnic German from Romania, became the focus of the rocket fad when he published *Die Rakete zu den Planetenraumen* (*The Rocket into Interplanetary Space*) in 1923. Unlike the more obscure works of the Russian

1 Ernst Stuhlinger and Frederick I. Ordway III, *Wernher von Braun: Crusader for Space* (Malabar, FL: Krieger Publishing Company, 1994), pp. 9–12.

2 Wernher von Braun, "Recollections of Childhood/Early Experiences in Rocketry" (1963), *http://history.msfc.nasa.gov/vonbraun/recall.html.*

Konstantin Tsiolkovsky and the more cautious publications of the American Robert Goddard, Oberth wrote in accessible prose, advocating liquid-fueled rockets for human spaceflight. An Austrian publicist and rocket enthusiast, Max Valier, publicized Oberth's ideas. Valier was among the charter members of an amateur rocket society founded by Johannes Winkler in 1927—the Society for Space Travel—that became known by its German acronym, VfR.[3] Oberth served as president of the society and also cooperated with the renowned film director Fritz Lang in the production of a science-fiction film, *Frau im Mond* (*Woman on the Moon*), that gave further publicity to the embryonic rocketry boom.[4]

Oberth's ideas stimulated the young von Braun. As a high school student at Ettersburg boarding school, Wernher sent Oberth a paper he had written on rockets. By the fall of 1929, having graduated from Ettersburg, von Braun had joined the VfR, which now had grown to 870 members. The following spring, he registered as an engineering student at the Technische Hochschule (Technical University) of Berlin Charlottenburg. There he met Oberth for the first time and helped him test a combustion chamber and nozzle that used gasoline and liquid oxygen as fuel.[5]

ROCKETRY IN THE GERMAN ARMY

By the late 1920s, the German army had developed interest in rockets. Lieutenant Colonel Dr. Karl Becker, chief of ballistics and ammunition for the Army Ordnance Office, was an artilleryman by trade, but the Versailles Treaty that ended WWI forbade the German army from developing heavy artillery, and rockets provided a possible alternative. In the winter of 1931–1932, Becker and two other army officers, Captain Dr. Walter Dornberger, who was responsible for powder rockets for the army, and Major Wolfram Ritter von Horstig, an ammunition expert, visited the amateur rocket experts of the VfR, including von Braun, who were experimenting at the Raketenflugplatz Reinickendorf, and invited them to Kummersdorf, where the army had begun experimenting with rockets. At Kummersdorf, the VfR amateurs set off a small rocket that flew 1,300 meters before crashing. Becker criticized the amateurish approach, particularly the lack of hard data, but offered von Braun a chance to work for the army.

Von Braun accepted, and, by early December 1932, he signed a contract to work on liquid-fueled rockets for the army at Kummersdorf. As von Braun's defenders point out,

3 Hermann Oberth, *Die Rakete zu den Planetenräumen* (Nuremburg: Uni-Verlog, 1960 (reprint)); and Michael J. Neufeld, *The Rocket and the Reich: Peenemünde and the Coming of the Ballistic Missile Era* (New York: Free Press, 1995), pp. 6–7.

4 Frederick I. Ordway III and Mitchell R. Sharpe, *The Rocket Team* (New York: Crowell, 1979), pp. 12–13.

5 Stuhlinger and Ordway, *Wernher von Braun*, pp. 15–17.

Photo occasioned by the certification of Hermann Oberth's liquid-fueled rocket engine in the 1930s. Left to right: Rudolf Nebel, Dr. Karl Ritter, Mr. Baermueller, Kurt Heinish, Klaus Riedel, Wernher von Braun, and unidentified person. (NASA Marshall Space Flight Center, negative number 6517791)

he went to work for the army two months before Hitler came to power. Von Braun was no admirer of the Nazis and indeed was frank (if perhaps unintentionally revealing) in explaining his decision to accept the army's offer: "Our feelings toward the army resembled those of the early aviation pioneers, who, in most countries, tried to milk the military purse for their own ends and who felt little moral scruples as to the possible future use of their brainchild." At the same time, the Technical University of Berlin accepted von Braun as a doctoral candidate. In a secret agreement with the army, he used the development of liquid-fueled rockets as the topic for his dissertation.[6]

Von Braun went to work under the military supervision of Dornberger, who assessed the young man as an energetic, shrewd, and temperamental student with an "astonishing" theoretical knowledge, whose ideas gushed forth in a "bubbling stream."[7] By 1934, von Braun and his team had designed their first rocket, the A-1 (Assembly-1 or Aggre-

6 Neufeld, *The Rocket and the Reich*, pp. 20–23.
7 Walter Dornberger, *V-2: The Nazi Rocket Weapon* (New York: Ballantine Books, 1954), pp. 33–34.

gate-1), a 4.6-foot-long tube 1 foot in diameter that developed 650 pounds of thrust. Its center of gravity was too far forward, however, and it blew up on the test stand. Von Braun modified its design, producing the A-2, and in December the group successfully fired its first liquid-propelled rocket.

While in Kummersdorf, Dornberger and von Braun never had unlimited funds and had to work with a small staff and be creative in getting needed materials. They began with only von Braun and a single mechanic. Dornberger and von Braun recruited others, and by the end of 1934 they had added Walter Riedel, a steady test engineer and designer, and Arthur Rudolph, who had already designed a liquid-propellant motor and who had become a member of the Nazi Party in 1931.

The solution to the financial problem came from an unexpected source. Herman Goering's recently established Luftwaffe threatened to disrupt the military missile program, but an alliance between the two services, fashioned by the Luftwaffe's Major Wolfram Freiher von Richthofen, brought them together. The Luftwaffe—brash, ambitious, and financially well heeled—challenged the hide-bound, bureaucratic, tightly budgeted army to act decisively. Indeed, when the Luftwaffe offered 5 million marks to initiate the alliance, General Becker more than matched the figure, offering 6 million—a 75-fold increase over the usual annual budget of only 80,000 marks![8]

PEENEMÜNDE

The operation began to outgrow its limited facilities, and Dornberger and von Braun began looking for a new location. Von Braun suggested Peenemünde on the wooded north end of the Baltic island Usedom, near where his grandfather had hunted ducks. Dornberger visited the site and agreed. The shift to Peenemünde took time, however, and during the transition operations were split between the new site and Kummersdorf. Thus when Hitler inspected the operation in 1939 for the first and only time, he came to Kummersdorf. Dornberger led the tour, but von Braun assisted and helped to present the technical progress to the Führer. Unlike most other visiting dignitaries, Hitler seemed strangely passive, although, he did ask about the range of the A-4, about how long it would take to make it operational, and whether steel could be used for the tank instead of aluminum. Without further ado, he said, "Well, that's grand," and departed.[9]

Peenemünde had several advantages, not the least of which was that test missiles could be fired over water. The remote location also offered secrecy, although nearby

8 Ordway and Sharpe, *The Rocket Team*, pp. 24–26.
9 Dornberger, *V-2*, p. 66.

summer beach resorts meant that isolation was limited. Furthermore, the site had room for expansion and allowed for the concentration of research, development, and production at one location, a concept that Dornberger called "everything under one roof." Von Braun at first resisted this approach, arguing that he lacked experience in production. Eventually he embraced it, and this "arsenal system" became one of the hallmarks of von Braun's approach to rocket development and a key to its success.[10]

At first, the imperative of secrecy allowed little cooperation with industry. Nor was there much incentive for industry to desire a role, since few had expertise in rocketry, and conventional arms contracts were more lucrative. Thus Dornberger and von Braun adopted the arsenal concept both by design and by default. As von Braun's colleague Ernst Stuhlinger recalled, Peenemünde used the arsenal system because nobody else could build rockets. "We had to develop it," he explained. "We did it in our Peenemünde laboratories and became the experts before anybody else was an expert."[11]

As operations at Peenemünde matured, von Braun sought closer relationships with industry and universities, but the in-house system was already in place. He contracted work to the universities and also recruited professors to Peenemünde, where many became lab directors. Although few of these recruits had worked in rocketry, they had expertise in disciplines such as physics, chemistry, and mechanical and electrical engineering, all of which had applications at Peenemünde. Many had advanced degrees, and many had worked in industry.[12] It was a substantial operation that, at its peak in 1943, employed 1,950 scientists, engineers, and technicians. At that time, Peenemünde had a budget of 112 million reichsmarks, or approximately $27 million.[13]

From 1938 to 1942, von Braun's research team conducted hundreds of test firings. They learned to profit from failure. Von Braun remembered that for a long time "our main objective was to make it more dangerous to be in the target area than to be with the launch crew."[14] They made progress in stability, propulsion, gas stream rudders used for steering, the wireless guidance communication system, and instruments to plot flight paths. Privately, among themselves, they discussed spaceflight.

The most memorable moment for veterans of rocket development at Peenemünde, and the pinnacle of von Braun's achievements in Germany, occurred on 3 October 1942. On

10 Michael J. Neufeld, "Peenemünde-Ost: The State, the Military, and Technological Change in the Third Reich," paper presented at the International Congress of the History of Science, Hamburg, West Germany (2 August 1989), p. 3. Neufeld cites von Braun team member Gerhard Reisig in attributing the phrase to Dornberger.

11 Ibid., pp. 4–6, 10–11; and Ernst Stuhlinger, interview by Andrew J. Dunar and Stephen P. Waring (24 April 1989), Huntsville, AL.

12 Ernst Stuhlinger, interview by Dunar and Waring (24 April 1989), Huntsville, AL.

13 Tom D. Crouch, *Aiming for the Stars: The Dreamers and Doers of the Space Age* (Washington: Smithsonian Institution Press, 1999), p. 79.

14 Major General John B. Medaris with Arthur Gordon, *Countdown for Decision* (New York: G. P. Putnam's Sons, 1960), pp. 37–38.

that date, an A-4 rocket became the first human-engineered creation to penetrate space. The A-4 achieved an altitude of 53 miles (85 kilometers) during a 5-minute flight, traveling 118 miles (190 kilometers) downrange. Von Braun remembered Dornberger's joy and his comment, "Do you realize what we accomplished today? Today the spaceship has been born! But I warn you: our headaches are by no means over—they are just beginning!"[15]

By May 1943, British intelligence had determined that Peenemünde was a center for rocket development. In August, the British struck with a bomber attack that killed 732 or 735 people (according to accounts by Dornberger) and destroyed test stands and transportation facilities. V-2 production facilities suffered little damage, but the raid prompted a decision that no production would take place at Peenemünde.[16]

THE NAZI PARTY

Labor for production had become a problem in any case by 1943, and the solution has influenced interpretations of von Braun's early career. Arthur Rudolph, who was the chief engineer of the Peenemünde factory, sought concentration camp prisoners as a source of labor, helped gain approval for their transfer, and served there as a technical director. V-2 production facilities at Nordhausen and the nearby concentration camp at Dora witnessed the death of approximately 20,000 people through execution, starvation, and disease. The major production facility, Mittelwerk, was in an abandoned gypsum mine that afforded interlocking tunnels, where slave labor built a factory that extended a mile into the hillside. There is no dispute that conditions at Mittelwerk were harsh beyond belief; even the high-ranking Nazi Albert Speer described conditions as "barbarous" and "scandalous."[17] Unlike Rudolph, von Braun never had direct supervisory responsibility over Mittelwerk's slave labor. He visited on several occasions, for periods ranging from a couple of hours to two days. On occasion he observed slave labor, and colleagues recall that he reported that he had never seen a dead person there and was deeply disturbed by what he saw, but that when he suggested that conditions ought to be improved, Stuhlinger asserts that he was told to mind his own business or he would find himself wearing the striped shirt of the prisoners.[18] The historian Michael Neufeld, who has conducted the most thorough review of the Peenemünde-Mittelwerk nexus, concluded that von Braun "essentially made a pact with the devil in order to build large rockets" and that "there is no evidence that he ever stuck his neck out for the concentra-

15 Stuhlinger and Ordway, *Wernher von Braun*, p. 29.
16 Neufeld, *The Rocket and the Reich*, pp. 198–200.
17 Stuhlinger and Ordway, *Wernher von Braun*, p. 42.
18 Ibid., pp. 41–46.

tion camp prisoners before his arrest, nor did he show any obvious pangs of conscience about their fate until the 1960s and 1970s, when protests by French prisoner survivors forced him to confront the issue more directly."[19]

Von Braun's relationship with the Nazi Party likewise is laden with ambiguities that give both his defenders and his critics evidence to debate. Many German academics, scientists, and technicians joined the Nazi Party. To do so offered the prospect of grants, promotions, and other preferential treatment. To refuse to do so risked untold consequences. In May 1940, an SS colonel brought von Braun an order from Heinrich Himmler, chief of the notorious SS, urging von Braun to join and accept the rank of lieutenant in the SS. Von Braun accepted, but only after conferring with his colleagues who agreed that refusal might provoke Himmler's wrath. Von Braun's colleagues recall this and a promotion offered in 1943 as part of an attempt to lure him from the army to the SS.[20]

In March 1944, one of the strangest events in von Braun's years under the Third Reich occurred when the Gestapo arrested him. The arrest came in part because of competition within the Nazi bureaucracy for control of the Peenemünde project. Himmler made a bid to wrest control of rocket development from Armaments Minister Albert Speer. Himmler summoned von Braun in February. After suggesting that von Braun must realize that the A-4 was no longer a toy and that the German people were awaiting its deployment, Himmler dangled his bait. He sympathized with von Braun's dilemma, being enmeshed in the army bureaucracy, and suggested that von Braun ought to come over to the SS, which had direct access to Hitler and could cut through red tape. Von Braun responded that he had the best chief he could hope for in Dornberger and that it was technical difficulty rather than the bureaucracy that was slowing development. Himmler dismissed von Braun but began compiling a dossier on him and other members of his team.

Early in March, von Braun relaxed with colleagues, discussing space travel as they often did in their off-duty hours. Among them was a woman who was an agent of the SD, the security arm of the SS; she reported the conversation to her superiors, although her report only added to the charges already assembled. At 2 a.m. on a March morning—the precise date is disputed, but was most likely March 22—the Gestapo awakened von Braun, arrested him, and took him to Stettin, nominally under protective custody. Gestapo officers also arrested other members of his team, including Klaus Riedel and von Braun's brother Magnus. Dornberger soon learned of the arrests and that the charges were so serious that it might cost the prisoners their lives. They stood accused of sabo-

---

19 Neufeld, *The Rocket and the Reich*, pp. 278–279.

20 Christopher Simpson, *Blowback: America's Recruitment of Nazies and Its Effects on the Cold War* (New York: Weidenfeld & Nicholson, 1988), pp. 32–36; and Neufeld, *The Rocket and the Reich*, pp. 178–179. The comments of von Braun's colleagues come from Ernst Stuhlinger; see notes on a draft chapter of Dunar and Waring, *Power to Explore: A History of Marshall Space Flight Center, 1960–1990* (Washington, DC: NASA SP 4313, 1999).

taging the A-4 program and diverting their attention to space travel rather than devoting their energies to weapons development. The charges alleged that von Braun had a plane at his disposal that he could use to flee to Britain with the A-4 plans. Von Braun indeed had a plane that he used for business flights in Germany, and the allegation was impossible to prove or disprove. Dornberger claimed that he defended von Braun and Riedel without reservation, saying they were not working on space rockets but on war missiles, and ultimately won their release. The incident, as historian Michael Neufeld observed, proved to be "one of the most fortunate things that ever happened" to von Braun in the Third Reich, since "after the war his defenders were able to credit him with an anti-Nazi record he never had."[21]

SURRENDERING TO THE AMERICANS

By early 1945, little doubt remained that the allies would win the war. With the Russians advancing toward Peenemünde, the group began to evacuate late in January, destroying material that the Russians might seize. SS General Hans Klammer, who had taken charge of missile development even as it was collapsing, directed von Braun and his colleagues to the Harz Mountains, near the notorious Mittlewerk site. On 1 April, as the Americans neared the Harz Mountains, Kammler directed von Braun and approximately 500 key people to move to the Bavarian Alps. As allied forces advanced in April, von Braun's men moved to villages in the vicinity of Oberammergau. They moved crates of documents estimated at 14 tons to an abandoned mine and detonated dynamite at the entrance to seal the treasure. Von Braun and his colleagues later claimed that they discussed their situation and agreed that their future would be more promising if they could surrender to the Americans, who had not suffered the physical damage that other combatant nations had endured and whose economy would be most able to support rocket development. Only at the end, however, did they have much control over their destiny. Early in May, as the Americans advanced toward Oberammergau, von Braun's brother Magnus rode out on a bicycle to meet the troops and surrendered.[22]

21 Neufeld, *The Rocket and the Reich*, pp. 214–220; Ordway and Sharpe, *The Rocket Team*, pp. 46–49; and Dornberger, *V-2*, pp. 178–184. Michael Neufeld elaborated on his investigation of von Braun's Nazi connections in an article in 2002, in which he concluded that "Wernher von Braun was neither an ideologically committed National Socialist nor an enthusiastic SS officer, but . . . like a great majority of Germans, he was enthusiastic about many of the 'accomplishments' of the 'Führer' during the late 1930s and early 1940s, and was correspondingly indifferent to the persecutions of the political opponents, Jews, and citizens of occupied countries. . . . Ultimately, it is not Wernher von Braun's membership in the SS nor his involvement in slave labor that is most bothersome. . . . It is his technocratic amorality, his single-minded obsession with his technical dreams, that is so disturbing." Michael J. Neufeld, "Wernher von Braun, the SS, and Concentration Camp Labor: Questions of Moral, Political, and Criminal Responsibility," *German Studies Review* 25/1 (2002): 72–73.

22 Ordway and Sharpe, *The Rocket Team*, pp. 261–267; Neufeld, *The Rocket and the Reich*, pp. 256–260, p. 265; Dunar and Waring, *Power to Explore*, p. 8; and Stuhlinger and Ordway, *Wernher von Braun*, pp. 58–61.

Von Braun (in cast) surrenders to U.S. Army counter-intelligence personnel of the 44th Infantry Division in Reutte, Tyrol, in April 1945. (NASA Marshall Space Flight Center, negative number 6517789)

The fate of the group now shifted out of the control of the dying Third Reich and into the crosscurrents of Soviet-American rivalry, the moral condemnation of Nazism, and the technological imperatives of the American military. Months before V-E Day, the Chief of the Rocket Branch of U.S. Army Ordnance, Colonel Gervais William Trichel, began taking steps that eventually led to the transfer of von Braun and more than 100 of his associates to the United States. In preparation for the U.S. Army's own rocket program, Trichel signed a contract with General Electric for the development of long-range missiles under Project Hermes. He hoped to use V-2 rockets in this research. In consultation with British intelligence, Major Robert Staver of Trichel's staff compiled a list of German rocket experts, ranked in order of significance, and von Braun's name was first on the list. Trichel directed Colonel Holger Toftoy, chief of Ordnance Technical Intelligence, to find 100 V-2 rockets and ship them to the Army's firing range in White Sands, New Mexico. When Toftoy learned about the allied discovery of Mittelwerk, he sent Major James P. Hamill to arrange for a shipment of V-2s to the United States. Staver meanwhile convinced von Braun's team to help locate the hidden Peenemünde documents and directed the shipment of 14 tons of documents just ahead of the British authorities who were closing access to the area.[23]

---

23 Dunar and Waring, *Power to Explore*, pp. 8–9.

In July 1945, the American Joint Chiefs of Staff established Project Paperclip (originally called Project Overcast), which gave authority to transfer German specialists who had expertise that might be of value to the military. Toftoy received permission to transfer about 120 members of the von Braun team to the United States, and, in September, von Braun and Major Hamill traveled to Fort Bliss in El Paso, Texas. There and in nearby White Sands, Toftoy planned Project Hermes, an effort to conduct rocketry research using V-2 rockets. By the spring of 1946, most of the Germans had arrived. In April, Project Hermes, assisted by the von Braun group, successfully fired the first V-2 to be launched from American soil. For the next several years, the von Braun group worked as consultants to the Army, Navy, and private contractors, including General Electric. Von Braun and his colleagues worked on a project of their own, designated Hermes B; it was a ramjet-powered second stage for the V-2. Among the accomplishments under Project Hermes was the launching of a Bumper-WAC (a modified V-2 first stage with a WAC Corporal second stage) from White Sands to an altitude of 250 miles.[24]

Perhaps as important as the rocket research conducted at Fort Bliss was the molding of a team under von Braun's leadership. The circumstances at Fort Bliss promoted the sense of group identity. Transferred to an unfamiliar country, separated from their families, united by professional interests, viewed with suspicion by citizens of El Paso (who had little interaction with them in any case), they naturally drew together. They hiked in the nearby mountains, played chess and ball games, and played pranks on one another. They were an elite group, and they knew it. One American described them as "a president and 124 vice presidents." There was no doubt who was the president; von Braun's leadership was never questioned. He negotiated for them and insisted on his prerogatives, sometimes sparring with Hamill, especially when Hamill made decisions affecting the group without working through him.

In October 1949, General Toftoy won approval to move Army rocket research to Huntsville, Alabama, site of Redstone Arsenal and the old Huntsville Arsenal, which the Army Chemical Corps wanted to sell. Toftoy moved the German team from Fort Bliss to Huntsville the following year, where they moved into leadership of the Ordnance Guided Missile Center. In 1952, the Army established the Ordnance Missile Laboratories at Redstone Arsenal, with von Braun as the Chief of the Guided Missile Development Division. Unlike their work at Fort Bliss, where they mainly worked as contractors for other groups' projects, in Huntsville von Braun's group had a project of their own: the development of the Redstone, a new surface-to-surface missile meant to augment the Army's Corporal and Hermes missiles.[25]

---

24 Ordway and Sharpe, *The Rocket Team*, pp. 310–314, 346–349; Major General H. N. Toftoy and Colonel J. P. Hamill, "Historical Summary on the Von Braun Missile Team" (29 September 1959), University of Alabama, Huntsville, Library Saturn Collection; "What We Have Learned from V-2 Firings," *Aviation Week* (26 November 1951): 23ff; Crouch, *Aiming for the Stars*, pp. 110–111; and Neufeld, telephone conversation with Dunar (11 December 2003).

25 Dunar and Waring, *Power to Explore*, pp. 12–14.

Von Braun employed the arsenal system in development of the Redstone, and it became the hallmark of his approach to rocket development from Redstone through the Saturn rockets that powered the American lunar program in the 1960s and 1970s. The approach was not uniquely German. In fact, the U.S. Army had used the arsenal system as early as the mid-19th century at its arsenal and armory at Harper's Ferry, West Virginia. By the late 1950s, however, when an interservice debate over in-house vs. contractor development took shape between the U.S. Army and Air Force, the von Braun group had come to epitomize the arsenal in-house approach. Indeed, their German training complemented the Army's approach because, as one of von Braun's lab chiefs explained, in Germany "you are not admitted to any technical college or university if you do not have some practical time."[26] Furthermore, during WWII Germany had followed the older statist tradition in which the building of arms and munitions was controlled by the state.[27] Von Braun's commitment to in-house development was also a response to funding constraints on the Army's missile program, since things could often be done more cheaply in house than by contracting the work to outside firms. After receiving a contractor's bid of $75 thousand for the construction of a static test stand, von Braun's team built their own for $1 thousand worth of materials and fondly named it the "Poor Man's Test Stand."[28]

PROPHET OF HUMAN SPACEFLIGHT

The Army's missile development program received little attention in the early 1950s, but von Braun nonetheless began to acquire a national reputation as a visionary prophet of human spaceflight. In a series of richly illustrated articles in 1952 in the popular magazine *Collier's*, von Braun discussed the prospects for space travel, advocating the development of a space station and even suggested that a Moon landing could occur within the next quarter-century. In another *Collier's* article the following year, von Braun advocated the development of an unmanned satellite, and, in 1954, he made a bold proposal for the exploration of Mars.[29] In 1955, he appeared on the enormously popu-

---

26 Karl Heimburg, oral history interview (OHI) by Dunar and Waring (2 April 1989), Huntsville, AL.

27 Neufeld, telephone conversation with Dunar (11 December 2003).

28 Heimburg OHI; David S. Akens, "Historical Origins of the George C. Marshall Space Flight Center," in *Marshall Space Flight Center Historical Monograph No. 1* (Huntsville: Marshall Space Flight Center (MSFC), 1960), p. 37; and Ordway and Sharpe, *The Rocket Team*, p. 372. Of course, as James Kingsbury, the former Director of Science and Engineering at MSFC, pointed out, the comparison is not entirely fair, since the $75,000 figure included labor costs, and the $1,000 figure did not. Kingsbury, undated note to Dunar and Waring on chapter draft of *Power to Explore.*

29 Wernher von Braun, "Man on the Moon: The Journey," *Collier's* (18 October 1952): 52–59. The space station article appeared on 27 June 1953. Other occasional articles by von Braun and other authorities on space travel appeared in 1953 and 1954.

lar television program *Disneyland*, where his authoritative German accent, charismatic enthusiasm, and ability to express complex concepts in understandable terms made him perhaps the most recognizable advocate for space exploration. A decade later, still one of the most effective spokespersons for the American space program, he began writing a monthly column in the magazine *Popular Science*, answering readers' questions about space.[30] He spoke to a wide range of audiences, from service clubs in Huntsville to industrial leaders around the nation.

Von Braun and his fellow Germans feared that they might encounter hostility in their new location, since, unlike their situation at Fort Bliss, they were living in the midst of a civilian population. The hostility never developed, and members of the team remarked that the green hills of Huntsville reminded them of Germany. The immigrants and their families had a dramatic impact on the city. The immigrants helped to start a symphony orchestra and contributed to the development of the city's public library. On 15 April 1955, von Braun, his wife Maria, and 40 members of his team became naturalized American citizens in a ceremony attended by 1,200 of their Huntsville neighbors and friends.[31]

Von Braun backed his public appeals with concrete proposals that might lead to the first steps toward spaceflight. In 1953, he argued that existing hardware could be used to launch a satellite into Earth orbit. The next year, the Army proposed an interservice satellite project, later the basis for the Army-Navy proposal Project Orbiter. The Air Force and the Naval Research Laboratories submitted similar proposals, and the Defense Department chose to support the Navy's Project Vanguard—in part, some suggested, because the department did not want to see the first American satellite launched by German rocket experts. Von Braun's dream, it appeared, would be executed by others.[32]

In 1956, the Army reorganized its missile development program, incorporating its Guided Missile Development Center and the Redstone project into the Army Ballistic Missile Agency (ABMA) at Redstone Arsenal under the command of General John B. Medaris. Von Braun's team received authorization to develop an Intermediate Range Ballistic Missile (IRBM) to be known as the Jupiter, a single-stage liquid-fuel rocket with a maximum range of 1,500 miles, a limitation designed to prevent competition with the Navy's Vanguard for the honor of launching the first artificial satellite. Von Braun chafed under the restriction, saying, "We at Huntsville knew that our rocket technology was fully capable of satellite applications and could quickly be implemented." The Defense

30 NASA MSFC Retiree Association, *50 Years of Rockets and Spacecraft in the Rocket City* (Paducah, KY: Turner Publishing Company, 2002), p. 69.

31 Stuhlinger and Ordway, *Wernher von Braun*, pp. 98–99; and Dunar and Waring, *Power to Explore*, pp. 14–16.

32 Wernher von Braun, "A Minimum Satellite Vehicle Based Upon Components Available from Missile Development of the Army Ordnance Corps"; Akens, "Historical Origins," pp. 38–39; and Ordway and Sharpe, *The Rocket Team*, p. 376.

Department even sent an observer to ensure that ABMA would not exceed its limits and orbit a booster by activating a dummy fourth stage.[33]

Despite the restrictions imposed on ABMA by the Defense Department, the greater altitude achieved by the new generation of missiles provided von Braun's team the opportunity to work on developments that had a bearing on spaceflight. One such challenge was the matter of how to deal with the tremendous heat developed during missile reentry into Earth's atmosphere. While the Air Force worked on a heat sink solution in which nosecone materials would absorb heat, the Huntsville team worked on an ablation system in which materials shielding the nosecone would melt and evaporate during reentry. Jupiter-C launches in 1956 and 1957 proved the viability of this system, and reentry studies gave ABMA's engineers experience in the technology of spaceflight.[34]

When Americans learned of the launch of the Russian satellite Sputnik on 4 October 1957, incoming Secretary of Defense Neil McElroy was visiting Redstone Arsenal. At dinner that evening, von Braun and Medaris sat on either side of McElroy and lobbied. Von Braun insisted that ABMA could launch a satellite into orbit in 60 days. Medaris, more cautious, said 90 days might be necessary. McElroy hesitated, but after the Soviet Union launched the 1,200-pound Sputnik II with the dog Laika aboard on 3 November, ABMA received approval, with 29 January designated as the launch date. ABMA worked in cooperation with the Jet Propulsion Laboratory (JPL) at the California Institute of Technology to develop the launch vehicle and its satellite. The launch vehicle combined a cluster of solid-propellant rockets designed and built by JPL with a Redstone rocket, which the von Braun team integrated into a new vehicle designated the Jupiter-C (sometimes called Juno I). Dr. William H. Pickering of JPL developed Explorer I, a 34-inch-long, 6-inch-diameter tube for that purpose. Weather delays postponed the launch until 31 January, but on that date Explorer I successfully achieved an orbit with an apogee of 1,594 miles.[35]

In the aftermath of the Sputnik launches, the Eisenhower administration conducted the first comprehensive review of American space policy. President Eisenhower, who wanted to avoid a space race with the Soviet Union, preferred a civilian space agency, and, on 29 July 1958, President Eisenhower signed the National Aeronautics and Space Act, which established the National Aeronautics and Space Administration (NASA).

33 Kurt H. Debus, "From A4 to Explorer I," paper presented at 24th International Astronautical Congress, Baku, USSR (8 October 1973), Debus/1973/Redstone/Pershing/Jupiter Folder, NASA Headquarters Historical Reference Collection, Washington, DC, p. 33; and Medaris and Gordon, *Countdown for Decision*, p. 72.

34 Debus, "From A4," p. 36; Medaris and Gordon, *Countdown for Decision*, pp. 142–144; William Lucas, OHI by Dunar and Waring (19 June 1989), Huntsville, AL.

35 Patricia Yingling White, "The United States Enters Space, 1945–1958: A Study of National Priorities and the Decision-Making Process in the Artificial Satellite Program" (master's thesis, Ohio State University, 1969), p. 95; Medaris and Gordon, *Countdown for Decision*, pp. 151–190, 207–226; Ordway and Sharpe, *The Rocket Team*, pp. 382–386; Akens, "Historical Origins," pp. 44–47; Debus, "From A4," pp. 52–54; and Stuhlinger and Ordway, *Wernher von Braun*, pp. 134–140.

During the months between the launch of Explorer I and the establishment of NASA, von Braun continued to pursue his dream of spaceflight. The Defense Department had responded to Sputnik by establishing the Advanced Research Projects Agency (ARPA), which had authority to sanction space projects for a one-year period, subject to presidential approval. Both ABMA and the Air Force submitted plans to put a man in space (Project Adam and Man-in-Space-Soonest). Von Braun also continued work on the remaining launches in the Juno series of missiles. Explorer II failed when the fourth stage did not ignite, but Explorer III went into orbit in March 1958. By October, when the Juno series came to an end, ABMA had recorded three successful launches and three failures.[36]

The Space Act assigned to NASA the 8,000 personnel and three laboratories of the National Advisory Committee for Aeronautics, the Navy's Vanguard project, and several Air Force projects. The status of the von Braun team was uncertain; three of ABMA's satellite projects and two of its lunar probes went to NASA, and the new NASA Administrator T. Keith Glennan requested transfer of more than half of von Braun's group. But General Medaris fought to retain von Braun and his German colleagues in ABMA, and for the time being they remained with the Army. Von Braun worried about possible dispersal of his team and NASA's opposition to in-house development; he wondered whether NASA would be able to support Saturn; and, in any case, he had little choice but to insist on his loyalty to Medaris.[37]

Despite von Braun's reservations, he could not ignore NASA. In December 1958, while still firmly attached to ABMA, von Braun and two of his lieutenants made a pitch to Glennan that looked beyond NASA's early plans to put a man in space. Von Braun had his eyes set on the Moon, and he told Glennan that he knew how to get there. He explained his concept of rocket clusters that could provide such power and suggested that the Saturn, already on the drawing boards in Huntsville, could reach the Moon, perhaps as early as the spring of 1967. Furthermore, the Saturn fit well into NASA's plans, since it could be developed even as NASA took its first steps into human spaceflight. As aerospace historian William Burrows observed, "It was right off the pages of the *Collier's* series, with one step locked into the next."[38] Von Braun the visionary had inspired von Braun the engineer, and the hardware to achieve the lunar dream was already in development.

---

36 Loyd S. Swenson, Jr., James M. Grimwood, and Charles C. Alexander, *This New Ocean: A History of Project Mercury* (Washington: NASA SP 4201, 1966), pp. 25–28; John M. Logsdon, *The Decision to Go to the Moon: Project Apollo and the National Interest* (Cambridge, MA: MIT Press, 1970), pp. 28–29; J. Boehm, H. J. Fichtner, and Otto A. Hoberg, "Explorer Satellites Launched by Juno and Juno 2 Vehicles," in *Peenemünde to Outer Space*, ed. Ernst Stuhlinger, et al. (Huntsville: Marshall Space Flight Center, 1962), pp. 163–165.

37 Richard L. Smoke, "Civil-Military Relations in the American Space Program, 1957–60" (bachelor's honors thesis, Harvard University, 1965), pp. 70–75; Medaris and Gordon, *Countdown for Decision*, pp. 242–245; Dunar and Waring, *Power to Explore*, p. 25; Jim G. Lucas, "Army Expects to Lose Von Braun," *New York World-Telegram & Sun* (31 October 1958); and "The Periscope," *Newsweek* (10 November 1958): 19.

38 William E. Burrows, *The New Ocean: The Story of the First Space Age* (New York: The Modern Library, 1999), pp. 286–287.

MANAGING AT MARSHALL

NASA's leaders wanted von Braun's expertise, but they had reservations about his way of doing business. From the start, suspicions about the German background were not far below the surface. Glennan's staff suggested that he should make it clear that he wanted "ABMA personnel and facilities, not the ABMA way of doing business."[39] NASA Deputy Administrator Hugh L. Dryden commented after reading an article by Walter Dornberger on the lessons of Peenemünde that "the general principles of the required management are well known; it seems difficult to get them adopted in a democracy."[40]

For a time, the relationship between ABMA and NASA was ambiguous. NASA contracted with ABMA to provide eight Redstone rockets for the early suborbital flights of Project Mercury, and ABMA continued the development work it had begun on the clustered Saturn booster, a powerful liquid-fueled vehicle that figured prominently in NASA's plans, but which promised to provide much more thrust than required for anything on the Army's drawing board. Indeed, the Saturn became the catalyst that enabled von Braun's contingent to transfer to NASA. To keep the Saturn in ABMA made little sense, despite Medaris's complaints about "project snatchers," and by October 1959—two years after Sputnik—the Army agreed to transfer von Braun's Development Operations Division of ABMA intact to NASA. The transfer required no physical move; instead, on 1 July 1960, a portion of Redstone Arsenal became the new George C. Marshall Space Flight Center with von Braun as Center Director.[41]

The year's delay in joining NASA had ramifications for von Braun's role in the new Agency. By the time the transfer took effect, NASA's culture had begun to form, shaped largely by a group of engineers from Langley Research Center that later transferred to Houston, where the group became the nucleus of the Manned Spacecraft Center (later renamed the Johnson Space Center). This group and the NASA Administrators in Washington would be at the center of planning for American human spaceflight and would harbor some suspicion of von Braun's approach at ABMA with its commitment to the arsenal system, engineering conservatism, and reliability testing, and its aversion to contracting out. Charles Murray and Catherine Bly Cox, in their account of the Apollo

---

39 Memorandum from Walter T. Bonney to T. Keith Glennan (30 September 1958), NASA-Army (ABMA) Folder, NASA Headquarters Historical Reference Collection, Washington, DC.

40 Letter from Dryden to Lieutenant General Arthur G. Trudeau (25 February 1959), Krafft Ehricke—The Peenemünde Rocket Center Folder (vertical file), NASA Johnson Space Center Historical Reference Collection, Houston, TX.

41 Wernher von Braun, "The Redstone, Jupiter, and Juno," *Technology and Culture* 4 (Fall 1963); Roger E. Bilstein, *Stages to Saturn: A Technological History of Apollo/Saturn Launch Vehicles* (Washington, DC: NASA SP 4206, 1980), pp. 35–37; Dunar and Waring, *Power to Explore*, pp. 26–30; and Medaris and Gordon, *Countdown for Decision*, p. 266.

program, argued that von Braun's group "had missed their chance to run the whole mission when they had stayed with the Army for the first year after NASA was founded."[42]

Von Braun became the first Director of Marshall Space Flight Center, and under his leadership the Huntsville Center became the heart of NASA's propulsion expertise. The early NASA test flights and manned suborbital and orbital flights of Project Mercury relied on four rockets, including the Little Joe, Redstone, and the Atlas. The von Braun team also brought from ABMA the Juno II, which was used for unmanned space science launches. Thus two of NASA's early launch vehicles, the Redstone and Juno II, were products of the von Braun team at ABMA. Whatever reservations NASA Administrators may have had, von Braun's success gave him enough immunity from criticism that he was able to carry out his program with little modification.

The apogee of von Braun's accomplishments at Marshall was the development of the Saturn V, the propulsion system of the Apollo program, and it is on this monumental achievement that his reputation rightly rests. Von Braun had the vision to conceive the development of a propulsion system of unprecedented complexity that was powerful enough to propel to the Moon the fuel, equipment, and life-support systems necessary to sustain a crew and return it safely to Earth. But he also had the hard-headed pragmatism of an engineer that leavened his visionary conceptual approach to scientific inquiry. He had acquired the engineering experience, assembled the personnel, and developed the managerial skills that enabled him to undertake such a daunting project. That he was able to do so in a political environment that dictated a demanding schedule and required working with managers, peers, and politicians who scrutinized his motives, resented his popularity, and questioned his loyalty makes his record all the more remarkable.

The success of the Saturn rested on the concept of clustering engines in order to achieve the thrust required for the lunar program. Concepts for the Saturn dated back to 1957. Von Braun recalled the concept of clustered engines developing out of ABMA's work with the Defense Department's ARPA in the late 1950s. "I don't know whether we came forth with drawings of clustering rockets, or whether ARPA came to us," von Braun reflected.[43] Saturn relied on the clustering of powerful rocket engines that used liquid fuel, and thus the project demanded the development and testing of complex engines and cryogenic tanks that could carry the enormous quantities of fuel consumed by these engines.

The statistics of these rockets are staggering, even decades after the last Saturn flight. The Saturn V was a three-stage rocket. Its first stage, the S-1C stage, had five clustered

42 Charles Murray and Catherine Bly Cox, *Apollo: The Race to the Moon* (New York: Simon and Schuster, 1989), p. 136.

43 Glen E. Swanson, ed., "Before This Decade Is Out . . ." in *Personal Reflections on the Apollo Program* (Washington, DC: NASA SP-4223, 1999), p. 49.

F-1 engines, each standing 18.5 feet in height. The F-1 used liquid oxygen and RP-1 (kerosene) for fuel, and each provided 1.5 million pounds of thrust, for a total of 7.5 million pounds of thrust during the first 2½ minutes of launch. In its original design, the first stage had four clustered rockets; von Braun had said that the "great big hole in the center is crying for a fifth engine," and the weight requirements added over the months of development made the fifth engine a fortunate decision. The second stage, the S-2 stage, clustered five J-2 engines, each powered by liquid oxygen and liquid hydrogen, and each providing 200,000 pounds of thrust for 500 seconds. Finally, the third stage employed one J-2 engine. A fully assembled Saturn V stood 364 feet in height and weighed 5.8 million pounds.[44] It provided power equivalent to 85 Hoover Dams.[45]

Development of the huge F-1 engine required dealing with issues of size rather than new technology, since the F-1 mainly used technology that was already understood. The J-2, however, was another matter, since the technology of dealing with liquid hydrogen was less well developed. At -423°F, liquid hydrogen is 130° colder than liquid oxygen, making more complex the technology of dealing with cryogenic propellants. Development of the Saturn's engines required coordination with NASA's Lewis Research Center (now the Glenn Research Center) in Cleveland, which had expertise in the use of liquid hydrogen, and with the contractors Rocketdyne and Pratt and Whitney.

Despite official concerns about how von Braun's team operated, Marshall's organization bore the mark of Peenemünde and ABMA. The Center had the capacity to design, test, and manufacture rockets from concept to completion. Marshall's matrix organization rested on the strength of its eight laboratories, each with a technical specialization, and each with its own facilities.[46] The laboratories gave Marshall expertise that exceeded its reputation as a propulsion center, and, after the Apollo program, this strength would allow the Center to diversify into other areas. Project offices—such as the Saturn I and Saturn V offices—would draw on the labs and form interdisciplinary teams to accomplish specific tasks.

Von Braun and his Center also remained deeply committed to the arsenal system. Von Braun believed that the system improved quality, contained costs, and allowed direct contact between engineers and technicians. When Marshall did contract work to industry, von Braun's engineers prided themselves on understanding the technology better than

---

44 Swanson, "Before This Decade," pp. 52–53; Roger E. Bilstein, *Stages to Saturn: A Technological History of the Apollo/Saturn Launch Vehicles* (Washington, DC: NASA SP-4206, 1980), pp. 110, 151; and William E. Burrows, *This New Ocean: The Story of the First Space Age* (New York: The Modern Library, 1998), p. 373.

45 *http://history.msfc.nasa.gov/rocketry/tl7.html.*

46 Marshall underwent periodic reorganizations, and the laboratories occasionally changed names, but the fundamental laboratory organization remained intact. The laboratory structure in 1963 included the following laboratories: Aeroastro-Dynamics, Astrionics, Computation, Manufacturing Engineering, Research Projects, Propulsion and Vehicle Engineering, Quality Assurance and Reliability, and Test.

the contractor; they believed they could "penetrate" the contractor, because the hands-on experience of Center engineers and technicians ensured that they could better monitor contractors and assist them in overcoming technical problems. They employed conservative engineering practices and tested beyond the usual requirements of industrial production. They remained committed to liquid fuel for propulsion, even as advocates of solid fuel stressed the cost savings of less-complex solid-rocket motors. Liquids suited the German research methodology; it allowed for component testing and provided a larger margin of safety since liquids could be shut down, whereas solids, once lit, could not.

While von Braun remained committed to conservative engineering practices, he came to appreciate the interaction with others—contractors, subcontractors, other NASA Centers, and NASA Administrators—inherent in the American system, which he once referred to as "a stock exchange of good ideas where we felt we picked the best things out."[47] Indeed the major contractors for the Saturn V (Boeing, McDonnell Douglas, IBM, North American Aviation Space Division, and North American Aviation Rocketdyne) used scores of subcontractors, spreading the space business around the nation, developing political support for the space program in the process.[48]

Von Braun insisted on open communication within the organization and devised managerial tools to ensure its practice. Among these practices were Marshall's "board meetings" and "weekly notes." The board meetings drew Center administrators, lab directors, project managers, and guests who provided outside expertise. Meetings included formal presentations, but participants remembered the freewheeling discussions and arguments over technical issues, policies, and problems. Two of von Braun's talents made these meetings particularly valuable. Participants often explained their point of view in complex scientific terminology. Von Braun would push experts to restate their argument in terms that everyone at the meeting could understand, and, if they were unable to do so, he would intervene and restate the issue in comprehensible terms himself. One participant recalled that specialists "would be talking almost like in unknown tongues," and that "finally von Braun would take over and explain what was being said in terms that everybody could understand." Von Braun also had the ability to summarize and to distill a consensus out of a contentious meeting. One of his engineers remembered the discussion of a technical point when von Braun interrupted, "Am I the only person at this meeting who doesn't understand this?" He looked around the room with a "quizzed look," stepped to the chalkboard, and made a diagram pertinent to the discussion.[49]

---

47 Swanson, "Before This Decade," p. 59.

48 A full listing of all subcontractors runs for nine pages in "Appendix A—Saturn V Subcontractors," in *Saturn V News Reference* (1968), *http://history.msfc.nasa.gov/saturnV/Subcontractors.pdf.*

49 Ralph A. Burns, "An Engineer Remembers," in *50 Years of Rockets and Spacecraft in the Rocket City*, comp. NASA MSFC Retiree Association (Paducah, KY: Turner Publishing Company, 2002), p. 108.

The weekly notes were a von Braun innovation that allowed open airing of difficult issues across the Center. Lab directors and project managers submitted weekly a one-page summary of their progress and problems of the previous week. Von Braun wrote marginal comments and circulated the notes among lead personnel. The resulting cross-fertilization of ideas kept key personnel aware of the overall status of Marshall's projects, prompted cross-disciplinary discussion of the Center's engineering challenges, and led to solutions that might otherwise have been overlooked. Managers began to require that their subordinates submit "Friday notes" in preparation for their own "Monday notes," bringing the benefits of the system to lower levels of the organization.[50]

The weekly notes demonstrated another aspect of Marshall's organization under von Braun. The Center's structure was hierarchical, disciplined, and conservative. One Marshall engineer described it as "a very conservative overview in management technique which went through the whole organization."[51] Those who worked closely with von Braun over the years considered it a creative system and believed that it fostered a team spirit that permeated the Center.[52] Those who were lower in the organization's hierarchy had a different perception. One assessment suggested that the weekly notes aired "problems and bad things—very few good things got surfaced. . . . Nobody at the bottom really felt free to do anything unless he got it approved from the next level up, the next level up, the next level up."[53] Another lower-ranking subordinate concluded that the weekly notes created "an almost iron-like discipline of organizational communication."[54]

The formality that characterized relations at Marshall was unusual in the freewheeling world of NASA, and it stemmed from von Braun and his German colleagues. Once when NASA Deputy Director Robert Seamans visited Marshall, he questioned von Braun's Deputy Director Eberhard Rees about this formality. Seamans asked Rees if he had always addressed von Braun as "Dr. von Braun." Rees turned to Seamans and replied that he used to call him "Herr Dr. von Braun."[55]

But Marshall's formality was only one side of von Braun. Stories are legion among Marshall veterans about memorable personal interactions with the Center Director, and many of the recollections refer to his charisma. There was a disarming boyishness and spontaneity to von Braun that captivated Marshall's workers and won their unstinting loyalty. One remembered an incident in 1962:

---

50 Dunar and Waring, *Power to Explore*, pp. 50–51.

51 Bob Marshall, OHI by Waring (29 August 1990), NASA Marshall Space Flight Center Historical Reference Collection, Huntsville, AL, p. 1.

52 Georg von Tiesenhausen, OHI by Dunar and Waring (29 November 1988), NASA Marshall Space Flight Center Historical Reference Collection, Huntsville, AL, p. 9.

53 Bob Marshall OHI, p. 1.

54 Phillip K. Tompkins, "Organization Metamorphosis in Space Research and Development," *Communication Monographs* 45 (June 1978): 116.

55 J. N. Foster, "Formality of the Von Braun Team," in *50 Years of Rockets*, pp. 129–130.

> I was sitting on the floor of the Recorder Room with my feet in a cable trench. Members of my crew were in the basement feeding long, black electrical cables up from below into the trench. The cables were stiff and heavy. I really needed someone to help with this end of the task. As I started to pull up another cable, I was aware of someone entering the room and taking hold of the end of the cable and pulling it across the room. The last cable seemed to be stuck. Without looking, I assumed that my helper was one of my crew, so I told him to get down in the trench and help. He dutifully complied. We broke the cable loose, and he pulled it across the floor. Still sitting in the trench trying to catch my breath, I heard my helper ask with a German accent, "Well, Mr. Weaver, what is the purpose of these cables?" I jumped out of the trench and faced the Center Director. As I searched for words, Dr. von Braun extended a hand, now very soiled by our task.[56]

TO THE MOON

Von Braun's central role in the American space program was well established when he became the Director of Marshall Space Flight Center, but the path that would lead to his primary achievement in aviation history began when President John F. Kennedy set the course for the American space program. Kennedy, in a speech in May 1961, announced a national goal of landing a man on the Moon and returning him safely to Earth by the end of the decade. Kennedy visited Marshall Space Flight Center twice. On his second visit, in May 1963, Kennedy asked von Braun about the prospects for achieving that goal. "Yes, Mr. President," von Braun replied, "we are going to meet your commitment of landing a man on the Moon, and we're going to do it within the time you set."[57]

Von Braun had a hand in many of the key decisions in the American lunar program in the 1960s. The decision to go to the Moon not only gave impetus and ample budget to Marshall's Saturn V program, it also touched off a debate among NASA leaders about the best way to reach the Moon, which in NASA parlance soon became known as the "mode decision." Each Center studied alternatives; Marshall examined "direct ascent," which would have required a rocket even more powerful than Saturn (called NOVA in the planning stages), and Earth Orbital Rendezvous (EOR), which stipulated launch to the Moon from a Saturn-launched vehicle in Earth orbit. EOR, which bore similarities to a von Braun concept in his *Collier's* articles, would have required two Saturn V vehicles; it became the Marshall favorite. Houston's Manned Spacecraft Center studied and preferred Lunar Orbital Rendezvous (LOR), which would have required one Saturn launch

56 Willie E. Weaver, "From Co-Op to Rocketeer," in *50 Years of Rockets*, p. 106.
57 Ed Buckbee, "JFK's Visit to Marshall," in *50 Years of Rockets*, p. 137.

Wernher von Braun, Marshall Space Flight Center Director, greets President John F. Kennedy during a visit to the Center on 12 October 1962. (NASA Marshall Space Flight Center, negative number 9806978)

of two spacecraft—one in lunar orbit, and a lunar lander that could take off and return to the lunar orbiter. Beyond practical applications, each Center had reason to back its own proposal; in Marshall's case, EOR would have taken the Center into new areas of future work in engineering and likely would have given Marshall new work at the end of the lunar program. The decision seemed to be one of numerous instances of Center rivalry that surfaced often between NASA's two principal Centers devoted to human spaceflight. The key meeting took place on 7 June 1962, with personnel from both Centers and Headquarters in attendance. At the end of the presentations, von Braun took the floor and announced, to the astonishment of all, that Marshall's position was to support LOR. Von Braun later explained that EOR was simply Marshall's study task, not its preference, and that LOR, with its single Saturn launch, offered the greatest chance for success within the decade. Von Braun's acceptance of the logic of LOR fostered the necessary cooperation for the success of the lunar program. Marshall, in what appears to have been something of a consolation prize, received designation as the Lead Center for a lunar rover.[58]

But there may have been more to it. In a curious way, the decision symbolized Marshall's culture. The Center under von Braun prided itself on teamwork, and, in confrontations with Houston, Marshall was more likely to be the team player, Houston the tenacious infighter. Von Braun was never reticent, always an eloquent spokesman for his Center and defender of its positions—but only to a point. He would back down rather than risk division, and he did so not only on EOR-LOR, but on other important decisions

58 Dunar and Waring, *Power to Explore*, pp. 54–58.

that affected the Center and its stake in NASA's programs. Other such moments included the decision to make Florida's Cape Canaveral, which had been Marshall's Launch Operations Directorate, an independent center and to accept the Air Force contracting system in a move away from Marshall's long commitment to the arsenal system.

The shadow of von Braun's German background affected his relationship with NASA Headquarters and other Centers. It was a subtle influence, seldom brought into the open, and it did decrease in frequency, but the issue came up too often to be incidental. Charles Sheldon, a senior staff member representing the White House on the National Aeronautics and Space Council in the early 1960s, remembered that people in Washington discounted rumors that von Braun might one day head NASA because "von Braun would never be given any political position. No one who had worked with Hitler and the Nazi government could be trusted." The Nazi issue came up often enough in public references to keep the matter alive. A film biography of von Braun produced in the early 1960s entitled *I Aim at the Stars* prompted one critic to add "but sometimes he hits London." Satirist Tom Lehrer, in one of his popular recordings in 1965, included a verse saying, "Once the rockets are up, who cares where they come down? 'That's not my department,' says Wernher von Braun."[59]

Jealousies at Headquarters compounded the problem. As one of von Braun's Huntsville associates noted, "When von Braun appeared at certain occasions—symposiums, meetings at Headquarters—he, rather than the upper Administrator, was the center of attention."[60] NASA Administrator James Webb, who served under Presidents Kennedy and Johnson, warned von Braun that his speeches contained overly optimistic projections of NASA's capabilities, creating unrealistic expectations. Webb, worried also about the propriety of von Braun making substantial profits from his speeches, restricted von Braun to four paid public appearances a year and required him to submit a list of intended speeches for approval. Thomas Paine, who followed Webb as NASA Administrator, said Webb wanted to keep von Braun out of Washington, saying, "I think Jim had the feeling that, well, the Jewish lobby would shoot him down or something—the feeling that basically you were dealing with the Nazi party here. And you could get away with it if he were a technician down in Huntsville building a rocket, but if you brought him up here . . . ."[61]

The issue lingered in relations with the Manned Spacecraft Center in Houston, too. Chris Kraft, who ran Houston's Mission Control during most of the Apollo flights, believed that "Wernher had a Teutonic arrogance that he'd honed to a fine edge. He saw himself as the number-one expert in the world on rockets and space travel, and had polished that self-image with magazine articles, books, lectures, and technical papers. He was famous. He was a NASA Center Director, equal to Bob Gilruth [Houston's Center

59 Ibid., pp. 154–155.
60 Georg von Tiesenhausen OHI.
61 Joseph J. Trento, *Prescription for Disaster* (New York: Crown Publishers, 1987), pp. 89–90.

Director], and probably trying to figure out ways to move Gilruth aside." Kraft remembered asking Gilruth over lunch one day what he thought of von Braun. "Gilruth looked up from his salad and gave me one of those looks that said 'this isn't a good subject.' But after a moment, he found the words to describe everything he felt about the German rocketmeister in one short sentence: 'Von Braun doesn't care what flag he fights for.'"[62]

In this atmosphere, von Braun, an extraordinarily self-confident person, realized that if he may have to give ground on occasion and exercise caution with outsiders, it would afford him the latitude to control his own domain in Huntsville. During the von Braun years, Marshall developed a reputation for secrecy that becomes understandable in this context. Engineer Bob Marshall remembered that von Braun's Center had a reputation as a "very good technical organization, but a poor management organization."[63] A 1968 study described von Braun as a model for the "reluctant supervisor" typical of the Huntsville Center, a man who wanted to keep his hands dirty and avoid red tape and committees.[64] Others saw it differently, and even von Braun's critics could not deny the record of success produced by the rocket team. NASA Administrator James Webb was not one to lavish praise on his subordinates, but when he visited Marshall Space Flight Center in 1965, he commented, "I saw here one of the most sophisticated forms of organized human effort that I have ever seen anywhere!"[65]

Furthermore, if questions arose about the management style at Marshall, von Braun had his own concerns about working with Headquarters. In the early 1960s, NASA's lunar program suffered the usual growing pangs of an organization experiencing rapid growth, and in 1963 von Braun complained to his associates that relations between the Centers and Headquarters had become "terribly complicated." He lamented "it is almost impossible to obtain a guideline, let alone a decision." Von Braun and the other Center Directors, Robert Gilruth in Houston and former von Braun team member Kurt Debus at Cape Canaveral, asked that one man at Headquarters be placed in charge of the Saturn-Apollo program. Webb concurred and appointed George Mueller as NASA's Director of the Office of Manned Space Flight.

The Mueller-von Braun relationship was critical and worked to the benefit of both men. Mueller came to his post sharing the prevalent Washington perspective that the NASA Centers preferred to operate as independent fiefdoms that desired contact with Headquarters only to receive money, but otherwise wanted to be left alone to do their work. Indeed, von Braun contributed to that perception, for he often quipped, "All we need is a rich uncle in Washington who sends the money but does not interfere with our

62 Chris Kraft, *Flight: My Life in Mission Control* (New York: Dutton, 2001), pp. 83, 103.
63 Bob Marshall OHI, p. 4.
64 Tompkins, "Organization Metamorphosis," p. 116.
65 Stuhlinger and Ordway, *Wernher von Braun*, p. 197.

work." Mueller told the Center Directors that things would not operate that way, and Debus and von Braun quickly came around. "Wernher turned out to be one of the strongest supporters," Mueller remembered. The relationship worked both ways, for Mueller became one of Marshall's protectors at Headquarters in the infighting among Centers and between Centers and Headquarters.[66]

At no point was the von Braun-Mueller collaboration more important than on another of the turning points in the Saturn program, a decision nearly as significant as the EOR-LOR debate. In November 1963, Mueller proposed that Saturn testing proceed with a compressed schedule in which the first Saturn IB flight and the first Saturn V flight would be conducted with all live stages rather than in incremental stage-by-stage tests—"all-up testing" in the NASA vernacular. The savings in money would be significant, and, in Mueller's view, it was the only way NASA could meet the schedule of landing on the Moon by the end of the decade. It meant that the first manned launch in each sequence would be on the third flight rather than the seventh. The concept went against Marshall conservative engineering practice, and von Braun's senior staff vehemently objected. Von Braun nonetheless decided to share the risk with Mueller and endorsed the concept despite continuing resistance from trusted subordinates. All-up testing proved to be a key to the success of Apollo.[67]

But all-up testing should not obscure the rigorous testing that went into the development of Saturn's components. Four years passed between the time von Braun accepted Mueller's proposal and the first Saturn V launch on 9 November 1967. During that crucial period, tests proceeded in the labs and on the test stands at Marshall Space Flight Center; without that program, Saturn would not have compiled its incredible record of reliability. Von Braun explained Saturn's dependability, insisting that

> Saturn V was not overdesigned in the sense that everything was made needlessly strong and heavy. But great care was devoted to identifying the real environment in which each part was to work—and "environment" included accelerations, vibrations, stresses, fatigue loads, pressures, temperatures, humidity, corrosion, and test cycles prior to launch. Test programs were then conducted against somewhat more severe conditions than were expected. A methodology was created to assess each part with a demonstrated reliability figure, such as 0.9999998. Total rocket reliability would then be the product of all these parts reliabilities and had to remain above the figure of 0.990, or 99 percent. Redundant parts were used whenever necessary to attain this reliability goal.[68]

---

66 Ibid., 196–197.

67 Bilstein, *Stages to Saturn*, pp. 348–351.

68 Wernher von Braun, "Saturn the Giant," in *Apollo Expeditions to the Moon*, ed. Edgar M. Cortright (NASA Scientific and Technical Information Office, NASA SP-350, 1975), *http://history.msfc.nasa.gov/special/pogo.html.*

Despite all of the testing, a serious problem developed on the second Saturn V flight. A longitudinal vibration developed in all three stages that von Braun described as similar to a concertina. As von Braun explained, the oscillation, which came to be known as the "Pogo effect," "was caused by resonance coupling between the springlike elastic structure of the tankage and the rocket engines' propellant-feed systems." Once the source of the problem was understood, a damping system minimized the Pogo effect, and Saturn flights continued with little interruption.

The next flight was one of the triumphant successes of the lunar program. Apollo 8, commanded by Frank Borman, reached lunar orbit during the Christmas season of 1968. The sixth flight of the Saturn V carried Apollo 11 to the Moon. In July 1969, Neil Armstrong became the first person to set foot on the Moon.

Altogether, NASA and its contractors produced 17 of the mammoth rockets. One was a dynamic test vehicle intended for testing rather than launch; 2 launched unmanned Apollo missions; 10 carried manned Apollo missions; 1 launched Skylab, the first American space station in 1973; and 3 became museum pieces, 1 in Huntsville, 1 in Houston, and 1 at Kennedy Space Center in Florida. "The Saturn V's track record of successful launches remains a marvel of technology," reflected Chris Kraft, often a von Braun critic. "In the 21st century, I still find it hard to believe that von Braun did so much in the 1960s. The world has nothing like a Saturn V today."[69]

AVIATION AND AEROSPACE PIONEER

Even as the peak achievements of the Apollo program captured the world's attention in the late 1960s, von Braun faced more mundane but nonetheless challenging problems in his role as Director of Marshall Space Flight Center. Pressured by the Kennedy and Johnson administrations to hire African Americans to counter the otherwise negative image of Alabama in civil rights, von Braun worked with local leaders in Huntsville to improve race relations in the city. Although he was never entirely successful—it was, after all, hard to convince African American engineers to come to Alabama—he helped to facilitate a more positive approach to race relations in north Alabama, and Huntsville never experienced the violent clashes that marked the civil rights struggle in Birmingham, Selma, or Montgomery. He was instrumental in establishing a university in Huntsville that became part of the University of Alabama system; it has collaborated with Marshall ever since.

One of his biggest challenges was managing the cutbacks at Marshall that began after the development of Saturn, particularly because these decisions were beyond his control. Cutbacks in spending and mandated personnel reductions in force came in waves, and

69 Kraft, *Flight: My Life*, p. 351.

Von Braun and Dr. Eberhard Rees during the launch of the Saturn A-6 on 28 May 1961. (NASA Marshall Space Flight Center, negative number 6673878)

the Center went through difficult readjustments even as its accomplishments in propulsion won worldwide praise.

The completion of Saturn V development posed questions about the future of Marshall Space Flight Center. Without a major propulsion project, and with the days of unlimited funding and abundant staffing fading, the future of the Center looked dim. Von Braun began to investigate other possible activities for the Center. Anticipating the decline, he formed a Future Projects Office in 1964 and directed the Research Projects Laboratory to conduct studies for space science projects. Between them, these two groups investigated possibilities that later became major NASA projects, including Skylab, the High Energy Astronomy Observatories, the Large Space Telescope, the Apollo Telescope Mount, lunar rover, and lunar science studies. These proposals became the basis for Marshall's diversification into fields other than propulsion, particularly space science. They also encouraged new space science for NASA.[70]

On 1 March 1970, von Braun left his position as Director of Marshall Space Flight Center to accept a position at NASA Headquarters as Deputy Associate Administrator for Planning, the fourth-ranking position in the Agency. Dr. Eberhard Rees, his long-time

70 Dunar and Waring, *Power to Explore*, pp. 137–138.

Deputy, succeeded him as Center Director. Five thousand Huntsville residents turned out on a drizzly day to bid "Huntsville's First Citizen" farewell. The *Huntsville Times* concluded "Dr. von Braun leaves this community bigger and better than he found it."[71]

In Washington, von Braun and his staff of 20 worked to develop long-term plans for NASA and to work out an approach for presenting the Space Shuttle to Congress.[72] Feeling marginalized and frustrated, von Braun left NASA after two years and four months at Headquarters. He accepted a position at Fairchild Industries in Germantown, Maryland. He died in Alexandria, Virginia, on 16 June 1977.[73]

Twenty-six years to the day after his death, the magazine *Aviation Week & Space Technology* announced at the Paris Air Show its list of the top 100 aviation pioneers. Von Braun's name was second on the list, behind only Orville and Wilbur Wright.[74] It was a fitting tribute to a man with the imagination to envision human space travel in the 1920s and the engineering expertise and managerial skills to make it a reality in the 1960s. His three milestone achievements—developing the first human-launched vehicle to reach space, the launch system that put the first American satellite in orbit around the Earth, and the Saturn V rocket that powered the American lunar program—represent a remarkable lifetime of achievement. That he was among the foremost individuals to popularize space travel adds luster to his record and demonstrates the range of his abilities. That his name is not free of its association with the Nazi Third Reich demonstrates that not even the loftiest of achievements can entirely escape the disturbing political undercurrents of the 20th century. He never escaped the charge of amoral opportunism or the stain of the concentration camps.

Not all of von Braun's ideas have reached fulfillment. NASA rejected the arsenal system of an all-in-one research and development organization. The Agency also rejected the conservative engineering approach that was the hallmark of the von Braun team, with its mission of step-by-step testing from component to subsystem, to system, to flight article, to test flight.

Other von Braun visionary concepts, such as a mission to Mars, remain unfulfilled. The dreams of a visionary are not always accomplished in one lifetime, and von Braun may yet speak to future generations.

ENDNOTE

Andrew J. Dunar wishes to thank Michael J. Neufeld and Stephen P. Waring for their helpful comments on a draft of this essay.

---

71 Ibid., pp. 152–153.

72 Jay Foster, "Dr. von Braun in Washington," in *The Marshall Retiree Report* (September 2003).

73 "Dr. Wernher von Braun: First Center Director, July 1, 1960–January 27, 1970," *http://history.msfc.nasa.gov/vonbraun/bio.html.*

74 Ed Buckbee, "Von Braun Honored," in *The Marshall Retiree Report* (September 2003).

11

Godfather to the Astronauts

Robert Gilruth and the Birth of Human Spaceflight

ROGER LAUNIUS

ROBERT R. GILRUTH WAS A LONGTIME NACA ENGINEER WORKING AT THE LANGLEY AERONAUTICAL LABORATORY WHEN HE BECAME ITS GURU OF SPACEFLIGHT. He had been Chief of Langley's Pilotless Aircraft Research Division from 1946 to 1952, developing the technologies necessary for reaching space, and his Space Task Group at Langley had been exploring the possibility of human spaceflight even before NASA was created in 1958. Because of this, he became the head of Project Mercury from 1959 to 1963 and served as the first Director of the Manned Spacecraft Center in Houston, Texas, retiring in 1972 after the successful completion of Project Apollo.

Gilruth, perhaps more than any other NASA official, served as the godfather of human spaceflight in the United States. Under his direction, NASA successfully completed Projects Mercury, Gemini, and Apollo. His organization recruited, trained, and oversaw the astronauts and the human spaceflight program throughout the heroic age of spaceflight. Yet, his name is much less well known than many others associated with these projects. He was a contemporary on a par with Wernher von Braun and a host of other NASA officials, and he certainly contributed as much to human spaceflight as any of them, yet his name is rarely invoked as a key person. He is a representative of the engineering entrepreneur, a developer and manager of complex technological and organizational systems, accomplishing remarkably difficult tasks through excellent oversight of the technical, fiscal, cultural, and social reins of the effort. Johnson Space Center Director George W. S. Abbey appropriately commented at the time of his death in 2000, "Robert Gilruth was a true pioneer in every sense of the word and the father of human spaceflight. His vision, energy, and dedication helped define the American space program. His leadership turned the fledgling Manned Spacecraft Center into what it is today, the leader in humanity's exploration of outer space."[1] This essay discusses the career of Robert Gilruth as an engineering entrepreneur who oversaw the vast human spaceflight effort for NASA during the "glory days" of the 1960s.

A MIDWESTERN CHILDHOOD

Robert Rowe Gilruth enjoyed a happy, tranquil boyhood in the small town of Nashwauk, Minnesota, a mining town in the Mesabi Range of northern Minnesota. Born on 8 October 1913, Gilruth's parents were both educators of Cornish/Scottish ancestry. His father, Henry Augustus Gilruth, was the Nashwauk superintendent of schools, and his mother, Francis Marian Rowe Gilruth, was an ex-school teacher. Gilruth recalled in an interview in 1986:

> My father was born in Davenport, Iowa, and my mother was born in Bessemer, Michigan. My mother was the daughter of a mining captain. They called them mining captains in those days if they became officials in the mine and had worked their way up from the pit, so to speak. He was born in England, was a Cornishman—you know there are a lot of mines in Cornwall—and he came to America because he heard they needed expert iron geologists. They didn't call them geologists; they called them mining captains. He was a self-educated geologist, and he could tell the men where to dig in order to get the rich iron ore.[2]

Gilruth and his older sister, Jean Marian Gilruth, enjoyed their experience in upper Minnesota.

---

1 JSC Press Release J00-49, "Statement By Johnson Space Center Director George W. S. Abbey Marking the Death of Dr. Robert R. Gilruth" (17 August 2000), NASA Headquarters Historical Reference Collection, Washington, DC.

2 Robert Gilruth Oral History Interview (OHI) No. 1 by David DeVorkin and Martin Collins (21 March 1986), Glennan-Webb-Seamans Project, National Air and Space Museum, Washington, DC.

His parents provided Gilruth not only with the necessities of life, but also encouraged his innate curiosity about the world around him. He started to carve wooden models of ships, grew intensely interested in the history and lore of the local Native American tribes, and gained an interest in airplanes and flight while in Nashwauk. At an early age, he began reading magazines such as *American Boy*, which reinforced his interest in making model airplanes from the sketches it carried. He soon moved on to *Popular Mechanics*, with its more advanced articles on model planes. Later he built telescopes to observe the planets and stars. Two constants remained in his life from his earliest years, his love of flight and his deep affection for the sea and sailing. "I've liked boats very much in my lifetime," he recalled, "and I've spent a lot of time building my own sail and power boats, and so on and so forth. I did also have a very good interest in a hydrofoil company with hydrofoil-fitted boats."[3]

Three of the four Apollo 13 flight directors applaud the successful splashdown of the Command Module Odyssey while Dr. Robert R. Gilruth, Director, Manned Spacecraft Center (MSC), and Dr. Christopher C. Kraft, Jr., MSC Deputy Director, light up cigars (upper left). The flight directors are (from left to right): Gerald D. Griffin, Eugene F. Kranz, and Glynn S. Lunney. Photo taken on 17 April 1970. (GRIN database number GPN-2000-001313)

3 Ibid.; and Pearce Wright, "Robert Gilruth: Rocket Engineer Who Put Americans into Space," in the *Guardian* (London, England: 21 August 2000).

As a young boy, Gilruth suffered from chronic bronchitis. He said, "I had missed so much school due to my bronchitis that I had as a younger chap that I just didn't do all that well." He never showed stellar grades in school until he reached the University of Minnesota as a junior in 1933.[4]

As a boy of almost eight years old, Gilruth moved with his family from Nashwauk to Hancock, Michigan, when his father accepted another education position. They did not stay long there, however, as his father ran afoul of the schoolboard over funding allocated to the system. Gilruth recalled:

> Hancock was a poor copper mining town in Michigan, and it was having tough going. The mines were just breaking even, and lots of men were out of work. They mined copper, and they had a refinery there called the Smelts. The copper came out of the ground as native copper. They had to crush the rock away from it and then melt it and put it in ingots. All those industries around there were just barely staying alive. It was hard for the school . . . . He led an effort to get a bond issue for a new school, and that was the end of his career with the Board of Education.[5]

By mutual agreement in 1922, Henry Gilruth left his position in Hancock and moved his family to Duluth, where he worked as a teacher in the local school system.

While in Duluth, about the age of 12, Gilruth entered a model airplane contest sponsored by the local newspaper, the *Duluth News Tribune*, which put the contest details on the front page to generate the greatest interest. "They gave people publicity who did well," Gilruth recalled, "and they got a lot of the boys there in Duluth to become interested in building models." He added, "I cared very little about whether it looked like a World War I airplane or Lindbergh's airplane, although I did build a model of Lindbergh's airplane. I did a little of both. When I made a model of Lindbergh's airplane, then I tried to make it look very much like his airplane." In addition to his model airplane building, Gilruth also built radios and other electronics. Indeed, his interest in things technical took root early and stayed with him throughout his life.[6]

While his parents thought Gilruth could spend his time more productively on other activities, he had caught the aviation bug by the time of Lindbergh's flight and decided to make it his life's work. Reading *American Boy, Popular Science,* and *Popular Mechanics* reinforced this decision.[7] So did reading the *Saturday Evening Post*, where he first heard

4 Gilruth OHI No. 1 by DeVorkin and Collins.

5 Ibid.

6 Ibid.

7 The importance of these periodicals for aspiring aerospace leaders before the dawn of the space age deserves sustained attention. Many of the first generation of NASA officials confess to reading these publications as children, and an analysis of their influence should be undertaken.

about the National Advisory Committee for Aeronautics (NACA), the government entity with laboratories dedicated to pursing the problems of flight, in the words of Hugh L. Dryden, "to separate the real from the imagined."[8] He sent away to the NACA for information about airfoils, an early research effort of the organization, and Gilruth used this data to help improve his model aircraft.[9]

Gilruth finished high school at Duluth in 1931, a bright but not particularly engaged student. In part because of this, he went to the local Duluth Junior College for two years. He also attended there because it was the height of the Great Depression, and his parents did not have the money to send him to the state university. He pursued his interest in aviation while at the junior college, but he also led the chess club on the campus and became an avid tennis player, both hobbies he pursued the rest of his life. At college he fell under the spell of Lewis A. Rodert, a recent graduate of the University of Minnesota and later a colleague of Gilruth's at the NACA. Gilruth recalled that Rodert "became a fast friend of mine," opening a world of discovery about aeronautics in his "Principles of Flight" course at the junior college. "Rodert was a good teacher," recalled Gilruth. "He was a good disciplinarian. There were only, I think, three of us that were taking the course," so Gilruth received considerable individual attention.[10]

When he transferred to the University of Minnesota for his junior year, his academic career, already taking off, accelerated even more. The university offered one of the few courses of study then available in aeronautical engineering, and Gilruth found a challenging and invigorating niche for his interests.[11] As he prepared for graduation in 1935, he had already decided that he wanted to work for the NACA. He commented, "I continued to take special note of anything I ever read about NACA, and my interest grew to such an extent that the only thing I wanted was to gain my aeronautical degree, hoping that I could then go to work with the group."[12]

With no positions available at the NACA because of the Great Depression, Gilruth decided to pursue graduate education and completed a master of science degree in aeronautical engineering in 1936, writing a thesis on "The Effect of Wing-Tip Propellers on

8 Michael H. Gorn, *Hugh L. Dryden's Career in Aviation and Space* (Washington, DC: NASA Monographs in Aerospace History No. 5, 1996).

9 Edgar S. Gorrell and H. S. Martin, *Aerofoils and Aerofoil Structural Combinations* (NACA Technical Report 13, 1918); Max M. Munk, *General Theory of Thin Wing Sections* (NACA Technical Report 142, 1922); Max M. Munk and Elton W. Miller, *Model Tests with a Systematic Series of 27 Wing Sections at Full Reynolds Number* (NACA Technical Report 221, 1927); and Virginius E. Clark, *Design Your Own Airfoils* (Langley Field, VA: Langley Memorial Aeronautical Laboratory, 1927).

10 Gilruth OHI No. 1 by DeVorkin and Collins; and Glenn E. Bugos, "Lew Rodert, Epistemological Liaison, and Thermal De-Icing at Ames," in *From Engineering Science to Big Science: The NACA and NASA Collier Trophy Research Project Winners*, ed. Pamela E. Mack (Washington, DC: NASA SP-4219, 1998), pp. 29–58.

11 Gilruth OHI No. 1 by DeVorkin and Collins.

12 Shirley Thomas, *Men of Space: Profiles of the Leaders in Space Research, Development, and Exploration*, vol. 4 (Philadelphia, PA: Chilton Company, 1962), p. 48.

the Aerodynamic Characteristics of a Low Aspect Ratio Wing."[13] Working as a graduate assistant at $50 a month, he helped the department chair build a hot air "barrage balloon" that used a ground-power generator to send electricity up a tether where it powered a space heater that kept the air in the balloon hot. The project proved ineffective and ended without a successful test. About this effort Gilruth recalled, "It was a job that I got paid for, so I did the best I could with it, but I certainly didn't think it was a good job to be doing. I was not interested in it. I didn't think it was a good thing for the university to do."[14] More interesting, Gilruth participated in a project with French high-altitude balloonist Jean Piccard, who had recently joined the University of Minnesota's faculty, to make a valve that could ensure constant pressure inside an aircraft's cockpit. Piccard told him that this work was necessary because the higher the aircraft could fly, the greater the possibilities for speed.[15]

Jean Piccard had an important early influence on Gilruth, but one that was more practical and less inspirational than Rodert's. "I learned many things from him," Gilruth commented, "ways of looking at problems. He had a way of simplifying things, in talking about [how components worked]." Gilruth said that he tried to remember how Piccard would break down an engineering problem into the smallest possible components and then tackle each in order, gradually working through an entire issue to achieve a meaningful solution. These might not be the most elegant resolutions of problems, but they worked, and a technical problem solved with a minimum of effort was often better than an elegant resolution requiring considerable expenditure of resources. Piccard practiced a form of KIS (keep it simple) before the name arose. Gilruth came back to it many times in his career as he worked to place Americans first in orbit and then on the Moon.[16]

Gilruth had some unique opportunities while attending graduate school. His advisor, John Akerman, received a contract from air showman Roscoe Turner to undertake technical work on an air racer of Turner's design, later christened the *Laird-Turner Meteor*.[17] Gilruth remarked:

> Roscoe Turner gave a contract to Akerman and Bud Barlow, assistant head of the department, to design this airplane. I think I had an input into just about every part of it. A small staff and I ran the wind tunnel tests, I designed the size of the tail, the wing,

13 Robert R. Gilruth, "The Effect of Wing-Tip Propellers on the Aerodynamic Characteristics of a Low Aspect Ratio Wing," (master's thesis, University of Minnesota, 1936).

14 Gilruth OHI No. 1 by DeVorkin and Collins.

15 David H. DeVorkin, *Race to the Stratosphere: Manned Scientific Ballooning in America* (New York: Springer-Verlag, 1989), pp. 234–246.

16 Gilruth OHI No. 2 by DeVorkin, Collins, and Linda Ezell (14 May 1986).

17 Carroll V. Glines, *Roscoe Turner: Aviation's Master Showman* (Washington, DC: Smithsonian Institution Press, 1995).

> did the stability and control work, and also a fair amount of the structural analysis. So, during the summer of 1935, I got a liberal course in airplane design by actually doing the work, which was invaluable because both Akerman and Barlow are very good aeronautical engineers. What we built was faster than anything else in the skies. I think it went a little over 400 miles an hour, which was at that time the world's speed record.[18]

This aircraft made it possible for Turner to win the National Air Races in 1938 and the Thompson Trophy Race in 1939.[19]

While in graduate school, Gilruth met and married Jean Barnhill, a fellow aeronautical engineering student and pilot who had flown in cross-country races. A friend of Amelia Earhart, Jean Gilruth claimed membership in the flying group she helped found, the 99s. They wed in the Episcopal Cathedral in Washington, DC. It proved a longlasting union, Jean outliving Gilruth. Jean gave up her flying when they had a daughter, but remained interested in aviation the rest of her life.[20]

REACHING THE NACA

Just as Gilruth finished his master of science degree in December 1936, the NACA offered him a position as an aeronautical engineer at its Langley Memorial Aeronautical Laboratory in Hampton, Virginia.[21] The NACA had been established in 1915 to foster aviation in the United States at a time when the nation lagged far behind the technological capabilities of Western Europe. Established via a rider to the Naval Appropriations Act of 1915, Congress established the NACA "to supervise and direct the scientific study of the problems of flight, with a view to their practical solution, and to determine the problems which should be experimentally attacked, and to discuss their solution and their application to practical questions."[22] This became an enormously important government research and development organization for the next half of a century, materially enhancing the development of aeronautics. All research projects undertaken by the NACA sought to pursue investigations that promised the compilation of fundamental aeronautical knowledge applicable to all flight, rather than working on a specific type of aircraft design because it smacked of catering to a particular aeronautical firm. Most

---

18 Thomas, *Men of Space*, p. 49.

19 Roger Huntington, *Thompson Trophy Races* (Osceola, WI: Motorbooks International, 1989).

20 Thomas, *Men of Space*, pp. 49–50.

21 Robert R. Gilruth Bio-Data Sheet (20 August 1970), NASA Headquarters Historical Reference Collection, Washington, DC.

22 From Public Law 271, 63rd Congress (approved 3 March 1915).

NACA research was accomplished "in house" by scientists or engineers on the federal payroll. The results of these activities appeared in more than 16,000 research reports of one type or another, which were distributed widely for the benefit of all. As a result of this work, the NACA received the coveted Robert J. Collier Trophy given annually for "great" achievement in aeronautics and astronautics in America a total of five times between 1929 and 1954.[23]

The NACA's research was conducted in government facilities, and its government scientists and engineers developed a strong technical competence, a commitment to collegial in-house research conducive to engineering innovation, and a definite apolitical perspective. While it never had more than about 8,000 employees and an annual budget of $100 million, the NACA maintained a small Washington Headquarters staff, three major research laboratories—the Langley Aeronautical Laboratory established in 1917, the Ames Aeronautical Laboratory activated near San Francisco in 1939, and the Lewis Flight Propulsion Laboratory built in Cleveland, Ohio in 1940—and two small test facilities at Muroc Dry Lake in the high desert of California and at Wallops Island, Virginia. This organization remained a significant entity until transformed into NASA in 1958.[24]

Gilruth reported to mother Langley, as its employees affectionately called it, in January 1937, newly graduated from the University of Minnesota. As he recalled:

> I reported for duty there, and after getting finger printed and everything like that, I went to see the head of the Aerodynamics Division. There were really three divisions at the NACA at that time. There was the Aerodynamics Division, which is sort of self-explanatory, there was the Wind Tunnel and Flight Research Section and so on, and there was the Hydro Division, which was the towing basin. Then there was the Engine Lab, which was what it says. I was obviously an aeronautical engineer with an aviation background, so I was sent to the Aerodynamics Division.

---

23 Alex Roland, *Model Research: The National Advisory Committee for Aeronautics, 1915–1958*, vol. 1 (Washington, DC: NASA SP-4103, 1985). For a discussion of the Collier trophy and the NACA/NASA research projects that received it, see Mack, ed., *From Engineering Science.*

24 On these centers, see Roland, *Model Research*, pp. 283–303; James R. Hansen, *Engineer in Charge: A History of the Langley Aeronautical Laboratory, 1917–1958* (Washington, DC: NASA SP-4305, 1987); Elizabeth A. Muenger, *Searching the Horizon: A History of Ames Research Center, 1940–1976* (Washington, DC: NASA SP-4304, 1985); Virginia P. Dawson, *Engines and Innovation: Lewis Laboratory and American Propulsion Technology* (Washington, DC: NASA SP-4306, 1991); Richard P. Hallion, *On the Frontier: Flight Research at Dryden, 1946–1981* (Washington, DC: NASA SP-4303, 1984); Lane E. Wallace, *Flights of Discovery: 50 Years at the NASA Dryden Flight Research Center* (Washington, DC: NASA SP-4309, 1996); and Glenn E. Bugos, *Atmosphere of Freedom: Sixty Years at NASA Ames Research Center* (Washington, DC: NASA SP-2000-4314, 2000).

The head of that division then looked at Gilruth's vita and said, "Well, your experience in stress analysis and airplane design will make you particularly valuable in flight research." He then sent him to Flight Research.[25]

Gilruth found a Center fabled both for its collegiality and cutting-edge aeronautical research. Almost with the opening of the Langley Center, the NACA had recruited Max Munk, a gifted student of Ludwig Prandtl. Munk—who had earned two doctorates, one in engineering and another in physics from Göttingen University—had reoriented the lab's efforts toward aerodynamics through the construction of a revolutionary Variable Density Tunnel that began operation in 1922. This instrument, and other later wind tunnels, transformed the Langley laboratory into a major research facility on par with the best of those anywhere in the world. As historian Deborah G. Douglas concluded, "By the late 1920s, the NACA's Langley Aeronautical Laboratory had begun to earn an international reputation, largely due to the construction of a trio of pioneering wind tunnels (the Variable Density Tunnel that became operational in 1922, the Propeller Research Tunnel in 1927, and the Full Scale Tunnel in 1931)."[26]

Because of this, Langley became a mecca for bright, young aeronautical engineers who wanted to make a difference in the progress of aircraft technology in the United States. As historian Roger E. Bilstein wrote, "Langley managed to attract the brightest young aeronautical engineers in the country because they knew that their training would continue to expand by close and comradely contact with many senior NACA engineers on the cutting edge of research."[27] This feature of the Langley Center and the whole of the NACA would become a hallmark of its technical success. Even though the organization grew more formal over the years, all agreed that it remained a place where uniquely creative individuals undertook remarkable research and made significant contributions to knowledge about the practical aspects of flight.[28]

While Gilruth found such an environment invigorating, he also found it competitive. The best aeronautical engineers in the world worked at Langley, with more arriving every day. Notwithstanding this, Gilruth rose to the occasion and soon became one of the premier researchers at the lab. His principal work revolved around the field of aircraft stability, control, and vehicle-handling qualities.[29] Throughout the war, the NACA aero-

---

25 Gilruth OHI No. 2 by DeVorkin, Collins, and Ezell.

26 Deborah G. Douglas, "Three-Miles-A-Minute: The National Advisory Committee for Aeronautics and the Development of the Modern Airliner," in *Innovation and the Development of Flight*, ed. Roger D. Launius (College Station: Texas A&M University Press, 1999), p. 156.

27 Roger E. Bilstein, *Orders of Magnitude: A History of the NACA and NASA, 1915–1980* (Washington, DC: NASA SP-4406, 1989), p. 5.

28 Ibid., pp. 1–14; and Michael H. Gorn, *Expanding the Envelope: Flight Research at NACA and NASA* (Lexington: University Press of Kentucky, 2001), pp. 9–96.

29 NASA Manned Spacecraft Center Release, "Robert R. Gilruth Biographical Data" (May 1968), NASA Headquarters Historical Reference Collection, Washington, DC.

nautical research program continued as it had in earlier years, but at a heightened pace. Requests for answers to specific problems came into the Committee and were then parceled out to laboratories for resolution. Researchers such as Gilruth worked closely with those seeking the information to ensure that they received what they needed on a timely basis. No fewer than 40 technical reports, notes, or other studies bore Gilruth's name as author or coauthor between his arrival at Langley and the end of WWII.[30]

The NACA expressed justifiable pride in its contributions in terms of both applied and fundamental research during WWII.[31] These related to research on the shape of wings and bodies, devices to improve engine power and propeller thrust, measures to safeguard stability and control, and apparatus to protect the planes against ice and other natural hazards. These involved all types of experiments at all of the NACA research institutions. The NACA periodically issued statements about its general work for the war. A January 1944 issue of *Aviation* described in proper patriotic fashion the Agency's efforts and urged support for it:

> How much is it worth to this country to make sure we won't find the Luftwaffe our superiors when we start that "Second Front"? We spend in one night over Berlin more than $20,000,000. The NACA requires—now—$17,546,700 for this year's work. These raids are prime factors in winning the war. How can we do more towards victory than by spending the price of one air raid in research which will keep our Air Forces in the position which the NACA has made possible?[32]

---

30 The following are representative NACA reports, which can be found in the NASA Langley Research Center Historical Reference Collection, Hampton, VA: Warren D. Reed and Robert R. Gilruth, "Results of Landing Tests with a Curtiss XF13C-3 Airplane" (12 August 1937); Gilruth, "Results of Landing Tests of Kellett TG-1 Autogiro" (15 November 1937); Gilruth, "Measurements of the Flying Qualities of the Martin B-10B Airplane" (11 January 1938); Gilruth, "An Investigation of the Lift and Thrust Theoretically Available from the Angular Momentum of a Propeller Wake" (19 February 1938); Gilruth and Melvin N. Gough, "Stalling Characteristics of a Douglas B-18 Airplane" (31 May 1938); Gilruth and Gough, "Stalling Characteristics of a Boeing B-17 Airplane" (14 October 1938); Gilruth and Gough, "Stalling Characteristics of the Boeing XB-15 Airplane" (14 November 1938); Gilruth and Gough, "Measurements of the Flying Qualities of a Douglas B-18 Airplane" (11 January 1939); Gilruth and Gough, "Measurements of the Fly Qualities of the Boeing B-17 Airplane" (14 April 1939); Gilruth and Gough, "Measurements of the Fly Qualities of the Boeing XB-15 Airplane" (31 July 1939); Gilruth, Gough, and William Gracey, "Measurements of the Flying Qualities of a Piper Cub Airplane (Model JSL-50)" (19 March 1940); Gilruth, Gough, and Gracey, "Measurements of the Fly Qualities of the Taylorcraft Airplane (Model 30-65)" (29 April 1940); Gilruth and Gough, "Modifications and Tests of Curtiss P-40 Airplane for Improvement of Ground Handling Characteristics" (1 July 1940); Gilruth and William N. Turner, "Longitudinal Stability and Maneuverability Characteristics of Republic XP-41 Airplane with Center of Gravity of 25 Percent M.A.C." (19 September 1940); Gilruth and Floyd L. Thompson, "Notes on the Stalling of Vertical Tail Surfaces and on Fin Design" (October 1940); Gilruth, "Requirements for Satisfactory Flying Qualities of Airplanes" (April 1941); Gilruth and Turner, "Lateral Control Required for Satisfactory Flying Qualities Based on Flight Tests of Numerous Airplanes" (1941); Gilruth, "Measurements of the Aileron Control Characteristics of a Republic P-47B Airplane" (22 March 1942); Gilruth, "Preliminary Trials of a Means for Testing Aerodynamic Bodies in the Transonic Speed Range" (September 1944); and Gilruth and Joseph R. Wetmore, "Preliminary Tests of Several Airfoil Models in the Transonic Speed Range" (May 1945).

31 Many of these activities have been detailed in George W. Gray, *Frontiers of Flight: The Story of NACA Research* (New York: Alfred A. Knopf, 1948). "Research and Air Supremacy," *New York Times* (3 April 1945).

32 "NACA: The Force Behind Our Air Supremacy," *Aviation* (January 1944): 22–23.

Committee Executive Director John F. Victory remarked: "The employees of the NACA have a big and important job to do. They are at war with similar research organizations in Germany, Japan, and Italy. It is their responsibility, and they are using their technical knowledge and skill to make sure that the airplanes that are given to American and allied flyers are better and more efficient instruments of war than those flown by enemy airmen."[33]

Gilruth soon earned a central role in the flight research efforts at Langley, demonstrated by the large number of reports he wrote and the increasing stature that he enjoyed at the laboratory. Working in flight research proved a real opportunity for Gilruth. "Although I didn't realize it at the time," he commented in 1986, "if I had gone to a wind tunnel or some other place, I probably would not have gotten the background that was to make it possible for me to do the things I did both in aviation and space. I was working with the actual airplanes, with test pilots, and, somehow or other, I found that the people that came out of Flight Research had a better chance for grasping the big picture than the people that were buried in wind tunnel work." Some pathbreaking studies emerged from his research, studies he chose to highlight more than 40 years later as the best of his early career, such as the following:

- "Notes on the Stalling of Vertical Tail Surfaces and on Fin Design," NACA Technical Report No. 778, 1940.
- "Analysis and Prediction of Longitudinal Stability of Airplanes," NACA Report No. 711, 1941.
- "Lateral Control Required for Satisfactory Flying Qualities, Based on Flight Tests of Numerous Airplanes," NACA Report No. 715, 1941.
- "Requirements for Satisfactory Flying Qualities in Airplanes," NACA Report No. 755, 1943.
- "Analysis of Vertical Tailoads and Rolling Pullout Maneuvers," NACA Confidential Bulletin L4H14, August 1944.[34]

Despite his productivity, these reports read like telephone books, and, despite a lifetime effort at communicating via the written word, he always wrote in the passive voice using dense engineering jargon. One example will suffice. In a 1942 report on ice detectors, he wrote:

33 John F. Victory, "National Advisory Committee for Aeronautics" (24 June 1942), pp. 2–3, John F. Victory Papers, Special Collection, United States Air Force Academy Library, Colorado Springs, CO. See also "NACA Research and the Nation's War Planes: A Brief History of the Efforts of the NACA To Improve the Performance of Military Airplanes" (9 September 1942), Record Group 255, National Archives and Records Administration, Archives II, College Park, MD.

34 Gilruth OHI No. 2 by DeVorkin, Ezell, and Collins.

> An ice detector, which served as a basis for a rate-of-icing indicator, has been developed and tested recently by the National Advisory Committee for Aeronautics . . . . The present investigation has disclosed two important characteristics of this instrument, either of which can be utilized in measuring the rate of icing. It has been found that a) the time required for the pressure to drop from any given level to another given level is inversely proportional to the icing rate, and b) the maximum rate of change of pressure or the average rate of change of pressure is proportional to the rate of icing.[35]

Gilruth will not win any writing awards for this, or other, similar studies. But the level of writing skill proved less important for engineering than the analysis, and there he excelled.

Gilruth quickly proved his capabilities, and, with the opportunities afforded by WWII, he soon found himself in charge of a number of other researchers. W. Hewitt Phillips, who would himself soon prove a leading aerodynamicist, recalled that "on starting work with NACA at Langley Field in July 1940, I was assigned to the Flight Research Division." He commented:

> The next few years, during the period of World War II, proved to be [an] exciting time. I was working under Dr. Robert R. Gilruth, who had undertaken the task of studying requirements for the flying qualities of airplanes. During that time, a new military airplane was produced practically every month, and many of these airplanes were assigned to Langley for study and improvement of their flying qualities.[36]

Phillips reported that in conducting this research Gilruth followed a set approach. An airplane would be fitted with recording instruments to measure "relevant quantities such as control positions and forces, angular velocities, linear accelerations, airspeed, altitude, etc." He then developed with his team of engineers and their research pilots, especially Melvin N. Gough, "a program of specified flight conditions and maneuvers . . . . After the flight, the data was transcribed from the flight records and plotted to show the relevant information, and the results were correlated with pilot opinion." They then undertook analysis of every aspect of the flight data, and Gilruth prepared reports on the individual studies. The ultimate study by Gilruth came in 1943, "Requirements for Satisfactory Flying Qualities of Airplanes." This major study involved research on tests of 16 different

35 Robert R. Gilruth, J. A. Zalovik, and A. R. Jones, "Flight Investigation of an NACA Ice-Detector Suitable for Use as a Rate-of-Icing Indicator," Wartime Report L-364 (November 1942), p. 1, NASA Langley Research Center Historical Reference Collection, Hampton, VA.

36 W. Hewitt Phillips, "Flying Qualities from Early Airplanes to the Space Shuttle," *Journal of Guidance* 12 (July–August 1989): 289.

airplanes of all types, ranging from light airplanes to the Boeing XB15.[37] As Phillips recalled, "this report formed the basis of subsequent military specifications for stability and control characteristics of airplanes."[38]

LEADING THE PILOTLESS AIRCRAFT RESEARCH DIVISION (PARD)

After nearly eight years as a "dirty hands" engineer at Langley, Gilruth made the most of a chance to lead his own organization, the Pilotless Aircraft Research Division (PARD). On 9 December 1944, Gilruth participated in a meeting at the Langley Memorial Aeronautical Laboratory to discuss the formation of an organization that would devote its efforts to the study of stability and maneuverability of high-speed weapons, especially guided missiles. From the outset, however, he understood that this work would point toward supersonic flight research, which was something he believed represented the cutting edge of flight activities in the postwar era. In early 1945, the NACA asked Congress for a supplemental appropriation to fund the activation of this unit, and a short time later the NACA opened the Auxiliary Flight Research Station (AFRS), soon redesignated the Pilotless Aircraft Research Division, with Gilruth as Director.[39]

The AFRS was established on 7 May 1945 at Wallops Island as a test-launching facility of Langley. Under Gilruth's direction on 4 July 1945, it launched its first test vehicle, a small two-stage, solid-fuel rocket to check out the installation's instrumentation. Also, by 1946 the PARD had begun testing rocket-launched X-2 models at Wallops to gather stability and control data. Additional tests helped NACA and Bell engineers design a pilot escape system for the X-2. Intended originally only to test rocket-powered models of aircraft and missiles at transonic and higher velocities to obtain aerodynamic data, under Gilruth's tutelage it also began pioneering work on supersonic inlets and ramjets. For example, Maxime E. Faget, one of PARD's staff, designed a compact (6½-inch diameter) ramjet engine and a supersonic flight-test vehicle that was powered by two of these ramjets.

---

37 Robert R. Gilruth, "Requirements for Satisfactory Flying Qualities of Airplanes," NACA Technical Report No. 755, 1943, NASA Langley Research Center Historical Reference Collection, Hampton, VA.

38 Phillips, "Flying Qualities," p. 294. The seminal nature of this report is attested to in "Specification for Stability and Control Characteristics of Airplanes," Bureau of Aeronautics, U.S. Navy, Washington, DC, SR-119A (April 1945); "Stability and Control Characteristics of Airplanes," Army Air Force Specification R-1815A (April 1945); and W. Hewitt Phillips, "Appreciation and Prediction of Flying Qualities," NACA Technical Report No. 927, 1948, NASA Langley Research Center Historical Reference Collection, Hampton, VA.

39 James M. Grimwood, *Project Mercury: A Chronology* (Washington, DC: NASA SP-4001, 1963), part 1A, p. 1; and Joseph Adams Shortall, *A New Dimension: Wallops Island Flight Test Range, the First Fifteen Years* (Washington, DC: NASA Reference Publication (RP)-1028, 1978). At first, only part of the land on Wallops Island was purchased; the rest was leased. In 1949, the NACA purchased the entire island.

During a flight test in 1950, this vehicle accelerated under ramjet power in a climbing flight achieving an altitude of 65,000 feet and a velocity of M=3.2, setting unofficial speed and altitude records for vehicles powered by air-breathing engines.[40]

Beyond a series of exploratory flight tests of rocket models, Gilruth's PARD advanced the knowledge of aerodynamics at transonic and later hypersonic speeds. They did so through exhaustive testing, which some at Langley considered excessive and overly expensive, launching between 1947 and 1949 at least 386 models, leading to the publication of Gilruth's first technical report on rocketry, "Aerodynamic Problems of Guided Missiles," in 1947. From this, Gilruth and the PARD filled in tremendous gaps in the knowledge of high-speed flight. As historian James R. Hansen writes, "The early years of the rocket-model program at Wallops (1945–1951) showed that Langley was able to tackle an enormously difficult new field of research with innovation and imagination."[41]

The NACA leadership promoted Gilruth to serve as the Assistant Director of Langley Memorial Aeronautical Laboratory in 1952. Thereafter, in addition to his administrative responsibilities, he worked on several of the ballistic missiles, but was deeply troubled by the advent of nuclear destruction. He said, "I felt that things had really gotten out of hand." He also was aware of the discussion of orbiting an artificial satellite, but at first paid little attention to its potential. On the other hand, he said, "When you think about putting a man up there, that's a different thing. That's a lot more exciting. There are a lot of things you can do with men up in orbit."[42]

Meantime, the PARD continued to advance its work on rockets and missiles. In 1952, the PARD started the development of multistage, hypersonic-speed, solid-fuel rocket vehicles. These vehicles were used primarily in aerodynamic heating tests at first and were then directed toward a reentry physics research program. On 14 October 1954, the first American four-stage rocket was launched by the PARD, and in August 1956 it launched a five-stage, solid-fuel rocket test vehicle, the world's first, that reached a speed of Mach 15.[43]

Also during 1956, Gilruth's engineers in PARD originated the idea of one of the most successful programs in NASA history—the small, inexpensive sounding rocket, Scout. Led by William E. Stoney, Jr., this team also included Max Faget, who would gain fame as the original designer of the Mercury capsule, Joseph G. Thibodaux, Jr., and Robert O.

40 Christopher C. Kraft, Jr., *Robert R. Gilruth, 1913–2000: A Biographical Memoir* (Washington, DC: National Academy of Sciences, 2000); Gilruth OHI No. 1 by DeVorkin and Collins; Robert L. Rosholt, *An Administrative History of NASA, 1958–1963* (Washington: NASA SP-4101, 1966), pp. 48, 81, fig. 3-1, and appendix B; and Glen Golightly, "Spaceflight Pioneer Maxime Faget Hospitalized," *http://www.space.com*.

41 Robert R. Gilruth, "Aerodynamic Problems of Guided Missiles," NACA Report (draft) (19 May 1947), Gilruth Papers, Special Collections, Carol M. Newman Library, Virginia Polytechnical University, Blacksburg, VA; and James R. Hansen, *Spaceflight Revolution: NASA Langley Research Center from Sputnik to Apollo* (Washington, DC: NASA SP-4308, 1995), p. 270.

42 Robert Gilruth OHI No. 3 by Linda Ezell, Howard Wolko, and Martin Collins (30 June 1986), pp. 19, 44.

43 Letter from NASA Space Task Group to NASA Headquarters (5 July 1960), NASA Headquarters Historical Reference Collection, Washington, DC; Eugene M. Emme, *Aeronautics and Astronautics: An American Chronology of Science and Technology in the Exploration of Space, 1915–1960* (Washington, DC: National Aeronautics and Space Administration, 1961), p. 76; and House Report 67, 87th Congress, 1st Session, p. 27.

Piland, who put together the first multistage rocket to reach the speed of Mach 10. Gilruth accepted that while the PARD had originated as an organization that collected data for transonic, supersonic, and eventually hypersonic speeds on aircraft models, it naturally evolved toward the design of rockets and missiles. Scout emerged as a logical outgrowth of the development of this expertise, as these individuals developed a multistage, solid-propellant rocket that could reach orbital speeds of Mach 18.[44]

In 1957, this group explored with the Aerojet Corporation how best to advance solid-rocket motor technology to achieve orbital velocity. Their most interesting attempt involved converting the Jupiter rocket, developed by Wernher von Braun's rocket team at the Army Ballistic Missile Agency in Huntsville, Alabama, to a solid-propellant missile for use aboard naval vessels. They called it the "Jupiter Senior," and its solid-propellant motor measured 30 feet long, 40 inches in diameter, and could provide a thrust of 100,000 pounds. From its first firing in March 1957, Jupiter Senior amassed a record of 13 static tests and 32 flights without a failure, proving the technology that made possible the Polaris and Minuteman intercontinental ballistic missiles that could be placed in a silo or on a submarine and launched reliably with a minimum amount of preparation. The PARD team then went on to argue for, but failed to receive, support to develop a four-stage launcher with the Jupiter Senior as the first stage; a plethora of launchers were developed in the latter 1950s in response to the need to create an intercontinental ballistic missile capability in the Cold War.[45]

While Gilruth ran PARD, he slowly became enamored with the prospects of human spaceflight. In 1952, German émigré Wernher von Braun burst on the broad public stage with a series of articles in *Collier's* magazine about the possibilities of spaceflight. The first issue of *Collier's* devoted to space appeared on 22 March 1952. In it readers were asked "What Are We Waiting For?" and were urged to support an aggressive space program. An editorial suggested that spaceflight was possible, not just science fiction, and that it was inevitable that humanity would venture outward. Von Braun led off the *Collier's* issue with an impressionistic article describing the overall features of an aggressive spaceflight program. He advocated the orbiting of an artificial satellite to learn more about spaceflight followed by the first orbital flights by humans, development of a reusable spacecraft for travel to and from Earth orbit, building a permanently inhabited space station, and finally human exploration of the Moon and planets by spacecraft launched from the space station. Willy Ley and several other writers then followed with elaborations on various aspects of spaceflight ranging from technological viability to

44 Hansen, *Spaceflight Revolution*, pp. 197–200, chapter 11; Shortal, *A New Dimension*, pp. vii, 533–534; T. Keith Glennan, *The Birth of NASA: The Diary of T. Keith Glennan*, ed. J. D. Hunley (Washington, DC: NASA SP-4105, 1993), p. 328; and Lloyd S. Swenson, Jr., James M. Grimwood, and Charles C. Alexander, *This New Ocean: A History of Project Mercury* (Washington, DC: NASA SP-4201, 1966), p. 65. On the history of Wallops, see Harold D. Wallace, Jr., *Wallops Station and the Creation of an American Space Program* (Washington, DC: NASA SP-4311, 1997), although it has far less material relevant to Scout than Shortal's older study.

45 Shortal, *A New Dimension*, pp. 706–707.

space law, to biomedicine.[46] The series concluded with a special issue devoted to Mars, in which von Braun and others described how to get there and predicted what might be found based on recent scientific data.[47]

Clearly the *Collier's* series helped to shape Gilruth's perception of spaceflight as something that was no longer fantasy. Gilruth recalled of von Braun and his ideas, "I thought that was fascinating. He was way ahead of all of us guys. . . . everybody was a space cadet in those days. I thought a space station was very interesting."[48] Gilruth confessed that he did not know von Braun well during that period:

> I had met him at the Pentagon one or two times when he was working for the Army. I think the first time I was with him was during some meetings that were held in connection with early reentry studies that they were doing using the Redstone rockets and some things that the Army was doing on reentry technology. I was there as one of the people in the government that was interested in that sort of thing and who was knowledgeable in the guided missile business.[49]

Gilruth claimed a working relationship with the spaceflight propagandist but not a friendship. They had similar objectives and worked diligently to achieve them.

Gilruth worked to close the gap between public perceptions of spaceflight and its near-term reality with the technological developments. The convincing of the American public that spaceflight was possible was one of the most critical components of the space policy debate of the 1950s. For realizable public policy to emerge in a democracy, people must both recognize the issue in real terms and develop confidence in the attainability of the goal. It was present by the mid-1950s, and without it NASA and the aggressive piloted programs of the 1960s could never have been approved.[50]

In a little more than 12 years, PARD made some significant strides in the development of the technology necessary to reach orbital flight above the atmosphere. Clearly, PARD held the lion's share of knowledge in the NACA above rocketry and the nascent field of astronautics. This organization enjoyed renewed attention and funding once the Soviet Union launched the world's first satellite, Sputnik, on 4 October 1957. "I can recall watching the sunlight reflect off of Sputnik as it passed over my home on the Chesapeake Bay in Virginia," Gilruth recalled in 1972. "It put a new sense of value and urgency on things we had been doing. When one month later the dog Laika was placed in orbit in Sputnik II, I was sure that the Russians were planning for man in space."[51]

---

46 "Man Will Conquer Space *Soon*" series, *Collier's* (22 March 1952): 23–76 ff.

47 Wernher von Braun with Cornelius Ryan, "Can We Get to Mars?" *Collier's* (30 April 1954): 22–28.

48 Robert Gilruth OHI No. 6 by David DeVorkin and John Mauer (2 March 1987).

49 Ibid.

50 Howard E. McCurdy, *Space and American Imagination* (Washington, DC: Smithsonian Institution Press, 1997), pp. 29–52.

51 NASA Press Release H00-127, "Dr. Robert Gilruth, an Architect of Manned Space Flight, Dies" (17 August 2000), NASA Headquarters Historical Reference Collection, Washington, DC.

THE SPACE TASK GROUP AND PROJECT MERCURY

The launch of Sputnik 1 in October 1957, and especially the successful flight of Sputnik 2 a month later, set the United States in crisis mode, kicking off an intensely competitive space race in which two superpowers locked in a Cold War sought to outdo each other for the world's accolades. Nothing seemed too much, no opportunity too small, to move toward that singular goal. At a fundamental level, NASA emerged because of pressures during the Cold War with the Soviet Union, a broad contest over the ideologies and allegiances of the nonaligned nations of the world in which space exploration emerged as a major area of contest. Sputnik had a "Pearl Harbor" effect on American public opinion, creating an illusion of a technological gap and providing the impetus for increased spending for aerospace endeavors, technical and scientific educational programs, and the chartering of new federal agencies to manage air and space research and development.[52]

Sputnik led directly to several critical efforts aimed at "catching up" to the Soviet Union's space achievements, which included the following:

- a wide-ranging review of the civil and military space programs of the United States (scientific satellite efforts and ballistic missile development);
- establishment of a Presidential Science Advisor in the White House who had responsibility for overseeing the activities of the federal government in science and technology;
- beginning the Advanced Research Projects Agency in the Department of Defense and consolidating several space activities under centralized management;
- creation of the National Aeronautics and Space Administration to manage civil space operations for the benefit "of all mankind" by means of the National Aeronautics and Space Act of 1958; and
- passage of the National Defense Education Act of 1958 to provide federal funding for education in scientific and technical disciplines.[53]

More immediately, the United States launched its first Earth satellite on 31 January 1958 when Explorer 1 documented the existence of radiation zones encircling Earth. Shaped by Earth's magnetic field, what came to be called the Van Allen Radiation Belt partially dictates the electrical charges in the atmosphere and the solar radiation that

52 See Rip Bulkeley, *The Sputniks Crisis and Early United States Space Policy: A Critique of the Historiography of Space* (Bloomington: Indiana University Press, 1991); Robert A. Divine, *The Sputnik Challenge: Eisenhower's Response to the Soviet Satellite* (New York: Oxford University Press, 1993); and Paul Dickson, *Sputnik: The Shock of the Century* (New York: Walker and Co., 2001).

53 Roger D. Launius, "National Aeronautics and Space Act of 1958," in Donald C. Bacon, et al., eds., *The Encyclopedia of the United States Congress* (New York: Simon and Schuster, 1994), pp. 1,437–1,438; and Alison Griffith, *The National Aeronautics and Space Act: A Study of the Development of Public Policy* (Washington, DC: Public Affairs Press, 1962).

reaches Earth. The United States also began a series of scientific missions to the Moon and planets in the latter 1950s and early 1960s.[54] As a direct result of this crisis, NASA began operations on 1 October 1958, absorbing the National Advisory Committee for Aeronautics intact—its 8,000 employees, an annual budget of $100 million, three major research laboratories (Langley Aeronautical Laboratory, Ames Aeronautical Laboratory, and Lewis Flight Propulsion Laboratory), and two smaller test facilities. NASA quickly incorporated other organizations into the new Agency. These included the space science group of the Naval Research Laboratory in Maryland, the Jet Propulsion Laboratory managed by the California Institute of Technology for the Army, and the Army Ballistic Missile Agency in Huntsville, Alabama, where Wernher von Braun's team of engineers was engaged in the development of large rockets. Eventually NASA created several other Centers; by the early 1960s, there were 10 located around the country.[55]

Just six days after the establishment of NASA, Gilruth's Space Task Group received approval from NASA Administrator T. Keith Glennan for a piloted satellite project to determine if it was possible for human spaceflight. On 8 October 1958, NASA established the Space Task Group at Langley Research Center. On 5 November 1958, Gilruth received his appointment as Project Manager, and Charles J. Donlan, Technical Assistant to the Director of the Langley Laboratory, became Assistant Project Manager. Thirty-five key staff members from Langley, many of whom had worked on a military man-in-space plan, were transferred to the new Space Task Group, as were 10 others from the Lewis Research Center in Cleveland, Ohio. These 45 persons formed the nucleus of the more than 1,000-person workforce that eventually became a part of Project Mercury. On 14 November, Gilruth requested the highest national priority procurement rating for this project, but it did not come until 27 April 1959. On 26 November 1958, NASA officially designated the program "Mercury," and by early 1959 it had contracted with the McDonnell Aircraft Corporation's bid to build the vehicle.[56] As Glennan recalled, "the philosophy of the project was to use known technologies, extending the state of the art as little as necessary, and relying on the unproven Atlas. As one looks back, it is clear that we did not know much about what we were doing. Yet the Mercury program was one of the best organized and managed of any I have been associated with."[57]

---

54 James A. Van Allen, *Origins of Magnetospheric Physics* (Washington, DC: Smithsonian Institution Press, 1983).

55 Roger D. Launius, *NASA: A History of the U.S. Civil Space Program* (Malabar, FL: Krieger Pub. Co., 1994), chapter 3.

56 Linda Neuman Ezell, *NASA Historical Data Book: Volume II: Programs and Projects, 1958–1968* (Washington, DC: NASA SP-4012, 1988), pp. 102, 139–140; and James M. Grimwood, *Project Mercury: A Chronology* (Washington, DC: NASA SP-4001, 1963), pp. 31–32.

57 T. Keith Glennan, *The Birth of NASA*, chapter 1. On Project Mercury, see Loyd S. Swenson, Jr., James M. Grimwood, and Charles C. Alexander, *This New Ocean: A History of Project Mercury* (Washington, DC: NASA SP-4201, 1966).

Gilruth received two responsibilities in one: 1) Project Manager of the Space Task Group and 2) Assistant Director of a new NASA "space projects center" to be located near Greenbelt, Maryland, which became the Goddard Space Flight Center. This facility, as Glennan believed, would serve as the operations control center for all NASA spaceflight activities. Until that new installation came on line, however, Gilruth's group would remain at Langley. This changed within three years, as it became apparent that the scope, size, and support for human spaceflight necessitated an entirely separate center, and politics decreed that it would be established in Houston, Texas.[58]

The Space Task Group, under Gilruth's tutelage, turned its attention to beginning the program as soon as Mercury received approval. From his mind sprang the astronaut corps needed to accomplish it. He firmly believed that humanity's future lay beyond this planet, and he intended to start society down that challenging and hopeful path. Concurrent with the decision to move forward with Project Mercury, NASA selected and trained the Mercury astronaut corps.[59] President Dwight D. Eisenhower directed that the astronauts be selected from among the armed services' test pilot force. Although this had not been the NASA leadership's first choice, this decision greatly simplified the selection procedure. The inherent risk of spaceflight, and the potential national security implications of the program, pointed toward the use of military personnel. It narrowed and refined the candidate pool, giving NASA a reasonable starting point for selection. It also made imminent sense in that NASA envisioned this astronaut corps first as pilots operating experimental flying machines and only later as scientists. As historian Margaret Weitekamp has concluded:

> From that military test-flying experience, the jet pilots also mastered valuable skills that NASA wanted its astronauts to possess. Test pilots were accustomed to flying high-performance aircraft, detecting a problem, diagnosing the cause, and communicating that analysis to the engineers and mechanics clearly. In addition, they were used to military discipline, rank, and order. They would be able to take orders. Selecting military jet test pilots as their potential astronauts allowed NASA to choose from a cadre of highly motivated, technically skilled, and extremely disciplined pilots.[60]

In addition, since most NASA personnel in Project Mercury came out of the aeronautical research and development arena anyway, it represented almost no stretch on the

58 On the formation of the Manned Spacecraft Center (renamed the Lyndon B. Johnson Space Center in 1973), see Henry C. Dethloff, *"Suddenly Tomorrow Came . . .": A History of the Johnson Space Center* (Washington, DC: NASA SP-4307, 1993).

59 Allan C. Fisher, Jr., "Exploring Tomorrow with the Space Agency," *National Geographic* (July 1960): 48, 52–89; and Kenneth F. Weaver, "Countdown for Space," *National Geographic* (May 1961): 702–734.

60 Margaret A. Weitekamp, "The Right Stuff, The Wrong Sex: The Science, Culture, and Politics of the Lovelace Woman in Space Program, 1959–1963" (Ph.D. dissertation, Cornell University, 2001), p. 98.

Agency's part to accept test pilots as the first astronauts. After all, they had been working with the likes of them for decades and knew and trusted their expertise. It also tapped into a highly disciplined and skilled group of individuals, most of whom were already aerospace engineers who had long ago agreed to risk their lives in experimental vehicles.[61]

From a total of 508 service records screened in January 1959 by NASA at the military personnel bureaus in Washington, they found 110 men that met the minimum standards established for Mercury. These standards were as follows:

1. Age—less than 40

2. Height—less than 5'11"

3. Excellent physical condition

4. Bachelor's degree or equivalent

5. Graduate of test pilot school

6. Total flying time of 1,500 hours

7. Qualified jet pilot

This list of names included 5 Marines, 47 Navy men, and 58 Air Force pilots. Several Army pilots' records had been screened earlier, but none was a graduate of a test pilot school.[62] The selection process began while the possibility of piloted Mercury/Redstone flights late in 1959 still existed, so time was a critical factor in the screening process; although, launch before the end of the year later proved impossible.[63]

A grueling selection process began in January 1959. Headed by the Assistant Director of the Space Task Group, Charles J. Donlan, the evaluation committee divided the list of 110 arbitrarily into three groups and issued invitations for the first group of 35 to come to Washington at the beginning of February for briefings and interviews. Donlan's team initially planned to select 12 astronauts, but as team member George M. Low reported:

> During the briefings and interviews it became apparent that the final number of pilots should be smaller than the 12 originally planned for. The high rate of interest in the project indicates that few, if any, of the men will drop out during the training program.

---

61 In some cases, this was literally the case. The best example is Neil A. Armstrong, who worked with the NACA and NASA as a civilian research pilot on the X-15 program at its Flight Research Center in the Mohave Desert prior to selection for astronaut training in 1962.

62 On this process, see Swenson, Grimwood, and Alexander, *This New Ocean*, pp. 155–165; and Joseph D. Atkinson, Jr., and Jay M. Shafritz, *The Real Stuff: A History of NASA's Astronaut Recruitment Program* (New York: Praeger Publishing, 1985), pp. 8–12.

63 Ibid., pp. 18, 43–45.

> It would, therefore, not be fair to the men to carry along some who would not be able to participate in the flight program. Consequently, a recommendation has been made to name only six finalists.[64]

Every one of the first 10 pilots interrogated on 2 February agreed to continue through the elimination process. The next week, a second third of the possible candidates arrived in Washington. The high rate of volunteering made it unnecessary to extend the invitations to the third group. By the first of March 1959, 32 pilots prepared to undergo a rigorous set of physical and mental examinations.

Thereafter, each candidate went to the Lovelace Clinic in Albuquerque, New Mexico to undergo individual medical evaluations. Phase four of the selection program involved passing an amazingly elaborate set of environmental studies, physical endurance tests, and psychiatric studies conducted at the Aeromedical Laboratory of the Wright Air Development Center in Dayton, Ohio. During March 1959, each of the candidates spent another week in pressure suit tests, acceleration tests, vibration tests, heat tests, and loud noise tests. Continuous psychiatric interviews, the necessity of living with two psychologists throughout the week, an extensive self-examination through a battery of 13 psychological tests for personality and motivation, and another dozen different tests on intellectual functions and special aptitudes were all part of the Dayton experience.[65]

Finally, without conclusive results from these tests, late in March 1959 Gilruth's Space Task Group began phase five of the selection, narrowing the candidates to 18. Thereafter, final criteria for selecting the candidates reverted to the technical qualifications of the men and the technical requirements of the program, as judged by Charles Donlan and his team members. "We looked for real men and valuable experience," said Donlan, and he pressed Gilruth to select the epitome of American masculinity.[66] Gilruth finally decided to select seven. The seven men became heroes in the eyes of the American public almost immediately, in part due to a deal they made with *Life* magazine for exclusive rights to their stories, and errantly became the personification of NASA to most Americans.[67]

---

64 Quoted in Swenson, Grimwood, and Alexander, *This New Ocean*, p. 161.

65 Although depicted as comic relief in the film version of *The Right Stuff* (1982), the battery of physiological tests were the most sophisticated designed up to that point in time. On these examinations, see W. Randall Lovelace II, "Duckings, Probings, Checks That Proved Fliers' Fitness," *Life* (20 April 1959); Mae Mills Link, *Space Medicine in Project Mercury* (Washington, DC: NASA SP-4003, 1965); and John A. Pitts, *The Human Factor: Biomedicine in the Manned Space Program to 1980* (Washington, DC: NASA SP-4213, 1985).

66 Quoted in Swenson, Grimwood, and Alexander, *This New Ocean*, p. 163. The masculine ideal has been analyzed in Susan Faludi, *Stiffed: The Betrayal of the American Male* (New York: HarperCollins, 1999), pp. 451–468.

67 See Tom Wolfe, "The Last American Hero," in *The Kandy-Kolored Tangerine-Flake Streamline Baby*, ed. Tom Wolfe (New York: Farrar, Straus, and Giroux, 1965); Atkinson and Shafritz, *The Real Stuff*, pp. 8–12; James L. Kauffman, *Selling Outer Space: Kennedy, the Media, and Funding for Project Apollo, 1961–1963* (Tuscaloosa: University of Alabama Press, 1994), pp. 68–92; and Mark E. Byrnes, *Politics and Space: Image Making by NASA* (Westport, CT: Praeger, 1994), pp. 25–46.

Despite the wishes of Gilruth and others within the NASA leadership, the fame of the astronauts quickly grew beyond all proportion to their assignments. Perhaps it was inevitable that the astronauts were destined for premature adulation considering the enormous public curiosity about them, the risk they took in spaceflight, and their exotic training activities, but the power of commercial competition for publicity and the pressure for political prestige in the space race also whetted an insatiable public appetite for this new kind of celebrity. Walt Bonney, a public information officer, foresaw the public and press attention, asked for an enlarged staff, and laid the guidelines for public affairs policy in close accord with that of other government agencies.[68]

Bonney's foresight proved itself in 1959 only a week before the cherry blossoms bloomed along the tidal basin in Washington, DC, drenching the city with spectacular spring colors. NASA had chosen to unveil the first Americans to fly in space on 9 April 1959. Excitement bristled in Washington at the prospect of learning who those space travelers might be. Surely they were the best the nation had to offer, modern versions of medieval knights of the round table whose honor and virtue was beyond reproach. Certainly, they carried on their shoulders all of the hopes and dreams and best wishes of a nation as they engaged in single combat with the ominous specter of communism. The fundamental purpose of Project Mercury was to determine whether or not humans could survive the rigors of liftoff and orbit in the harsh environment of space. From this perspective, and it was the central one for men like Gilruth, the astronauts were not comparable to earlier explorers who directed their own exploits. Comparisons between them and Christopher Columbus, Admiral Richard Byrd, and Sir Edmund Hillary left the astronauts standing in the shadows.[69]

At the same time, Gilruth had enormous respect for the astronauts. He was genuinely impressed with all of them and enjoyed working with them. These individuals, he realized, embodied the deepest virtues of the United States. They strode the Earth as latter-day saviors whose purity coupled with noble deeds would purge this land of the evils of communism by besting the Soviet Union on the world stage. John Glenn, perhaps intu-

68 Walter T. Bonney (1909–1975) was NASA's first Director of the Office of Public Information (1958–1960). From 1949 to 1958, he had worked for the National Advisory Committee for Aeronautics. For more information on Bonney, see Walter T. Bonney Biographical File, NASA Headquarters Historical Reference Collection, Washington, DC.

69 On this dynamic, see Roger D. Launius, "Project Apollo in American Memory and Myth," in *Space 2000: Proceedings of the Seventh International Conference and Exposition on Engineering, Construction, Operations, and Business in Space*, ed. Stewart W. Johnson, Koon Meng Chua, Rodney G. Galloway, and Philip J. Richter (Reston, VA: American Society of Civil Engineers, 2000), pp. 1–13; Harvey Brooks, "Motivations for the Space Program: Past and Future," in *The First 25 Years in Space: A Symposium*, ed. Allan A. Needell (Washington, DC: Smithsonian Institution Press, 1983), pp. 3–26; Perry Miller, "The Responsibility of a Mind in a Civilization of Machines" *The American Scholar* 31 (Winter 1961–1962): 51–69; and Thomas Park Hughes, *American Genesis: A Century of Invention and Technological Enthusiasm, 1870–1970* (New York: Viking, 1989), p. 2.

itively or perhaps through sheer zest and innocence, understood this better than most others and delivered on numerous occasions ringing sermons on God, country, and family that melted the souls of all who heard him. He called upon the memory of how Wilbur and Orville Wright flipped a coin at Kitty Hawk in 1903 to see who would fly the first airplane and how far we had come since. "I think we would be most remiss in our duty," he said at the press conference where the Mercury Seven were unveiled by NASA in 1959, "if we didn't make the fullest use of our talents in volunteering for something that is as important as this is to our country and to the world in general right now. This can mean an awful lot to this country, of course." Other astronauts proved just as eloquent and spoke of their sense of duty and destiny as the first Americans to fly in space. The astronauts emerged as noble champions who would carry the nation's manifest destiny beyond its shores and into space. James Reston of the *New York Times*, a newspaper with a history of pooh-poohing spaceflight going back to a criticism of Robert Goddard in 1920, exulted the astronaut team. He said he felt profoundly moved by the press conference, and even reading the transcript of it made one's heart beat a little faster and step a little livelier. "What made them so exciting," he wrote, "was not that they said anything new, but that they said all the old things with such fierce convictions . . . . They spoke of 'duty' and 'faith' and 'country' like Walt Whitman's pioneers . . . . This is a pretty cynical town, but nobody went away from these young men scoffing at their courage and idealism."[70]

The astronauts put a very human face on the grandest technological endeavor in history, and the myth of the virtuous astronaut was born at that moment in 1959. In some respects, it was a natural occurrence. The Mercury Seven were, as Gilruth perceived, surrogates for each of us. None were either aristocratic in bearing or elitist in sentiment. They came from everywhere in the nation, excelled in the public schools, trained at their local state university, served their country in war and peace, married and tried to make lives for themselves and their families, and ultimately rose to their places on the basis of merit. They represented the best we had to offer, and, most importantly, they expressed at every opportunity the virtues ensconced in the democratic principles of the republic.

The astronauts, of course, were the "main architects" of their image.[71] But they appeared at a time when NASA desperately needed to inspire public trust in its ability to

---

70 John H. Glenn, "A New Era: May God Grant Us the Wisdom and Guidance to Use It Wisely," *Vital Speeches of the Day* (15 March 1962): 324–326; Dora Jane Hamblin, "Applause, Tears, and Laughter and the Emotions of a Long-Ago Fourth of July," *Life* (9 March 1962): 34; and Roger D. Launius and Bertram Ulrich, *NASA and the Exploration of Space* (New York: Stewart, Tabori, and Chang, 1998), chapter 2.

71 Letter from Don A. Schanche to P. Michael Whye (28 December 1976), NASA Headquarters Historical Reference Collection, Washington, DC.

carry out the nation's goals in space. Rockets might explode, but the astronauts shined. The astronauts seemed to embody the personal qualities in which Americans of that era wanted to believe: bravery, honesty, love of God and country, and family devotion. How could anyone distrust a government agency epitomized by such people? The trust that the public placed in the astronauts spread through NASA and to the government as a whole. As one of the *Life* reporters summarized, "*Life* treated the men and their families with kid gloves. So did most of the rest of the press. These guys were heroes, most of them were very smooth, canny operators with all of the press. They felt that they had to live up to a public image of good, clean, all-American guys, and NASA knocked itself out to preserve that image."[72]

Gilruth understood that the astronauts were critical to the success of the program. Early on he made astronauts an important part of the organizational structure, inviting them into the inner councils of NASA and into the decision-making process. Gilruth recalled in 1987, "They certainly had every right to sit in and listen to things that were going on in the design of the spacecraft. They certainly had every right to make an input." In essence, Gilruth put the astronauts to work for him, co-opting them on behalf of his larger ideals.

Despite the success of motivating the public with Gilruth's astronauts, stubborn problems arose with Project Mercury at seemingly every turn. The first spaceflight of an astronaut, made by Alan B. Shepard, had been postponed for weeks so NASA engineers could resolve numerous details; it finally took place on 5 May 1961, less than three weeks before the Apollo announcement. The second flight, a suborbital mission like Shepard's, launched on 21 July 1961, had problems as well. The hatch blew off prematurely from the Mercury capsule, *Liberty Bell 7*, and it sank into the Atlantic Ocean before it could be recovered. In the process, astronaut "Gus" Grissom nearly drowned before being hoisted to safety in a helicopter. These suborbital flights, however, proved valuable for NASA technicians who found ways to solve or work around literally thousands of obstacles to successful spaceflight.[73]

As these issues were being resolved, NASA engineers began final preparations for the orbital aspects of Project Mercury. In this phase, NASA planned to use a Mercury capsule capable of supporting a human in space for not just minutes, but eventually for as much as three days. As a launch vehicle for this Mercury capsule, NASA used the more powerful Atlas instead of the Redstone, but this decision was not without controversy. There were technical difficulties to be overcome in mating it to the Mercury

---

72 Letter from Dora Jane Hamblin to P. Michael Whye (18 January 1977), NASA Headquarters Historical Reference Collection, Washington, DC.

73 Swenson, Grimwood, and Alexander, *This New Ocean*, pp. 341–379.

capsule, but the biggest complication was a debate among NASA engineers over its propriety for human spaceflight.[74]

When Atlas was first conceived in the 1950s, many believed that it was a high-risk proposition because to reduce its weight Convair engineers, under the direction of Karel J. Bossart, a pre-WWII immigrant from Belgium, designed the booster with a very thin, internally pressurized fuselage instead of massive struts and a thick metal skin. The "steel balloon," as it was sometimes called, employed engineering techniques that ran counter to a conservative engineering approach used by Wernher von Braun for the V-2 and the Redstone at Huntsville, Alabama.[75] Von Braun, according to Bossart, needlessly designed his boosters like "bridges" to withstand any possible shock. For his part, von Braun thought the Atlas was too flimsy to hold up during launch. He considered Bossart's approach much too dangerous for human spaceflight, remarking that the astronaut using the "contraption," as he called the Atlas booster, "should be getting a medal just for sitting on top of it before he takes off!"[76] The reservations began to melt away, however, when Bossart's team pressurized one of the boosters and dared one of von Braun's engineers to knock a hole in it with a sledge hammer. The blow left the booster unharmed, but the recoil from the hammer nearly clubbed the engineer.[77]

Most of the differences had been resolved by the first successful orbital flight of an unoccupied Mercury-Atlas combination in September 1961. On 29 November, the final test flight took place with the chimpanzee Enos occupying the capsule for a two-orbit ride before being successfully recovered in an ocean landing. Not until 20 February 1962, however, could NASA get ready for an orbital flight with an astronaut. On that date, John Glenn became the first American to circle Earth, making three orbits in his *Friendship 7* Mercury spacecraft. The flight was not without problems, however; Glenn flew parts of the last two orbits manually because of an autopilot failure and left his normally jettisoned retrorocket pack attached to his capsule during reentry because of a loose heatshield.

Glenn's flight provided a healthy increase in national pride, making up for at least some of the earlier Soviet successes. The public, more than celebrating the technological success, embraced Glenn as a personification of heroism and dignity. Hundreds of requests for personal appearances by Glenn poured into NASA Headquarters, and NASA

74 Wernher von Braun, "The Redstone, Jupiter, and Juno," in *The History of Rocket Technology: Essays on Research, Development, and Utility*, ed. Eugene M. Emme (Detroit: Wayne State University Press, 1964), pp. 107–122.

75 Richard E. Martin, *The Atlas and Centaur "Steel Balloon" Tanks: A Legacy of Karel Bossart* (San Diego, CA: General Dynamics Space Systems Division, 1989).

76 Interview with Karel J. Bossart by John L. Sloop (27 April 1974), quoted in John L. Sloop, *Liquid Hydrogen as a Propulsion Fuel, 1945–1959* (Washington, DC: NASA SP-4404, 1978), pp. 176–177.

77 Martin, *The Atlas and Centaur*, p. 5.

learned much about the power the astronauts had in swaying public opinion. The NASA leadership made Glenn available to speak at some events, but he often substituted other astronauts and declined many other invitations. Among other engagements, Glenn did address a joint session of Congress and participated in several ticker-tape parades around the country. NASA discovered in the process of this hoopla a powerful public relations tool that it has employed ever since.[78]

Three more successful Mercury flights took place during 1962 and 1963. Scott Carpenter made three orbits on 20 May 1962, and, on 3 October 1962, Walter Schirra flew six orbits. The capstone of Project Mercury was the 15–16 May 1963 flight of Gordon Cooper, who circled Earth 22 times in 34 hours. The program had succeeded in accomplishing its purpose: to successfully orbit a human in space, explore aspects of tracking and control, and learn about microgravity and other biomedical issues associated with spaceflight.[79]

MANAGING PROJECT APOLLO

In May 1961, Robert Gilruth's life changed forever. After President John F. Kennedy announced the decision on 25 May that the United States would land an American on the Moon by the end of the decade, Apollo consumed NASA's every effort. It required significant expenditures, costing $24.5 billion in 1960s dollars over the life of the program (more than $110 billion in 2004 dollars) to make it a reality. Only the building of the Panama Canal rivaled the Apollo program's size as the largest nonmilitary technological endeavor ever undertaken; only the Manhattan Project was comparable in a wartime setting. Even NASA leaders expressed concern that it might prove too daunting a challenge. When Kennedy made his speech, Gilruth was flying to a meeting in Tulsa. He recalled that he was "aghast" at the lunar landing goal and what it would portend for the future. After all, he reasoned, his organization now had to accomplish it. Rising to the challenge, project participants exhibited single-minded devotion to it for a decade.[80]

In 1986, Gilruth talked about the decision-making process leading up to Kennedy's speech on 25 May. He commented on the intense technical and political review and how government officials reached closure on the initiative. He recalled:

---

78 Swenson, Grimwood, and Alexander, *This New Ocean*, pp. 422–436.

79 Ibid., pp. 446–503.

80 By far the most influential study making this case is the seminal work of John M. Logsdon, *The Decision to Go to the Moon: The Space Program and the National Interest* (Cambridge, MA: MIT Press, 1970). There are several ways to view the decision. These have been analyzed in Stephen J. Garber, "Multiple Means to an End: A Reexamination of President Kennedy's Decision to Go to the Moon," *Quest: The History of Spaceflight Quarterly* 7 (Summer 1999): 5–17.

I told President Kennedy when he said, "I want to go to the Moon," I said, "Well, that's very hard to do." "But," I said, "I don't know that you can't." So that was fair and square. I didn't know that you couldn't. And it turned out, it was pretty straightforward. But how we ever did it, and all those things worked, with all those single-point failures in the sequence—there were some people who wanted to keep on flying those things, you know. A lot more of them—I said "Not me, you get another boy. You'll have to get another guy to handle it. You'll have to get another boy because I'm not going to stay around for it if you're going to keep doing it."[81]

President John F. Kennedy presents Dr. Robert R. Gilruth with the Medal for Distinguished Federal Civil Service. The ceremony took place on the White House lawn. In attendance were (foreground, left to right): astronauts Alan Shepard and John Glenn, Gilruth, NASA Administrator James Webb, and President John F. Kennedy. Photo taken on 1 August 1962. (GRIN database number GPN-2000-001681)

Gilruth agreed to take on the responsibility for managing the human element of the program, reluctant though he may have been.

Of course, Gilruth fully supported President Kennedy's decision, recognizing that he had correctly gauged the mood of the nation. This commitment captured the American imagination and enjoyed strong support during the days and weeks following the announcement. No one seemed concerned either about the difficulty or about the expense at the time. Congressional debate was perfunctory, and NASA found itself literally pressing to expend the funds committed to it during the early 1960s. Like most political decisions, at least in the U.S. experience, the decision to carry out Project Apollo was an effort to deal with an unsatisfactory situation (world perception of Soviet leadership in space and technology). As such, Apollo was a remedial action

81 Gilruth OHI No. 6 by DeVorkin and Mauer.

ministering to a variety of political and emotional needs floating in the ether of world opinion. Apollo addressed these problems very well and was a worthwhile action if measured only in those terms. In announcing Project Apollo, Kennedy put the world on notice that the United States would not take a back seat to its superpower rival. John Logsdon commented, "By entering the race with such a visible and dramatic commitment, the United States effectively undercut Soviet space spectaculars without doing much except announcing its intention to join the contest."[82] It was an effective symbol, just as Kennedy had intended.

Without question, Kennedy gave Gilruth an opportunity to shine by approving Apollo. He fully recognized that the lunar landing was so far beyond the capabilities of either the United States or the Soviet Union in 1961 that the early lead in space activities taken by the Soviets would not predetermine the outcome. It gave the United States a reasonable chance of overtaking the Soviet Union in space activities and recovering a measure of lost status. Gilruth recalled telling Kennedy the following:

> Well, you've got to pick a job that's so difficult that it's new, that they'll have to start from scratch. They just can't take their old rocket and put another gimmick on it and do something we can't do. It's got to be something that requires a great big rocket, like going to the Moon. Going to the Moon will take new rockets, new technology, and, if you want to do that, I think our country could probably win because we'd both have to start from scratch.[83]

Even though Kennedy's political objectives were essentially achieved with the decision to go to the Moon, there were other aspects of the Apollo commitment that required assessment. Those who wanted to see a vigorous space program, a group led by NASA scientists and engineers, obtained their wish with Kennedy's announcement. An opening was present to this group in 1961 that had not existed at any time during the Eisenhower administration, and they made the most of it. They inserted into the overall package supporting Apollo programs that they believed would greatly strengthen the scientific and technological return on the investment to go to the Moon. In addition to seeking international prestige, this group proposed an accelerated and integrated national space effort incorporating both scientific and commercial components.

The first challenge Gilruth and other NASA leaders faced in meeting the presidential mandate was securing funding. While Congress enthusiastically appropriated funding for Apollo immediately after the President's announcement, NASA leaders rightly ques-

82 John M. Logsdon, "An Apollo Perspective," *Astronautics & Aeronautics* (December 1979): 112–117.

83 Quoted in Glen E. Swanson, ed., *"Before This Decade is Out . . ." Personal Reflections on the Apollo Program* (Washington, DC: NASA SP-4223, 1999), p. 68.

tioned if the momentary sense of crisis would subside and if the political consensus present for Apollo in 1961 would abate. They worked, albeit without much success, to lock the presidency and Congress into a long-term obligation to support the program. While they had made an intellectual commitment, NASA's leadership was concerned that they might renege on the economic part of the bargain at some future date. But they did err on the side of caution. While Apollo never enjoyed unlimited funding, there was always enough for the program because of this strategy. Additionally, after the assassination of Kennedy in November 1963, Gilruth and other NASA leaders used the slain President's commitment to the program as a means of convincing—some would say shaming—Congress into continuing to support the program with considerable public resources. Accordingly, the space agency's annual budget increased from $500 million in 1960 to a high point of $5.2 billion in 1965. The NASA funding level represented 3.3 percent of the federal budget in 1965.[84]

Out of the budgets appropriated for NASA each year, approximately 50 percent went directly to human spaceflight, most of it directly under the control of Gilruth and his leadership team at the Manned Spacecraft Center in Houston. For 11 years after Kennedy's Apollo decision, through the flight of Apollo 17 in December 1972, Robert Gilruth politicked, coaxed, cajoled, and maneuvered for the program. After Kennedy's assassination in 1963, moreover, he often appealed for continued political support for Apollo because it represented a fitting tribute to the fallen leader. In the end, through a variety of methods, he and Administrator James Webb built a seamless web of political liaisons that brought continued support for and resources to accomplish the Apollo Moon landing on the schedule Kennedy had announced.

In the immediate aftermath of the Apollo decision, NASA created the Manned Spacecraft Center (renamed the Lyndon B. Johnson Space Center in 1973) near Houston, Texas. It moved Gilruth's team from Hampton, Virginia, where Gilruth had lived since the latter 1930s, to this new facility. Gilruth hated that prospect. He loved the Virginia peninsula and had become an avid sailor. He had built a sailboat in his basement—which some of his underlings thought silly since he had to disassemble it to get it out of his basement—and was determined to sail it around the continent to Galveston. He took several weeks out of his work in 1962 to accomplish this voyage, traveling down the intercoastal waterway around Florida and across the Gulf. Later he built the first successful sailing hydrofoil system and participated in many hydrofoil projects. Upon reaching Houston, Gilruth set his team to work not only in settling into their new facility, but also in completing the design and development of the Apollo spacecraft and the launch plat-

84 As an example, see the 1963 letter in defense of Apollo by Vice President Lyndon B. Johnson to the President (13 May 1963), with attached report, John F. Kennedy Presidential Files, NASA Headquarters Historical Reference Collection, Washington, DC; Ezell, *NASA Historical Data Book*, pp. 122–123; and *Aeronautics and Space Report of the President, 1988 Activities* (Washington, DC: NASA Annual Report, 1990), p. 185.

form for the lunar lander. His Center also became the home of NASA's astronauts and the site of Mission Control.[85] The cost of the expansion of NASA's facilities, not only for the Manned Spacecraft Center, but also for other Apollo infrastructure, was great—more than $2.2 billion over the decade, with 90 percent of it expended before 1966.

Within its first few months in Houston, said Gilruth in June 1962, "the Manned Spacecraft Center has doubled in size; accomplished a major relocation of facilities and personnel; pushed ahead in two new major programs; and accomplished Project Mercury's design goal of manned orbital flights twice with highly gratifying results."[86]

The mobilization of resources was not the only challenge facing those charged with meeting President Kennedy's goal. NASA had to meld disparate institutional cultures and approaches into an inclusive organization moving along a single unified path. Each NASA installation, university, contractor, and research facility had differing perspectives on how to go about the task of accomplishing Apollo.[87] To bring a semblance of order to the program, Gilruth employed a systems management concept borrowed from the military/industrial complex to oversee the Apollo capsule development effort. One of the fundamental tenets of the program management concept was that three critical factors—cost, schedule, and reliability—were interrelated and had to be managed as a group. Many also recognized these factors' constancy; if program managers held cost to a specific level, then one of the other two factors, or both of them to a somewhat lesser degree, would be adversely affected. This held true for the Apollo program. The schedule, dictated by the President, was firm. Since humans were involved in the flights, and since the President had directed that the lunar landing be conducted safely, the program managers placed a heavy emphasis on reliability. Accordingly, Apollo used redundant systems extensively so that failures would be both predictable and minor in result. The significance of both of these factors forced the third factor, cost, much higher than might have been the case with a more leisurely lunar program such as had been conceptualized in the latter 1950s. As it was, this was the price paid for success under the Kennedy mandate, and program managers made conscious decisions based on the knowledge of these factors.[88]

---

85 On the creation of this center see, Henry C. Dethloff, *"Suddenly Tomorrow Came . . .": A History of the Johnson Space Center* (Washington, DC: NASA SP-4307).

86 *Space News Roundup* (11 July 1962), Lyndon B. Johnson Space Center, Houston, TX.

87 On the NASA organizational culture and Apollo, see Howard E. McCurdy, *Inside NASA: High Technology and Organizational Change in the U.S. Space Program* (Baltimore, MD: Johns Hopkins University Press, 1993); John M. Logsdon, moderator, *Managing the Moon Program: Lessons Learned from Project Apollo* (Washington, DC: NASA, Monographs in Aerospace History No. 14, 1999); and Stephen B. Johnson, *The Secret of Apollo: Systems Management in American and European Space Programs* (Baltimore, MD: Johns Hopkins University Press, 2002).

88 Aaron Cohen, "Project Management: JSC's Heritage and Challenge," in *Issues in NASA Program and Project Management*, ed. Francis T. Hoban (Washington, DC: NASA SP-6101, 1989), pp. 7–16; C. Thomas Newman, "Controlling Resources in the Apollo Program," in *Issues in NASA Program*, pp. 23–26; and Eberhard Rees, "Project and Systems Management in the Apollo Program," in *Issues in NASA Program*, issue 2, pp. 24–34. For a full explication of the approach taken, see Johnson, *Secret of Apollo*.

Gilruth oversaw every aspect of this effort under his domain. He had excellent people working for him, and they read like a who's who of space history. Chris Kraft, Gene Kranz, John Aaron, and Glynn Lunney served in Mission Control. Joseph Shea and later George Low oversaw the Apollo spacecraft development effort. Aaron Cohen, Barouk el-Faz, Max Faget, Wendell Mendell, and others trained astronauts, developed subsystems, and worked scientific aspects of the program. These were the people who made the dreams of spaceflight real. There is a moving ballad in the filk (science-fiction folk song) community that captures the essence of this group. Written by Mary Jean Holmes in 1992, "Everyman" begins with the lament that these individuals will never leave the ground, but they enabled the astronauts to do so. The chorus states:

> For I'm the man who took up tools and laid out the designs.
> Of starships, I'm the one who built their sleek and burnished lines.
> I'm everyman who ever fashioned cold refined steel.
> Into the dreams of spaceflight, I'm the one who made them real.[89]

The program management concept was recognized as a critical component of Project Apollo's success in November 1968, when *Science* magazine, the publication of the American Association for the Advancement of Science, observed:

> In terms of numbers of dollars or of men, NASA has not been our largest national undertaking, but in terms of complexity, rate of growth, and technological sophistication, it has been unique . . . . It may turn out that [the space program's] most valuable spin-off of all will be human rather than technological: better knowledge of how to plan, coordinate, and monitor the multitudinous and varied activities of the organizations required to accomplish great social undertakings.[90]

Understanding the management of complex structures for the successful completion of a multifarious task was an important outgrowth of the Apollo effort.

Gilruth's organization orchestrated more than 200 contractors working on both large and small aspects of Apollo. These prime contractors, with more than 150 subcontractors, provided millions of parts and components for use in the Apollo spacecraft, all meeting exacting specifications for performance and reliability. Getting all of the personnel elements to work together challenged the program managers, regardless of whether or not

89 Mary Jean Holmes's "Everyman" lyrics are available online at *http://216.109.117.135/search/cache?va=filk+everyman&ei=UTF-8&n=20&fl=0&u=www.speculations.com/rumormill/index.php%3Fshow_all_topics%3D0%26t%3D211&w=filk+everyman&d=E96872EAB2&c=482&yc=6089&icp=1* (accessed 3 April 2004).

90 Dael Wolfe, Executive Officer, American Association for the Advancement of Science, editorial for *Science* (15 November 1968).

they were civil service, industry, or university personnel. There were various communities within NASA that differed over priorities and competed for resources. The two most identifiable groups were the engineers and the scientists. As ideal types, engineers usually worked in teams to build hardware that could carry out the missions necessary for a successful Moon landing by the end of the decade. Their primary goal involved building vehicles that would function reliably within the fiscal resources allocated to Apollo. Again as ideal types, space scientists engaged in pure research and were more concerned with designing experiments that would expand scientific knowledge about the Moon. They also tended to be individualists, unaccustomed to regimentation and unwilling to concede gladly the direction of projects to outside entities. The two groups contended with each other over a great variety of issues associated with Apollo. For instance, the scientists disliked having to configure payloads so that they could meet time, money, or launch vehicle constraints. The engineers, likewise, resented changes to scientific packages added after project definition because these threw their hardware efforts out of kilter. Both had valid complaints and had to maintain an uneasy cooperation to accomplish Project Apollo.[91]

The scientific and engineering communities within NASA, additionally, were not monolithic, and differences among them thrived. Add to these groups representatives from industry, universities, and research facilities, and competition on all levels to further their own scientific and technical areas was the result. The NASA leadership generally viewed this pluralism as a positive force within the space program, for it ensured that all sides aired their views and emphasized the honing of positions to a fine edge. Competition, most people concluded, made for a more precise and viable space exploration effort. There were winners and losers in this strife, however, and sometimes ill-will was harbored for years. Moreover, if the conflict became too great and spilled into areas where it was misunderstood, it could be devastating to the conduct of the lunar program. The head of the Apollo program worked hard to keep these factors balanced and to promote order so that NASA could accomplish the presidential directive.[92]

BRIDGING THE TECHNOLOGICAL GAP: FROM GEMINI TO APOLLO

Even as the Mercury program was underway and work took place developing Apollo hardware, Gilruth and his colleagues in the NASA leadership perceived a huge gap in the capability for human spaceflight between that acquired with Mercury and what would

91 To discipline the system, Gilruth had a sophisticated system of information flow to enable all to keep informed on the status of the program. As an example, every couple of days George M. Low sent him "Apollo Notes for Dr. Gilruth" (1 April 1967–5 November 1969), located in the Robert R. Gilruth Papers, Virginia Tech Special Collections, Blacksburg, VA.

92 McCurdy, *Inside NASA*, pp. 11–98.

be required for a lunar landing. They closed most of the gap by experimenting and training on the ground, but some issues required experience in space. Three major areas immediately arose where this was the case. The first was the ability in space to locate, maneuver toward, and rendezvous and dock with another spacecraft. The second was closely related—the ability of astronauts to work outside a spacecraft. The third involved the collection of more sophisticated physiological data about the human response to extended spaceflight.[93]

To gain experience in these areas before Apollo could be readied for flight, NASA devised Project Gemini. Hatched in the fall of 1961 by engineers at Gilruth's Space Task Group in cooperation with McDonnell Aircraft Corp. technicians, builders of the Mercury spacecraft, Gemini started as a larger Mercury Mark II capsule; it soon became a totally different proposition. It could accommodate two astronauts for extended flights of more than two weeks. It pioneered the use of fuel cells instead of batteries to power the ship and incorporated a series of modifications to hardware. Its designers also toyed with the possibility of using a paraglider being developed at Langley Research Center for "dry" landings instead of a "splashdown" in water and recovery by the Navy. The whole system was to be powered by the newly developed Titan II launch vehicle, another ballistic missile developed for the Air Force. A central reason for this program was to perfect techniques for rendezvous and docking, so NASA appropriated from the military some Agena rocket upper stages and fitted them with docking adapters.

Problems with the Gemini program abounded from the start. The Titan II had longitudinal oscillations, called the "pogo" effect because it resembled the behavior of a child on a pogo stick. Overcoming this problem required engineering imagination and long hours of overtime to stabilize fuel flow and maintain vehicle control. The fuel cells leaked and had to be redesigned, and the Agena reconfiguration also suffered costly delays. NASA engineers never did get the paraglider to work properly and eventually dropped it from the program in favor of a parachute system like the one used for Mercury. All of these difficulties shot an estimated $350-million program to over $1 billion. The overruns were successfully justified by the space agency, however, as necessities to meet the Apollo landing commitment.[94]

By the end of 1963, most of the difficulties with Gemini had been resolved, albeit at great expense, and the program was ready for flight. Following two unoccupied orbital test flights, the first operational mission took place on 23 March 1965. Mercury astronaut

93 Barton C. Hacker, "The Idea of Rendezvous: From Space Station to Orbital Operations in Space-Travel Thought, 1895–1951," *Technology and Culture* 15 (July 1974): 373–388; Barton C. Hacker, "The Genesis of Project Apollo: The Idea of Rendezvous, 1929–1961," in *Actes 10: Historic des techniques* (Paris: Congress of the History of Science, 1971), pp. 41–46; and Barton C. Hacker and James M. Grimwood, *On the Shoulders of Titans: A History of Project Gemini* (Washington, DC: NASA SP-4203, 1977), pp. 1–26.

94 James M. Grimwood and Ivan D. Ertal, "Project Gemini," *Southwestern Historical Quarterly* 81 (January 1968): 393–418; James M. Grimwood, Barton C. Hacker, and Peter J. Vorzimmer, *Project Gemini Technology and Operations* (Washington, DC: NASA SP-4002, 1969); and Robert N. Lindley, "Discussing Gemini: A 'Flight' Interview with Robert Lindley of McDonnell," *Flight International* (24 March 1966): 488–489.

Grissom commanded the mission, with John W. Young, a Naval aviator chosen as an astronaut in 1962, accompanying him. The next mission, flown in June 1965, stayed aloft for four days, and astronaut Edward H. White II performed the first extravehicular activity (EVA) or spacewalk. Eight more missions followed through November 1966. Despite problems great and small encountered on virtually all of them, the program achieved its goals. Additionally, as a technological learning program, Gemini had been a success, with 52 different experiments performed on the 10 missions. The bank of data acquired from Gemini helped to bridge the gap between Mercury and what would be required to complete Apollo within the time constraints directed by the President. Gilruth always believed that Gemini represented the fundamental point at which NASA demonstrated its abilities to surpass the Soviets in human spaceflight.[95]

THE APOLLO SPACECRAFT

Gilruth had his Space Task Group working to develop a spacecraft capable of taking astronauts to the Moon even before the announcement by JFK. What they came up with was a three-person command module capable of sustaining human life for two weeks or more in either Earth orbit or in a lunar trajectory; a service module holding oxygen, fuel, maneuvering rockets, fuel cells, and other expendable and life-support equipment that could be jettisoned upon reentry to Earth; a retrorocket package attached to the service module for slowing to prepare for reentry; and finally a launch escape system that was discarded upon achieving orbit. The teardrop-shaped command module had two hatches—one on the side for entry and exit of the crew at the beginning and end of the flight, and one in the nose with a docking collar for use in moving to and from the lunar landing vehicle.[96]

Production work on the Apollo spacecraft stretched from 28 November 1961, when the prime contract for its development was let to North American Aviation, to 22 October 1968, when the last test flight took place. In between there were various efforts to design, build, and test the spacecraft both on the ground and in suborbital and orbital flights. For instance, on 13 May 1964, Gilruth's team tested a boilerplate model of the

95 Reginald M. Machell, ed., *Summary of Gemini Extravehicular Activity* (Washington, DC: NASA SP-149, 1968); *Gemini Summary Conference* (Washington, DC: NASA SP-138, 1967); Ezell, *NASA Historical Data Book*, pp. 149–170; and Gilruth OHI No. 6 by DeVorkin and Mauer.

96 A lengthy discussion of the development of the Apollo spacecraft can be found in Ivan D. Ertal and Mary Louise Morse, *The Apollo Spacecraft: A Chronology, Volume I, Through November 7, 1962* (Washington, DC: NASA SP-4009, 1969); Mary Louise Morse and Jean Kernahan Bays, *The Apollo Spacecraft: A Chronology, Volume II, November 8, 1962–September 30, 1964* (Washington, DC: NASA SP-4009, 1973); Courtney G. Brooks and Ivan D. Ertal, *The Apollo Spacecraft: A Chronology, Volume III, October 1, 1964–January 20, 1966* (Washington, DC: NASA SP-4009, 1973); and Ivan D. Ertal and Roland W. Newkirk, with Courtney G. Brooks, *The Apollo Spacecraft: A Chronology, Volume IV, January 21, 1966–July 13, 1974* (Washington, DC: NASA SP-4009, 1978). A short developmental history is in Ezell, *NASA Historical Data Book*, pp. 171–185.

Apollo capsule atop a stubby Little Joe II military booster, and another Apollo capsule actually achieved orbit on 18 September 1964 when it was launched atop a Saturn I. By the end of 1966, NASA leaders declared the Apollo command module ready for human occupancy. The final flight checkout of the spacecraft prior to the lunar flight took place on 11–22 October 1968 with three astronauts.[97]

As these development activities were taking place, tragedy struck the Apollo program. On 27 January 1967, Apollo-Saturn (AS) 204, scheduled to be the first spaceflight with astronauts aboard the capsule, was on the launch pad at Kennedy Space Center, Florida moving through simulation tests. The three astronauts to fly on this mission—"Gus" Grissom, Edward White, and Roger B. Chaffee—were aboard running through a mock launch sequence. At 6:31 p.m., after several hours of work, a fire broke out in the spacecraft, and the pure oxygen atmosphere intended for the flight helped it burn with intensity. In a flash, flames engulfed the capsule, and the astronauts died of asphyxiation. It took the ground crew 5 minutes to open the hatch. When they did so, they found three bodies. Although three other astronauts had been killed before this time—all in plane crashes—these were the first deaths directly attributable to the U.S. space program.[98]

Shock gripped NASA and the nation during the days that followed. As the nation mourned, NASA appointed an eight-member investigation board, chaired by longtime NASA official and Director of the Langley Research Center, Floyd L. Thompson. It set out to discover the details of the tragedy: what happened, why it happened, could it happen again, what was at fault, and how could NASA recover? The members of the board learned that the fire had been caused by a short circuit in the electrical system that ignited combustible materials in the spacecraft fed by the oxygen atmosphere. They also found that it could have been prevented and called for several modifications to the spacecraft, including a move to a less oxygen-rich environment. Changes to the capsule followed quickly, thanks to the efforts of a dedicated team of engineers under Gilruth's and others' direction. Within a little more than a year, it was ready for flight.[99]

---

97 Ezell, *NASA Historical Data Book,* pp. 182–185.

98 On this subject, see "The Ten Desperate Minutes," *Life* (21 April 1967): 113–114; Erik Bergaust, *Murder on Pad 34* (New York: G. P. Putnam's Sons, 1968); Mike Gray, *Angle of Attack: Harrison Storms and the Race to the Moon* (New York: W. W. Norton and Co., 1992); Erlend A. Kennan and Edmund H. Harvey, Jr., *Mission to the Moon: A Critical Examination of NASA and the Space Program* (New York: William Morrow and Co., 1969); Hugo Young, Bryan Silcock, and Peter Dunn, *Journey to Tranquillity: The History of Man's Assault on the Moon* (Garden City, NY: Doubleday, 1970); and Brooks, Grimwood, and Swenson, *Chariots for Apollo,* pp. 213–236.

99 United States House, Committee on Science and Astronautics, Subcommittee on NASA Oversight, *Investigation into Apollo 204 Accident, Hearings, Ninetieth Congress, First Session* (Washington, DC: Government Printing Office, 1967); United States House, Committee on Science and Astronautics, *Apollo Program Pace and Progress; Staff Study for the Subcommittee on NASA Oversight, Ninetieth Congress, First Session* (Washington, DC: Government Printing Office, 1967); United States House, Committee on Science and Aeronautics, *Apollo and Apollo Applications: Staff Study for the Subcommittee on NASA Oversight of the Committee on Science and Astronautics, U.S. House of Representatives, Ninetieth Congress, Second Session* (Washington, DC: Government Printing Office, 1968); and Robert C. Seamans, Jr., and Frederick I. Ordway III, "Lessons of Apollo for Large-Scale Technology," in *Between Sputnik and the Shuttle: New Perspectives on American Astronautics,* ed. Frederick C. Durant III (San Diego: Univelt, 1981), pp. 241–287.

GODFATHER TO THE ASTRONAUTS

The Apollo fire shook Gilruth personally, for he had served since the beginning of the Mercury program as leader and mentor to the young men selected to fly the Apollo missions, whether they set foot on the Moon or performed a less visible but no less significant role. Gilruth included the astronauts in the decision-making process at the Manned Spacecraft Center, not only for crew systems, but also for the larger Apollo technical requirements. Gilruth held weekly meetings with his senior staff, including the astronaut corps. "We'd spend the morning talking about all of our problems," he recalled, "and this was a pretty effective thing to do. Sometimes it would maybe take 1 or 2 hours, and sometimes it would take all morning." Gilruth added, "It was obvious they should report to the Director's office, to me. They shouldn't report to the people designing the spacecraft, because if they disliked something the spacecraft designers were doing they ought to be able to bitch about it."[100] Gilruth had enormous respect for them, sometimes calling them his boys. This seems fully in keeping with this philosophy of leadership, with an older visionary overseeing the stupendous accomplishments of young and virile heroes. It is one of the most powerful conceptions in myth and human history—Merlin/King Arthur, Obi-Wan Kenobi/Luke Skywalker, Gandalf/Frodo Baggins, Lincoln/Union Army, FDR/G.I. generation—with the greybeard prophet teaching and motivating the young civic-minded heroes who accomplish great tasks under that guidance. That certainly took place in the context of the relationship between the older Gilruth and the young astronauts who went to the Moon. Gilruth was a prophet, possessing vision, values, and ideals concerning a future for America in space. As heroes, the astronauts possessed community, affluence, and technology. They made a powerful team, accomplishing the task required. NASA eventually landed six sets of astronauts on the Moon between 1969 and 1972. The first landing mission, Apollo 11, succeeded on 20 July 1969 when astronaut Neil Armstrong first set foot on the lunar surface, telling millions of listeners that it was "one small step for [a] man—one giant leap for mankind." Five more landing missions followed Apollo 11 at approximately six-month intervals through December 1972.[101]

The astronauts were in too many instances rambunctious boys, as Gilruth called them. They roughhoused and drank and drove fast and got into sexual peccadilloes. Rumors swirled around several of the Apollo astronauts, especially Gus Grissom, whom Gilruth considered a consummate professional in the cockpit and an incorrigible adolescent whenever off duty. Everyone laughed, including Gilruth, when Grissom said, "There's a certain

100 Gilruth OHI No. 6 by DeVorkin and Mauer.

101 For a stimulating discussion of this dynamic in American history, see William Strauss and Neil Howe, *The Fourth Turning* (New York: Broadway Books, 1996); and William Strauss and Neil Howe, *Generations: The History of America's Future, 1584 to 2069* (New York: William Morrow, 1992).

Dr. Robert R. Gilruth's official portrait while he was serving as Director of the Manned Spacecraft Center in Houston, TX. 1965. (Center number S87-26820)

kind of small black fly that hatches in the spring around the Space Center south of Houston. Swarms of the bugs can splatter [on] windshields, but their real distinction is that male and female catch each other in midair and fly along happily mated." Grissom told a *Life* magazine reporter that he envied those insects. "They do the two things I like best in life," he said, "flying and fucking—and they do them at the same time." For years thereafter, the insects were known as "Grissom Bugs" to local residents.[102] Several memoirs have recounted these and other anecdotes of the astronauts, many of which are the stuff of legend. It should come as no surprise to anyone that many astronauts had a wild, devil-may-care side to their personalities, certainly Gilruth understood it.[103]

Sometimes the astronauts caused Gilruth grief, and he could rule with an authoritarian hand. More often, however, he was benevolent and patriarchal toward the astronauts. Often this had to do with what rules they needed to follow and the lack of well-understood guidelines for their ethical conduct. For example, when the Space Task Group moved to Houston in 1962, several local developers offered the astronauts free houses. This caused a furor that reached the White House and prompted the involvement of Vice President Lyndon B. Johnson. In this case, Gilruth had to disallow an outright gift to the astronauts.[104] Gilruth's boys also got into trouble over what they could and could not do to make additional money on the outside. NASA had facilitated the Mercury Seven in selling their stories to *Life* magazine. This had raised a furor, and NASA policies were changed thereafter. However, in 1963 Forrest Moore complained to LBJ that the new astronauts were seeking to do essentially the same thing. Gilruth had to intervene and explain that any deals for "personal stories" would be worked through the NASA General Counsel and would only take place in a completely open and legal manner.[105] Gilruth also defended the astronauts to the NASA leadership when they accepted tickets to see the Houston Astros season opener baseball game in the

102 James Schefter, *The Race: The Uncensored Story of How America Beat Russia to the Moon* (Garden City, NY: Doubleday and Co., 1999), p. 72. LBJ also had great confidence in Grissom. See the letter from Lyndon B. Johnson to Gus Grissom (21 July 1961), LBJ Papers, Vice Presidential Papers, Box 116, January–July 1961, Space and Aero File, LBJ Library, University of Texas, Austin, TX.

103 See Guenter Wendt and Russell Still, *The Unbroken Chain* (Burlington, Ontario: Apogee Books, an imprint of Collector's Guide Publishing Ltd., 2001).

104 Letter from Edward Welsh to Lyndon B. Johnson, "Gift of Houses to Astronauts" (2 April 1962), VP Papers, LBJ Library, Box 182; and Gilruth OHI No. 6 by DeVorkin and Mauer.

105 Letter from LBJ to Forrest Moore, President, Rominger Advertising Agency (14 June 1963), VP Papers, LBJ Library, Box 237.

new Astrodome in 1965, although he reprimanded several for poor judgment. While he told his superiors that he saw no reason why they should not enjoy the experience, he ensured that this type of media problem did not repeat itself. He also disliked and privately chastised, but publicly defended, Gus Grissom over the famous corned beef sandwich episode during Gemini III. He took licks such as the following for these actions from the NASA Administrator:

> If this were a military operation and this kind of flagrant disregard of responsibility and of orders were involved, would not at least a reprimand be put in the record? The only way I know to run a tight ship is to run a tight ship, and I think it essential that you and your associates give the fullest *advance* consideration to these matters, rather than to have them come up in a form of public criticism which takes a great deal of time to answer and which make the job of all of us more difficult.[106]

None of this suggests that Gilruth let the astronauts run amuck. He tried to maintain order through more patriarchal means than military ones, but on occasion—as in the case of the Apollo 15 stamp cover sales by the crew—he could be enormously stern. Gilruth said he tried to keep issues in perspective. These men put their lives on the line and deserved some leniency when minor problems arose. After all, they rose to the challenge repeatedly in conducting Mercury, Gemini, and Apollo.

At a fundamental level, moreover, he basked in the reflected glory of the astronauts. He enjoyed being with them and loved their humor and passion for life. He said, "People used to tell me that I had no control over the astronauts. I'll tell you, those boys were wonderful." He took seriously the selection of the best crews and worked with them personally to ensure success. He worked with Deke Slayton to establish the most effective crews. Slayton "would just tell me who he thought would be the best for the next crew. I'd ask him the questions I had to ask about why, what he based it on. If there was something that I knew about this young man, where he might have had some problem, physical problems or something, was he all better from those? Of course he wouldn't have been picked if he hadn't been. As you say, they were all very high-caliber people, and all would have done a good job."[107] These men were his charges. They carried with them his and the entire nation's best wishes. Gilruth agreed with the sentiments expressed by Cynthia McQuillin in her 1983 poem "Star Fire" about the Apollo astronauts:

---

106 Letter from James E. Webb to Robert R. Gilruth (15 April 1965), James E. Webb Papers, Box 113, NASA-Astronaut Notes, Truman Library, Independence, MO.

107 Gilruth OHI No. 6 by DeVorkin and Mauer.

Ten thousand hands to build the shining shell.
    It took a dozen years and love to build it well.

Everyone who touched its birth,
    Though they be bound on Earth,
Will be with the astronauts that in her dwell.[108]

Gilruth, and everyone who met them, was fascinated by the apparent willingness of these young men to risk their lives for the good of a national cause. Tom Wolfe captured the method of this imagery some 20 years later in *The Right Stuff*. He wrote: "The astronauts were not brave in a stupid, unknowing way. Any fool could throw his or her life away. No, the idea here . . . seemed to be that a man should have the ability to go up in a hurling piece of machinery and put his hide on the line and then have the moxie, the reflexes, the experience, the coolness to pull it back in the last yawning moment—and then to go up again *the next day*."[109] The bravery of the astronauts touched emotions deeply seated in the American experience of the 20th century, felt by Gilruth perhaps most of all. He remained friends with them—but also in awe of them—his entire life.[110]

GILRUTH AND THE LEGACY OF APOLLO

Between 1968 and 1972, Apollo achieved its goals. Gilruth, approaching retirement, recognized that with the end of the Apollo program nothing would be the same. He spent 10 years as the Director of the Manned Spacecraft Center (1961–1971) directing 25 human spaceflights, including Alan Shepard's first Mercury flight in May 1961, the first lunar landing by Apollo 11 in July 1969, the dramatic rescue of Apollo 13 in 1970, and the Apollo 15 mission in July 1971. In January 1972, as Apollo was wrapping up, he served as Director of Key Personnel Development at NASA Headquarters in Washington, DC, retiring from federal service in December 1973. At that point, he devoted himself full time to building the boat of his dreams, one in which he could circumnavigate the world. He completed a 52-foot multi-hull boat, sailed it to his retirement home in Virginia, but never felt his nautical skills matched his astronautical ones; so he never sailed around the world. His close associate in Houston, George Low, said of him, "There

108 Cynthia McQuillin, "Star Fire," on *To Touch the Stars* (Sunnyvale, CA: Prometheus Music, 2004).
109 Tom Wolfe, *The Right Stuff* (New York: Farrar, Straus, and Giroux, 1979), p. 24.
110 Richard Slotkin, *Gunfighter Nation* (New York: Atheneum, 1992).

is no question that without Bob Gilruth there would not have been a Mercury, Gemini, or an Apollo program. He built in terms of what he felt was needed to run a manned spaceflight program . . . it is clear to all who have been associated with him that he has been the leader of all that is manned spaceflight in this country."[111]

Gilruth was justifiably proud of his role in Project Apollo. He commented on several key legacies from it. First, the Apollo program was successful in accomplishing the political goals for which it had been created. Kennedy had been dealing with a Cold War crisis in 1961 brought on by several separate factors—the Soviet orbiting of Yuri Gagarin and the disastrous Bay of Pigs invasion only two of them—that Apollo was designed to combat. At the time of the Apollo 11 landing, Mission Control in Houston flashed the words of President Kennedy announcing the Apollo commitment on its big screen. Those phrases were followed with these: "TASK ACCOMPLISHED, July 1969." When Gilruth lit up his characteristic cigar in Mission Control when Apollo 11 touched down, he understood that no greater understatement could probably have been made. It became more obvious with every passing year that any assessment of Apollo that did not recognize the accomplishment of landing an American on the Moon and safely returning before the end of the 1960s was incomplete and inaccurate, for that was the primary goal of the undertaking, and Gilruth and the other individuals who worked on the effort deserved recognition for making this dream a reality.[112]

Second, Project Apollo was a triumph of management in meeting the enormously difficult systems engineering and technological integration requirements. Gilruth always believed that Apollo was much more a management exercise than anything else and that the technological challenge, while sophisticated and impressive, was also within grasp. More difficult was ensuring that those technological skills were properly managed and used.[113] Through the decade of the 1960s, the space program provided one of the leading examples of a government technological program that worked. It inspired public confidence in the ability of government to accomplish great feats. Even as other government initiatives failed, civilian spaceflights continued to succeed. Actor Carroll O'Connor perhaps said it best in an episode of *All in the Family* in 1971. Portraying the character of Archie Bunker, the bigoted working-class American whose perspectives were more common in our society than many observers were comfortable with, O'Connor summarized well how most Americans responded to the culture of competence that Apollo engendered. He observed that he had "a genuine facsimile of the Apollo 14 insignia. That's

111 NASA Press Release H00-127, "Dr. Robert Gilruth, An Architect of Manned Space Flight, Dies" (17 August 2000), NASA Headquarters Historical Reference Collection, Washington, DC; and Pearce Wright, "Robert Gilruth: Rocket Engineer Who Put Americans into Space," *Guardian* (London, U.K., 20 August 2000).

112 John Pike, "Apollo—Perspectives and Provocations," Cold War History Symposium (11 May 1994), Ripley Center, Smithsonian Institution, Washington, DC.

113 See Arnold S. Levine, *Managing NASA in the Apollo Era* (Washington, DC: NASA SP-4102, 1982); and Sylvia D. Fries, *NASA Engineers and the Age of Apollo* (Washington, DC: NASA SP-4104, 1992).

the thing that sets the U.S. of A. apart from . . . all them other losers."[114] In very specific terms, Archie Bunker encapsulated for everyone what set the United States apart from every other nation in the world—success in spaceflight. At a basic level, Gilruth and Apollo provided the impetus for the perception of NASA as a culture of competence, one of the great myths emerging from the lunar landing program. This contention was true beyond all bounds, and subsequent tragedies have failed to demonstrate this level of continuing excellence shown throughout the Apollo era.

Third, by sheer serendipity Apollo taught humanity about itself and in the process altered our perception of the world in which we live. Apollo 8 was critical to this fundamental change, as it treated the world to the first pictures of Earth from afar. For example, Anne Morrow Lindbergh suggested in *Earthshine* (1969) that humanity gained a "new sense of awe and mystery in the face of the vast marvels of the solar system." She added, "Man had to free himself from Earth to perceive both its diminutive place in the solar system and its inestimable value as a life-fostering planet." The poet Archibald MacLeish said it this way: "To see the Earth as it truly is, small and blue and beautiful in that eternal silence where it floats, is to see ourselves as riders on the Earth together, brothers on that bright loveliness in the eternal cold—brothers who know now that they are truly brothers."[115] The modern environmental movement was galvanized in part by this new perception of the planet and the need to protect it and the life that it supports.[116]

Finally, Gilruth would agree that the Apollo program, while an enormous achievement, left a divided legacy for NASA and the aerospace community. The perceived "golden age" of Apollo created for the Agency an expectation that the direction of any major space goal from the President would always bring NASA a broad consensus of support and provide it with the resources and license to dispense them as it saw fit. Something most NASA officials did not understand at the time of the Moon landing in 1969, however, was that Apollo had not been a normal situation and would not be repeated. The Apollo decision was, therefore, an anomaly in the national decision-making process. The dilemma of the "golden age" of Apollo has been difficult to overcome, but moving beyond the Apollo program to embrace future opportunities has been an important goal of the Agency's leadership in the recent past. Exploration of the solar system and the universe remains as enticing a goal and as important an objective for humanity as it ever has been. Project Apollo was an important early step in that ongoing process of exploration.[117]

---

114 "Carroll O'Connor Obituary," *Morning Edition*, National Public Radio (22 June 2001). This report by Andy Bowers is available online at *http://www.npr.org* (accessed 2 July 2001).

115 Oran Nicks, ed., *This Island Earth* (Washington, DC: NASA SP-250, 1970), p. 3.

116 R. Cargill Hall, "Project Apollo in Retrospect" (20 June 1990), pp. 25–26, R. Cargill Hall Biographical File, NASA Headquarters Historical Reference Collection, Washington, DC.

117 As an example, see the argument made in George M. Low, Team Leader, to Mr. Richard Fairbanks, Director, Transition Resources and Development Group, "Report of the NASA Transition Team" (19 December 1980), NASA Headquarters Historical Reference Collection, Washington, DC, which advocated for strong presidential leadership to make everything right with the U.S. space program.

CONCLUSION

Gilruth wrote several reflective essays on the method of accomplishment and the meaning of Apollo.[118] He commented, "In thinking back over the flights of Apollo, I am impressed at the intrinsic excellence of the plan that had evolved. I have, of course, somewhat oversimplified its evolution, and there were times when we became discouraged, and when it seemed that the sheer scope of the task would overwhelm us in some areas there were surprises, and other areas proceeded quite naturally and smoothly." He added:

> The most cruel surprise in the program was the loss of three astronauts in the Apollo fire, which occurred before our first manned flight. It was difficult for the country to understand how this could have occurred, and it seemed for a time that the program might not survive. I believe that the self-imposed discipline that resulted, and the ever-greater efforts on quality, enhanced our chances for success, coming as they did while the spacecraft was being rebuilt and final plans formulated . . . .
>
> The flights came off almost routinely following Apollo 8 on through the first lunar landing and the flight to the Surveyor crater. But Apollo 13 was to see our first major inflight emergency when an explosion in the service module cut off the oxygen supply to the command module. Fortunately, the LM was docked to the CSM, and its oxygen and electric power, as well as its propulsion rocket, were available. During the four-day ordeal of Apollo 13, the world watched breathlessly while the LM pushed the stricken command module around the Moon and back to Earth. Precarious though it was, Apollo 13 showed the merit of having separate spacecraft modules, and of training of flight and ground crews to adapt to emergency. The ability of the flight directors on the ground to read out the status of flight equipment, and the training of astronauts to meet emergencies, paid off on this mission.
>
> Apollo surely is a prototype for explorations of the future when we again send men into space to build a base on the Moon or to explore even farther away from Earth.[119]

His last sentence is an understatement—something he fully realized as the years passed and America did not return to the Moon.

Robert Gilruth died on 17 August 2000, and the spaceflight community mourned. He was 86 years old by then and had suffered from Alzheimer's disease for a number of years. He never enjoyed the public stature of Wernher von Braun or James Webb. He certainly

118 See, for example, Robert R. Gilruth, "To the Moon and Beyond," *The Aeronautical Journal* 75 (January 1971): 1–17.

119 Robert R. Gilruth, "I Believe We Should Go to the Moon," in *Apollo Expeditions to the Moon*, ed. Edgar M. Cortright (Washington, DC: NASA SP-350, 1975), p. 39.

Wearing special germ-free clothing, Dr. Robert R. Gilruth, right, inspects lunar samples collected during the Apollo 17 mission, NASA's final Apollo flight in 1973. (Center number S73-34103)

never enjoyed the adulation of the astronauts. But he may be seen in innumerable film and photographs standing beside—or more often slightly behind—those people at some public event. He was decorated for his efforts by Kennedy and Johnson, and he specifically was singled out for recognition with a Robert J. Collier Trophy in 1971 "as representative of the engineering genius of the manned spaceflight team culminating in Apollo 15—man's most prolonged and scientifically productive lunar mission."[120] He was an unsung leader in the race to the Moon.

I believe Gilruth would agree that Apollo represented the unique creation and operation of technological systems that accounted for the perceptions of the U.S. as a people who master machines. We are a technological and an organizational people, and the structures we created to carry out Apollo are not just the so-called hardware of the system, but also the management structure, the organizations, the processes and procedures of operation, the people assigned, and the transportation and information networks that interconnect everything. In NASA, these technological systems included not just the spacecraft, but also the organizations, people, communications, manufacturing components, and even the political structure. It is the task of the modern makers of systems to direct the values of order, system, and control that are embedded in the machines of modern technology. The real legacy of Apollo for the modern American should probably be to understand the system and how to control it so that order will emerge sufficiently to make the success of the endeavor a reality. That, coupled with the dreams of the possible and the unwillingness to believe the experts when they say it cannot be done, might be the greatest lesson of Apollo for the modern

120 NASA Press Release H00-127, "Dr. Robert Gilruth"; JSC Press Release J00-49, "Statement by Johnson Space Center Director George W. S. Abbey Marking the Death of Dr. Robert R. Gilruth" (17 August 2000); and W. Henry Lambright, "Managing America to the Moon: A Coalition Analysis," in *From Engineering Science*, ed. Mack, pp. 193–211.

nation. Who knows what might happen when we fail to listen to those who say, "It can't be done." We might be able, as we once were, to promise the Moon and to deliver.

He would not have enjoyed this lament for what the United States has lost with the end of the Apollo program captured in the following poem entitled "Legends," by Bill Roper:

Once upon a time,
You could hear the Saturn's roar
As it rose upon its fiery tail to space.
And once upon a time, the men that we sent out
Landed in a strange and alien place.

And as I watched them walk upon the Moon,
I remembered Icarus,
Who flew too close to the Sun.

Once upon a time, they tore the gantries down
And the rockets flew no longer to the Moon.
And once upon a time,
We swore that we'd return,
But it doesn't look like we'll be back there soon.

And as the Moon shines down
On the shattered launching ground,
I remember Apollo,
Who flew the chariot of the Sun.
And I wonder of the legends they will tell
A thousand years from now.[121]

He would have applauded the announcement made by President George W. Bush on 14 January 2004 that the United States would return to the Moon between 2015 and 2020. With sufficient diligence and resources, Gilruth always believed that virtually anything humans can imagine in spaceflight may be achieved. He would have been concerned, however, that neither sufficient diligence nor resources would be available for this great initiative.[122]

121 Bill Roper, "Legends" (1980), *To Touch the Stars* (Sunnyvale, CA: Prometheus Music, 2004).

122 White House Press Release, "Fact Sheet: A Renewed Spirit of Discovery" (14 January 2004), available online at *http://www.whitehouse.gov/news/releases/2004/01/20040114-1.html* (accessed 4 April 2004).

12

Celebrating the Invention of Flight

in a Hands-On Way: Replicating the 1902 Experimental Glider Flights of the Wright Brothers

EDWARD J. PERSHEY

IN 1913, IN REFLECTING ON THE IMPORTANCE OF THE 1902 GLIDER TEST FLIGHTS, WILBUR WRIGHT WROTE THE FOLLOWING:

> The flights of the 1902 glider had demonstrated the efficiency of our system of maintaining equilibrium, and also the accuracy of our laboratory work upon which the design of the glider was based .... Before leaving camp in 1902 we were already at work on the general design of a new machine which we proposed to propel with a motor ... the general construction and operation of the machine was to be similar to that of the 1902 glider.[1]

---

1 Wilbur Wright, "How We Made the First Flight," *Flying* (December 1913), pp. 10–12, 35–36, as reprinted in Peter L. Jakab and Rich Young, eds., *The Published Writings of Wilbur and Orville Wright* (Washington, DC: Smithsonian Institution, 2000).

The Wright brothers applied for a U.S. Patent for "New and Useful Improvements in Flying Machines" in March of 1903, prior to the flights of their powered flyer in December of that year. The underlying claims in the patent covered the means of controlling a flying machine while in flight. The patent illustrations, which show an aircraft without a motor, were based on the Wrights' successful glider of 1902. In 2003, as the world celebrated 100 years of powered flight, the historical record showed that as far as the Wright brothers were concerned, they had found the essential piece in the puzzle of human flight in 1902 in the design and testing of their glider.

In anticipation of the celebration of the centennial of flight, the Crawford Auto-Aviation Museum in Cleveland in the spring of 2000 contracted with Wright researcher and master woodworker Nick Engler of West Milford, Ohio to produce for the museum an accurate, full-scale replica of the Wright 1902 glider.[2] This decision was based on a variety of circumstances, but ultimately was made for two important reasons. First, the replication work of Nick Engler was clearly superb, both from the standpoint of craftsmanship and from the excellence of his research and understanding of the work of the Wright brothers. Second was the important historical role that the 1902 glider played in aviation history, a role that was largely unknown to the general public. The museum felt that a replica of this crucial aircraft would afford opportunities to interpret the invention of the airplane in ways that would be unexpected but yet clear for a general museum audience. This decision was to prove correct in some unexpected ways when the glider was flown in the fall of 2002 on the dunes of North Carolina's Outer Banks.

This narrative will place the replication of the 1902 glider in the dual context of the history of the invention of the airplane and the role that museums can play in recreating events as a way of enhancing public interest in and understanding of historical developments.

THE ORIGINAL 1902 GLIDER

The Wrights had become interested in the possibility of human flight at the end of the 19th century, joining a community of European and American researchers pursuing that dream. They began their first experiments in flight in 1899 with the construction of a maneuverable kite. The story of how they came to be interested in flight, these early

2 The museum is a unit of the Western Reserve Historical Society, Cleveland's oldest cultural institution founded in 1867. Nick Engler is a private researcher who formed the Wright Aeroplane Company and is replicating all of the aircraft built by the Wright brothers, starting with the 1899 experimental kite. Engler has a working and interpretive shop at the Boonshoft Museum of Discovery in Dayton. The Crawford Museum wishes to acknowledge the value of the work that Engler did under contract to build the 1902 glider replica.

Wright Brothers' glider tests in 1902. (GRIN database number GPN-2002-000125)

experiments, and the crucial work in 1902–1903 that led to the successful powered flights at the very end of 1903 has been told many times.[3] The Wrights pursued two important areas of research during the period of 1900–1903 that are at the very heart of the design of the 1902 glider: control of aircraft in flight using a system that they referred to as "wing warping" and the correct curvature for the wings of an aircraft that would provide the optimal combination of lift and drag.

The Wrights concentrated their efforts on solving the problems of human flight on the issues of how a human operator would and could control the aircraft while it was flying. While other researchers designed machines that were meant to be stable in flight, with minimal involvement by the operator, the Wrights took their cue from observing the flight of birds, especially large, soaring buzzards. They noted how birds twisted the ends of their wings, one side up and one side down, to correct imbalances and change course as they flew. In 1900, Wilbur wrote a letter to Octave Chanute, a leading aviation researcher in the United States, in which he described these observations of birds. The Wrights commented on these observations in several of their published writings that reviewed their experiments on flight. Their August 1903 article in the *Journal of the Western Society of Engineers,*

3 On the Wrights' invention of the airplane, see Tom D. Crouch, *A Dream of Wings: Americans and the Airplane 1875–1905* (NY: W. W. Norton, 1981); and Peter L. Jakab, *Visions of a Flying Machine: The Wright Brothers and the Process of Invention* (Washington, DC: Smithsonian Institution Press, 1990).

which has an excellent account of the 1902 glider flights, devotes pages to their observations of the flight of birds along the dunes at Kill Devil Hills. They devised a mechanical way, using wires and pulleys, for the operator to twist the wings of a biplane glider to mimic this behavior of the birds. "Wing warping" was incorporated into all of their experimental gliders in 1900, 1901, and 1902, as well as into the 1903 powered flyer. It was brought to final form and use in the 1902 glider.

In the first three years of experiments in flight, the Wrights used tables of aerodynamic data that had been developed by Otto Lilienthal (European glider researcher) and others at the end of the 19th century. The wing surfaces derived from this data and incorporated into the gliders of 1900 and 1901 performed poorly. The revised glider of 1901 was an especially poor performer, leading the brothers in the fall of 1901 to conduct a series of tests, ultimately using a wind tunnel of their own design, to recalculate the research data. Then, from this improved data, they designed new lifting surfaces for another larger, improved glider. It was this model which they transported to Kill Devil Hills, North Carolina in the late summer of 1902. One year later, Wilbur Wright described this glider:

> The 1902 pattern was a double-deck machine having two surfaces, each 32 feet from tip to tip and 5 feet from front to rear. The total area of the main surfaces was about 305 square feet. The front rudder spread 15 square feet additional and the vertical tail about 12 square feet, which was subsequently reduced to 6 square feet. The weight was 116½ lbs. Including the operator, the total weight was from 250 to 260 lbs. It was built to withstand hard usage, and in nearly a thousand glides was injured but once. It repeatedly withstood without damage the immense strains arising from landing at full speed in a slight hollow where only the tips of the wings touched the earth, the entire weight of machine and operator being suspended between.[4]

The structural members of the glider were constructed of fine-grained spruce. The fabric covering of the wings was made of 200-plus-count cotton cloth. Wire and wrought and cast iron fittings were used for tensioning the two lifting surfaces. Wire and bicycle chain were used for the wing-warping mechanism.[5] As with their previous experimental craft, the brothers first flew the 1902 glider as a tethered kite. A famous photo of the 1902 glider shows it sailing almost directly overhead, very unkitelike.[6]

They began gliding in the 1902, expecting it to outperform the 1901. It certainly had much better lifting characteristics, but the improved 1902 version still had problems. The

4 Wilbur Wright, "Experiments and Observations in Soaring Flight," *Journal of the Western Society of Engineers* (August 1902): 400–417.

5 Ibid., see Wilbur Wright's answer to a query in a presentation to an engineering group.

6 Photo 15-6-56; N336 in the Wright Brothers Collection at Wright State University. You can search this collection online at *http://www.libraries.wright.edu/special/wright_brothers/dmc.html.*

1901, which did not have a tail assembly, had suffered from lateral instability, often slip-sliding out of a glide into the ground. They had added a fixed, double vane tail rudder to the 1902 glider. In the first series of glides in September 1902, the glider proved slightly more stable than the 1901, but, in certain banked turns actuated by wing warping, the new tail sometimes overcorrected the problem, leading to further instability. In late September, the brothers made an important and critical change to the glider. They reduced the tail to a single surface and made it moveable. They coupled this moveable rudder to the wing-warping mechanism so that when the operator warped the wings to bank to the right, the rear rudder would move accordingly. This new interconnected control system proved extremely successful. In early October, the brothers began a series of flights in the 1902 glider that demonstrated the validity of their design. The first flights with the interconnected moveable tail rudder probably occurred on 8 October 1902.

In October of 1902, the Wright brothers flew the glider 1,000 to 1,400 times on the dunes at Kill Devil Hills. They reckoned that this provided them with more than 4 hours of time in the air, far more than any other researcher had ever achieved. The brothers kept no specific record of each flight, though they claimed that the longest flights were about 622 feet long with a time-in-the-air of almost 30 seconds. While they made measurements of the angle of the incidence of the wings to the air and the angle of the glides relative to the horizon and ground, they used their time in the air in the 1902 glider primarily to gain skill as pilots.[7]

According to the Wrights, the glider needed winds of about 12 mph in order to lift off the ground and flew best in winds in excess of 18 mph. Winds exceeding 25 to 30 mph, which occurred occasionally in the fall of 1902, allowed the glider to sometimes soar over the dune. It would fly "in place" over the dune, the wind moving over the wing creating enough lift to counter the total weight of pilot and aircraft.

The glider used gravity to create forward motion. It was launched by two assistants who would lift the glider off the sand dune, the moving wind providing most of the "muscle work." Then the two assistants would tow the glider forward, propelling it off the edge of the dune. Gravity did the rest, as the dune fell away in a gentle slope. While the glider rose to over 40 to 50 feet on some glides, the brothers reported that most flights were only a few feet off the ground and that the glider seemed to perform more predictably at lower altitudes.

The Wrights' written descriptions of the glider in flight were mostly about the angle of descent and the length of flights/time in air for various angles. They did not provide much in the way of verbal imagery. It was clear, though, from their descriptions that they were able to fly the glider in a wide range of wind speeds and over dunes with heights of 30 to 100 feet. And while they became the most skillful flyers in the world by the end of October 1902, the design of the machine itself exceeded their skill as pilots.

7 Wilbur Wright, "Experiments and Observations."

The photographs of the 1902 glider demonstrated the performance of this aircraft. Whether skimming low along the dune or rising high above the sand, the glider could rightfully be called beautiful and graceful. Especially noteworthy were the photos of the glider in banked turns as shot by the photographer from behind the aircraft.

At the end of October 1902, the brothers had made enough flights in the improved glider that they felt confident of their ability and capacity to build a powered aircraft. They left the 1902 in a protective shed on a dune at Kill Devil Hills and returned to Dayton. When they went back to North Carolina in September of 1903, they found the 1902 in "pretty fair condition" and made practice flights in it as they assembled the 1903 flyer.[8] After the four powered flights in mid-December 1903, the Wrights returned to Dayton. They took the flyer with them, but did not bother to pack up the 1902. As far as we know, it was taken apart, and the pieces were used by the local people who had assisted the Wrights in a variety of ways. The world's first fully controllable airplane disappeared.

But the design of the glider survived. The Wrights clearly based the overall design of the 1903 powered flyer on that of the 1902. Their skill as pilots survived as evidence of the quality, the "rightness" of the 1902 design. The design of the 1902 was recorded for posterity in the patent illustrations in the 1903 application to the U.S. Patent Office.

One of the more interesting ways that the 1902 glider "survived" was the influence that it had on aviation in Europe. Octave Chanute, an American glider experimenter who had observed the flights of the 1902 glider at Kill Devil Hills in 1902, traveled to Europe in 1903. He gave public lectures on the experimental work of the Wrights in North Carolina and distributed scale drawings of the 1902 glider. This design was used several times in France in the form of replica gliders and eventually powered airplanes. As with the U.S. patent, a French patent of 1904 was based openly on this design, including the Wright system of wing warping.[9] In 1909, a glider designed on the model of the 1902 was built in England as an aid to training pilots. At least in terms of its overall design, the influence of the 1902 glider exceeded in many ways that of the 1903 flyer.

THE 1902 GLIDER AS MUSEUM EXHIBIT

The Crawford Auto-Aviation Museum in Cleveland, Ohio dates to the late 1930s when Frederick C. Crawford, then President and CEO of Thompson Products, Inc., began collecting antique automobiles. Thompson Products produced both automotive and aviation components, and was a major sponsor of the world-famous Cleveland National Air Races. The winner of the daredevil pylon air race was awarded the Thompson Trophy

---

8 Wilbur Wright, "How We Made the First Flight."

9 Letter to the editor from Wilbur Wright, *Scientific American* (16 July 1910): 47, in Jakab and Young, eds., *The Published Writings of Wilbur and Orville Wright.*

by Fred Crawford each year. So Crawford's collection of automobiles also included early Ohio-related and air-racing aircraft. In the middle of WWII, the company opened the Thompson Auto Album in downtown Cleveland, creating a private museum and public venue in which to display the rich collection of autos and aircraft rescued from wartime scrap drives. In 1963, this corporate collection was given to the Western Reserve Historical Society (WRHS), forming the museum named after its founder. The collection was moved from its downtown location into an addition built onto the WRHS headquarters in Cleveland's cultural center, University Circle, on the city's east side.

The collection grew to about 200 autos and a dozen aircraft, with an exceedingly rich archival collection of owners' manuals, repair and parts literature, and advertisements. Anticipating the centennial of flight celebration in 2003, staff at WRHS observed the several attempts already underway by a number of researchers to replicate the 1903 Wright flyer as a flying exhibit. While this was enticing, it was also clear that such an undertaking would be expensive and difficult. Demonstrating the powered flyer posed some complicated planning. The written descriptions of previous flying replicas also indicated that it was an extremely unstable and difficult aircraft to fly. The challenge of a 1903 replication seemed too great, and the budget seemed too small.

In 1999, the team first heard about the work of a Wright researcher, Nick Engler, who was a master woodworker in the process of replicating all the various Wright aircraft and had worked his way to the 1902 glider. Visiting with Nick Engler at his workshop in West Milford, just north of Dayton, the Crawford team was impressed with his partially constructed 1902 glider. The team was especially impressed with Engler's determination to make the glider a rugged, flyable, hands-on teaching artifact. His idea was that since the original 1902 glider was a very rugged piece of machinery, a well-built replica could be handled by museum visitors, flown on dunes, and used in school programs for students. As a way of funding his ongoing work and looking ahead to the centennial-of-flight celebrations in 2003, he was offering to build 1902 replicas at a set price. He also had made a complete set of drawings available free to any craftsman or Wright enthusiast who would choose to build their own. In April 2000, the Crawford team contracted with Engler to build a 1902 glider. The contract included a working replica of the Wright brothers' wind tunnel.

THE REPLICA 1902 AS ARTIFACT

When the staff of the Crawford museum first visited Nick Engler at his workshop in West Milford, Ohio in the summer of 1999, they were immediately impressed with the elegance of the 1902 glider replica that he had under construction. Engler allowed the staff to assist him in partially assembling the glider. He talked about his research into all of the Wright aircraft and his career as a master woodworker. He also discussed his

Captain James Alexander, USAF, pilots a replica of the 1902 Wright glider at Jockey's Ridge State Park near Kitty Hawk, NC in October 2002. (Photograph by Scott Carpenter, courtesy of the Western Reserve Historical Society)

intense interest in building a series of replica Wright aircraft as an educational project to get children interested in and to learn about Wright aviation history. The Crawford design team returned to Cleveland that summer, and, over the next several months, they weighed the merits of constructing a replica Wright aircraft built from Engler's plans or contracting with Engler himself.

The decision was relatively easy. The museum wanted a replica Wright aircraft that was as accurate as possible and that was well researched. On 5 April 2000, the museum entered into an agreement with Nick Engler to provide the Crawford museum with an accurate, flyable replica of the Wrights' 1902 glider.

The construction of the glider was detailed on the Wright Aeroplane Company Web site. Based on photos of the original and an existing replica built by the U.S. Army in 1934 (with the help of Orville Wright), Engler had developed a replica design that incorporated his own research into the way that the Wrights built their aircraft. He also brought to the replication his skills as a woodworker and his acquired skills in tying knots; the Wrights used wax-covered twine to lash the metal hardware for the tenioning wires and struts to the wooden framework. The finished replica represented a wealth of knowledge and experience.

The Crawford glider was Engler's second 1902 replica. He had taken one to Jockey's Ridge State Park in 2000 to test-fly it. Based on the performance of that replica, he made some changes to the Crawford model. First, he used a slightly different cloth, with a thread count closer to that reported by the Wrights for the wing surfaces. But he also made another change to the wing surfaces, one that he surmised from his research but was *not* documented by the Wrights.

The 1999 replica flew well enough on the dunes in North Carolina in 2000. But even though Engler had carefully matched the construction of his replica to that of the original, the performance of the replica did not match that reported by the Wrights. The replica seemed to need stronger winds to fly, and the flights were shorter in both duration and distance. Engler became convinced that the Wrights also had covered the wings of their glider with paraffin. He reasoned that they would have done this in the tents that they used on the beach dunes at Kitty Hawk. He decided to coat the Crawford glider with wax, leaving his first replica uncoated. The final form of the replica contained all of the information available from the archival and craftsmanship data, as well as the Wrights' written description and photographic journal of their original experimental glider.

Engler was working on the replica late in the summer of 2002, just prior to the Cleveland Air Show held each Labor Day weekend at Cleveland's downtown Burke Airport. He brought it to the air show, finishing some of the work in his tent hangar in front of an audience. After the show, he returned to Dayton to finish the glider. The Crawford staff made arrangements with Engler to meet him on the Outer Banks of North Carolina in early October to watch the replica fly. Engler had selected 8 October as the centennial date for the 1902 glider. From his research and reading of the Wrights' journals, it was on that date that the 1902 with its modified movable rudder made its first successful flights.

One of the visitors to the Cleveland Air Show that weekend was Captain Tanya Markow, a U.S. Army helicopter pilot, on leave visiting her family in the Cleveland area. That was a fortunate happenstance for Captain Markow, Engler, and the Crawford team. Engler had been working with the U.S. military to arrange for pilots for the flights in October. Markow maintained contact with Engler to see about getting assigned by the Army to fly the glider.

THE 2002 FLIGHTS OF THE 1902 REPLICA

By late September 2002, Engler notified the staff at the Crawford that he had lined up four pilots for the 1902 glider replica, one each from the Army and Navy, and two from the Air Force. He had made arrangements to transport the finished glider to North Carolina at the end of the first week in October. Jockey's Ridge State Park in Nags Head, North Carolina is about 10 miles south of Kitty Hawk and Kill Devil Hills. The dunes at Kitty Hawk are now overgrown with vegetation and no longer resemble the wind-swept dunes on which the Wrights flew from 1900 to 1903. Those at Jockey's Ridge, however, closely mimic the historic dunes of Kitty Hawk. North Carolina's park system preserves and interprets this natural environment. A hang-gliding school operates at the state park on the dunes, and permits can be obtained from the park for hang-gliding.

The Crawford team arrived at Jockey's Ridge State Park on Saturday, 5 October 2002. Engler and two of the pilots had already made some short flights on Friday, but the winds had been light; so the full flight capability of the glider had not been tested. The pilots on loan from the U.S. military were Lieutenant Commander Klaus Ohman, a Navy fighter pilot from the carrier USS *Kitty Hawk*, stationed in Japan; Captain Tanya Markow, an Army helicopter pilot stationed nearby in North Carolina and scheduled to teach the following year at West Point; Captain Jim Alexander, an Air Force C-130 pilot who had just returned from a stint in Afghanistan; and Captain Dawn Dunlop, an Air Force pilot stationed at the Pentagon. These four pilots combined had many years of experience on a variety of aircraft, but the 1902 glider promised to present new challenges quite unlike an F-18 or C-130. The four pilots, along with Allan Unrein, Collection Manager of the Crawford Auto-Aviation Museum who was also scheduled to fly the glider, took hang-gliding lessons at Jockey's Ridge State Park gliding school.

The weather on Saturday was gorgeous, with blue skies and warm temperatures, but disappointingly no wind. Engler, the pilots, and the Crawford team waited all day Saturday in vain for any breath of breeze. Also patiently waiting were several documentary film crews from the History Channel, PBS at the University of North Carolina, an independent film production company, an individual researcher from Germany, and a North Carolina TV station. While the filmmakers and videographers interviewed Engler and the pilots, the rest of the crew tried to stay cool in Engler's large hangar tent. In the tent was Engler's 1999 glider, and arrayed around the outside were the Crawford 1902 and Engler's 1901 and 1900 Wright replicas. The setting was impressive, but, alas, also without any blowing sand.

Sunday morning, 6 October 2002, greeted the pilots, staff, Engler, and film crews with beautiful weather once again and 15- to 18-mph winds on the crest of the dunes! Five or six of the assembled group lifted the 1902 off the ground and carried it the few hundred yards up to the top of the nearest dune with enough height and a good, gentle slope.[10]

Launching the glider was a relatively simple procedure, but one that took a dozen or so flights to understand. Four "ground crew" lifted the glider by holding onto the four corners of the lower wing. Once the glider was about 2 feet off the ground and facing into the wind, the 15–18-mph breeze provided ample force to allow the glider to begin flying in place. It was no longer necessary for the ground crew to lift the glider, but rather they needed to hold the glider down and keep it from flying backward.

With the ground crew holding the now flying glider in place, each pilot climbed onto the glider (separately), stretching him/herself prone in the center of the lower wing. The pilots' hips were positioned in the cradle connected to the wing-warping mechanism.

10 The descriptions of the flights made in the fall of 2002 that follow are based on notes and visual memories by Edward J. Pershey, Director of Education and Research, Western Reserve Historical Society.

Lieutenant Commander Klaus Ohman, U.S. Navy, pilots the 1902 replica on 6 October 2002. At an altitude of 12 to 15 feet, the glider will stay in the air 12 to 15 seconds, traveling 125 to 150 feet down the slope of the dune on this flight. (Photograph by Scott Carpenter, courtesy of the Western Reserve Historical Society)

Extending the arms straight ahead and slightly down, the pilots braced themselves in a manner similar to doing a pushup, grasping the bar that controlled the front elevator. Up/down movement of the elevator was actuated by twisting this bar, rotating it forward and backward. Banked turns were actuated by the pilots shifting their hips in the cradle left/right to warp the wings via cables and bicycle chains. These controls were very much not similar to any aircraft controls familiar to the pilots. Nor were these controls quite like hang-gliding, which essentially uses the shifting of the body weight of the hang-glider to establish control.

With the pilot situated at the controls, he/she gave the signal to launch. While the pilot attempted to hold the elevator in a position slightly up from neutral to provide minimal climb, the two ground crewmembers on the leading edge of the wing began to pull the glider forward, giving it forward momentum. Running with it about 15 to 20 feet and then at the shout of the pilot to "RELEASE!", the two launch crewmembers would propel the glider forward with as much energy as possible. That release off the ridge of the dune put the glider into a gravity well down the slope of the dune. Forward speed was attained by lowering the elevator and allowing the glider to nose down into the slope of the dune. With speed increase, the pilot then returned the elevator to a slightly up position to gain altitude. The length of the glide in distance and seconds depended on the wind speed, wind direction, slope of the dune, and initial forward momentum of the crew. At the end of the glide, the pilots tried to bring the elevator to the neutral position, allowing the glider to settle flat onto the sand.

The initial flights were barely glides, lasting just a few seconds with the glider traveling 50 feet or so. This corresponded to the Wrights' own statements about early attempts to glide with the 1902. However, after only a few such attempts, Jim Alexander on Sunday morning successfully flew the glider well over 100 feet and stayed in the air about 15 seconds. This first of many lengthier flights was beautiful to observe. Launching from the top of the dune, Alexander caught a good, steady breeze and was able to use the momentum of the launch to climb to about 12 to 15 feet over the dune. Silhouetted against the blue Outer Banks sky, the tan and wood glider soared elegantly through the air, recreating many of the famous historic photographs taken by the Wright brothers of the original glider flying over the dunes at Kitty Hawk.

Klaus Ohman, the Navy representative on the piloting team, stood at 5 feet, 9 inches, and weighed in at about 140 pounds—about the same as either of the Wrights. Not surprisingly, Klaus made some of the longest flights in the glider on Sunday and again later in the week. On Sunday, he flew the glider well over 200 feet and stayed in the air 20 seconds or so.

Captains Markow and Dunlop flew the glider many successful times. In the case of Markow, who weighs about 115 pounds or about the same as the glider itself, it was necessary to add about 20 pounds of lead shot to the glider in order to bring the glider into proper balance. Dunlop, the tallest of the four pilots, had the most trouble adjusting to the length between elevator control and the hip cradle for wing warping.

The glider proved to be laterally unstable. In flight it would tend to slip-slide to either the right or left. The pilots then attempted to correct this by warping the wings to bank the glider into a correcting turn. These corrections did not seem to have much affect. The pilots commented that the wing-warping control cradle was difficult to use. It tended to stick a bit as it slid across the wooden pilot cradle. Since even the best flights lasted only about 10 to 12 seconds, this delayed the corrective wing warping just long enough to allow the glider to continue to slip-slide into the ground.

The glider also proved to be extremely unstable in pitch control. The elevator oscillated wildly in the wind, requiring the pilots to concentrate on holding the elevator in the neutral or slightly up position. Any small corrective control by the pilot up or down was overcompensated by the elevator movement, requiring a strong corrective movement in the opposite direction. The glider flew in an erratic up/down "roller coaster" fashion. The elevator mounts had been built by Engler with tensioning springs that were meant to dampen out this oscillation (as on the original glider), but, in the light winds of Friday's test glides, the pilots had thought that the springs restricted the elevator control too much. The springs had been removed.

Attempts to dampen the elevator using lead weights were ineffective. Toward midafternoon on Sunday, Engler reinstalled the dampening springs. But by then the tension in the control wires on the elevator had loosened considerably. There was no easy way to tighten those wires. The glider continued to pitch erratically up/down, although the springs modified the movement somewhat. About midafternoon, a short flight was completed with a rather rough landing. The structural members in the pilot's cradle cracked. With the winds dying down and the afternoon coming to a close, it was time to suspend flying anyway.

Monday's weather was essentially a repeat of Saturday's—a calm, sunny day. No flying was done on Monday, but this gave Engler the time to repair the pilot cradle. In doing so, he added a piece of flat metal under the wing-warping cradle, allowing it to slide more freely. He also installed turnbuckles on the elevator control cables, providing a means for adjusting tension. These improvements were to be tested the next day.

Tuesday, 8 October, had been selected by Engler, from his reading of the Wrights' journals, as the probable day that the 1902 was first flown with the wing-warping mechanism coupled to a movable rudder, creating the world's first truly controllable aircraft. Winds were a minimum of 18 mph, gusting to more than 22 mph. The adjustments and improvements to the 1902 glider proved to be solid decisions by Engler. All of the pilots were able to fly the glider in excess of 100 feet with much more control. Once again, Klaus Ohman had the longest flights, one reaching 325 feet and almost 25 seconds in the air. Banked turns were more common. By midday, the pilots were more familiar with the improved controls and began to feel like they were becoming successful 1902 pilots! The ground crew was also "seasoned" by mid-Sunday morning. Smooth launches and careful positioning of the glider along the ridge and into the wind gave the pilots optimal flying conditions. The glider continued to enthrall all those who had assembled to watch it fly.

Around 11 a.m., as Dunlop brought the glider down after a medium-length flight, a gust of wind caught the elevator, pushing it into the extreme up position. The wind then cartwheeled the glider backward, end over end, with Major Dunlop still in the pilot's cradle. Fortunately Dunlop was not hurt, but the rudder broke off from the tail of the glider. This was undoubtedly the most serious accident of the two days of flying. Engler was able to repair the glider on the dune in about an hour. The glider even seemed to fly better after these repairs.

More flights were made, including several by Al Unrein. These were important for the museum, since any future demonstration flights would need to be done by museum staff. By the end of the day on Tuesday, hang-glider instructors who had been volunteering as ground crew also were flying the 1902 successfully. In all, over 120 flights of the 1902 were made over two days of flying on 6 and 8 October.

After his last flight of the day on Tuesday, Lieutenant Commander Klaus Ohman provided this assessment of the 1902 glider:

> The glider is behaving like a true airplane. You can start to feel the controls feeding back on you. When we get enough experience, we can anticipate the movements and actually make the airplane go where we want it to go. Fantastic flyer. Great job by Nick building it. It's a solid machine.[11]

Captain Markow, when interviewed about her experiences with the 1902, reported:

> The thing about the accidents really . . . is that in some cases the Wright brothers learned more from their mishaps than from their successful flights. Those mishaps taught us, for instance, that this airplane likes to be low. It is not an aircraft designed for a great deal of altitude. We found out that if we have too much up elevator or catch a bad wind, it will go end over end.[12]

Her overall assessment of the 1902 glider expressed sentiments that had been shared by all the pilots and the observers on the ground: "Oh yeah, it's beautiful. It's pure aerodynamics. You put enough thrust behind anything, and it will fly. Not this!"[13]

The experiences with the replica 1902 glider were clearly similar to those described by the Wright brothers in both 1902 and 1903. The Wrights described smooth flights down the dune and understood that the glider flew best at low altitudes. However, they also seemed to fly it much higher at times, contrary to what the 2002 pilots experienced. The Wrights also flew their 1902 glider 600 to 1,000 times, far more than the 120 flights in 2002. And those experiences were shared by two pilots who communicated in important nonverbal ways. The 120 flights in 2002 were accomplished by four military pilots, a museum curator, and several hang-glider experts. No one person flew more than 20 or 25 times. The military pilots were convinced that had they enough time in the air with the 1902, they would have achieved the kinds of flights of which the Wrights wrote in their accounts—600 feet and more and a full minute in the air!

The 2002 experiment also revealed something about the Wright brothers that is not often highlighted in the histories of aviation. The military pilots came away from the glider flights with well-bruised hips. Controlling the 1902 took physical strength and dexterity. Captain Markow summed this up: "They [Wrights] look like normal Victorian guys, but they had to be just extremely athletic just to do this."[14] Flying in 1902 was as much a physical sport as it was an engineering experiment.

---

11 Videotape interview of Lt. Cmdr. Klaus Ohman by Pershey (8 October 2002).

12 Videotape interview of Capt. Markow (7 October 2002) by History Channel crew in conference room at hang-gliding school at Jockey's Ridge State Park.

13 Ibid.

14 Ibid.

THE LEGACY OF THE 1902 AND 1902 REPLICA

On Wednesday, 9 October, Engler began packing up the 1902 replica, and the Crawford team packed up for the trip back to Cleveland. The documentary filmmakers began reviewing their film and tapes. The replica glider had brought the invention of the airplane alive for a small group of people in North Carolina in the fall of 2002. Not until well into their second and third iterations of their powered flyers did the Wright brothers spend more time in the air than they did in the 1902 glider. They had learned to fly, to become the world's first pilots in the 1902 glider. In 2002, a group of historians, military pilots, museum curators, hang-glider pilots, filmmakers, aviation enthusiasts, and onlookers discovered something for themselves by recreating that history in a hands-on way. This little aircraft which became the initial pattern for what an airplane ought to look like was indeed a good and graceful flyer. Its story needs to be made part of the popular historical record about how two brothers from Dayton accomplished one of the great engineering feats of all time. The 1902 glider was the world's first controllable airplane, and its passengers became the world's first airplane pilots.

About the Contributors

TAMI DAVIS BIDDLE is an associate professor in the Department of National Security and Strategy at the U.S. Army War College, the Army's senior-level staff college, where she teaches strategic studies and military history. She was the 2001–2002 Harold K. Johnson Visiting Professor of Military History at the U.S. Army's Military History Institute. She received her Ph.D. from Yale University. Her research focus has been warfare in the 20th century, especially fighting wars and diplomacy during the two World Wars and the early Cold War period. In particular, she has concentrated on the history of air warfare and the history of the Cold War. She has written many articles and book chapters on these subjects, and has most recently published *Rhetoric and Reality in Air Warfare: The Evolution of British and American Ideas about Strategic Bombing, 1914–1945* (Princeton, NJ: Princeton University Press, 2002), which was a *Choice* outstanding academic book in 2002.

ROGER E. BILSTEIN is a professor of history, emeritus, at the University of Houston—Clear Lake. He is the author or editor of nine books and monographs, including *Flight in America: From the Wrights to the Astronauts* 3d ed. (Baltimore, MD: Johns Hopkins University Press, 2000) and *Enterprise of Flight: The American Aviation and Aerospace Industry* 2d ed. (Washington, DC: Smithsonian Institution Press, 2001). His honors and awards include recognition by the Aviation/Space Writers Association, the National Space Council, and the American Institute of Aeronautics and Astronautics (AIAA). He has held the Lindbergh Chair of Aerospace History at the Smithsonian Institution's National Air and Space Museum (1992–1993) and was visiting professor at the Air War College of the U.S. Air Force (1995–1996).

MARK D. BOWLES received a B.A. in psychology (1991) and an M.A. in history (1993) from the University of Akron. He earned his Ph.D. in the history of technology and science (1999) from Case Western Reserve University and an M.B.A. in technology management (2005) from the University of Phoenix. He has been a Tomash Fellow from the Charles Babbage Institute at the University of Minnesota. He has written several books on the history of medicine, and his work on space science includes coauthoring *Taming Liquid Hydrogen: The Centaur Upper Stage Rocket* (Washington, DC: NASA SP-2004-4230, 2004) with Virginia Dawson and *NASA's Nuclear Frontier: The Plum Brook*

Reactor Facility (Washington, DC: NASA SP-2004-4533, 2004) with Robert Arrighi. His forthcoming monograph on the history of the Plum Brook nuclear test reactor was awarded the American Institute of Aeronautics and Astronautics History Manuscript Award in 2005. He is currently an independent scholar and director at Tech Pro, Inc.

AMY SUE BIX is an associate professor in the History Department at Iowa State University and a faculty member of university's program in the history of technology and science. Her book *Inventing Ourselves Out of Jobs?: America's Debate over Technological Unemployment, 1929–1981* was published by Johns Hopkins University Press (Baltimore, MD) in 2000. She also has published on the history of breast cancer and AIDS research, on the history of eugenics, on the history of home economics, on the history of alternative medicine, and on post-WWII physics and engineering, among other subjects. She is currently finishing a book entitled *Engineering Education for American Women: An Intellectual, Institutional, and Social History.*

TOM D. CROUCH is the Senior Curator of the Division of Aeronautics at the National Air and Space Museum. He is the author or editor of a number of books and many articles for both popular magazines and scholarly journals. Most of his work has been on aspects of the history of flight technology. Crouch's leading books include: *The Bishop's Boys: A Life of Wilbur and Orville Wright* (New York: W. W. Norton, 1989); *Aiming for the Stars: The Dreamers and Doers of the Space Age* (Washington, DC: Smithsonian Institution Press, 1999); *Eagle Aloft: Two Centuries of the Balloon in America* (Washington, DC: Smithsonian Institution Press, 1983); *Bleriot XI: The Story of a Classic Airplane* (Washington, DC: Smithsonian Institution Press, 1982); and *A Dream of Wings: Americans and the Airplane, 1875–1905* (New York: W. W. Norton, Inc., 1981). Dr. Crouch received a Christopher Award, a literary prize recognizing "significant artistic achievement in support of the highest values of the human spirit," for *The Bishop's Boys.* In the fall of 2000, President Clinton appointed Dr. Crouch to the chairmanship of the First Flight Centennial Federal Advisory Board, an organization created to advise the Centennial of Flight Commission on activities planned to commemorate the 100th anniversary of powered flight.

VIRGINIA P. DAWSON is President of History Enterprises, Inc., a history-consulting firm. She is the author of books and articles on the history of science, technology, and medicine, including *Engines and Innovation: Lewis Laboratory and American Propulsion Technology* (Washington, DC: NASA SP-1991-4306, 1991), *Taming Liquid Hydrogen: The Centaur Upper Stage Rocket* (coauthor with Mark Bowles, 2004), and *Ideas Into Hardware: A History of NASA's Rocket Engine Test Facility* (Cleveland, OH: NASA Glenn Research Center, 2004). Dr. Dawson is currently working on a history for the Cleveland

Clinic. She is an adjunct associate professor of history at Case Western Reserve University in Cleveland, Ohio.

ANDREW J. DUNAR is a professor of history at the University of Alabama in Huntsville. A graduate of Northwestern University, he has degrees from UCLA (M.A.) and the University of Southern California (Ph.D., 1981). He is the author of *The Truman Scandals and the Politics of Morality* (Columbia, MO: University of Missouri Press, 1984); coauthor (with Dennis McBride) of *Building Hoover Dam: An Oral History of the Great Depression* (New York: Twayne Publishers, 1993); and coauthor (with Stephen P. Waring) of *Power to Explore: A History of Marshall Space Flight Center, 1960–1990* (Washington, DC: NASA Headquarters, 1999), which won the History Book Award from the AIAA in 2000. He is currently working on an oral history of "The Farm," a hippie commune in Summertown, Tennessee. He is the editor of the *Oral History Review*, the journal of the Oral History Association.

MICHAEL GORN received his doctorate in history in 1978 from the University of Southern California. He has served as deputy Air Force historian and later as the first historian of the U.S. Environmental Protection Agency in Washington, DC. Since 1996, he has been employed by the National Aeronautics and Space Administration and is presently the Chief Historian of the NASA Dryden Flight Research Center, Edwards, California. Gorn is the author of a number of books about aeronautics and spaceflight, most recently (with Richard P. Hallion) *On the Frontier: Experimental Flight at NASA Dryden* (Washington, DC: Smithsonian Institution Press, 2003). He also has published *Expanding the Envelope: Flight Research at NACA and NASA* (Lexington, KY: The University Press of Kentucky, 2001) and *The Universal Man: Theodore von Kármán's Life in Aeronautics* (Washington, DC: Smithsonian Institution Press, 1992). He is the former co-editor of the Smithsonian's *History of Aviation* book series. Dr. Gorn serves presently on the History and Education Committee of the U.S. Centennial of Flight Commission.

ALAN L. GROPMAN is a distinguished professor of national security policy at the Industrial College of the Armed Forces, a graduate-degree-granting war college for senior military and civilian officials. From 1996 until 2002, he was chairman of the Department of Grand Strategy and Mobilization at the Industrial College of the Armed Forces. He is also an adjunct professor of security studies at Georgetown University. A retired colonel in the United States Air Force, he flew over 670 missions in Vietnam, later serving as the Deputy Director of Air Force Plans for Planning Integration at the United States Air Force Headquarters. He earned, among other awards, the Defense Superior Service Medal, Legion of Merit, Distinguished Flying Cross, Air Medal with Five Oak Leaf Clusters, and Vietnam Cross of Gallantry with Palm. Dr. Gropman received his

Ph.D. in history from Tufts in 1975. Dr. Gropman's publications include *The Air Force Integrates, 1945–1964*, 2d ed. (Washington, DC: Smithsonian Institution Press, 1998), and *Mobilizing U.S. Industry in World War II* (Washington, DC: National Defense University Press, 1996).

ROGER D. LAUNIUS is chair of the Division of Space History at the Smithsonian Institution's National Air and Space Museum in Washington, DC. Between 1990 and 2002, he served as Chief Historian of the National Aeronautics and Space Administration. He received his Ph.D. from Louisiana State University, Baton Rouge, in 1982. Among the many books on aerospace history he has written or edited are: *Space Stations: Base Camps to the Stars* (Washington, DC: Smithsonian Institution Press, 2003), which received the AIAA's history manuscript prize; *Flight: A Celebration of 100 Years in Art and Literature* (New York: Welcome Books, 2003), edited with Anne Collins Goodyear, Anthony M. Springer, and Bertram Ulrich; *To Reach the High Frontier: A History of U.S. Launch Vehicles* (Lexington, KY: University Press of Kentucky, 2002), with Dennis R. Jenkins; *Imagining Space: Achievements, Possibilities, Projections, 1950–2050* (San Francisco, CA: Chronicle Books, 2001), with Howard E. McCurdy; *Reconsidering Sputnik: Forty Years Since the Soviet Satellite* (Amsterdam, The Netherlands: Harwood Academic Publishers, 2000), with John M. Logsdon and Robert W. Smith; and *Innovation and the Development of Flight* (College Station, TX: Texas A&M University Press, 1999).

WILLIAM M. LEARY is the E. Merton Coulter Professor of History at the University of Georgia. After serving in the U.S. Air Force during the Korean War, he worked for Eastern Airlines, KLM Royal Dutch Airlines, and the Air Transport Section of General Motors Corporation. He received his doctorate from Princeton University in 1966 and taught at Princeton, California State University at San Diego, and the University of Victoria in British Columbia before joining the faculty of the University of Georgia. He is the author of *We Freeze to Please* (Washington, DC: NASA SP-2002-4226, 2002), a history of NASA's Icing Research Tunnel; *Under Ice: Waldo Lyon and the Development of the Arctic Submarine* (College Station, TX: Texas A&M University Press, 1999); *Perilous Missions: Civil Air Transport and CIA Covert Operations in Asia* (Tuscaloosa, AL: University of Alabama Press, 1984); and numerous other books and articles on civil transport, aircraft safety, and the U.S. Air Mail Service.

W. DAVID LEWIS holds the title of distinguished university professor in the History Department at Auburn University. In addition to books on the history of steel making, he is the author of many books and articles on the history of aviation, including *The Airway to Everywhere: The History of All-American Aviation* (Pittsburgh, PA: Pittsburgh University Press, 1988) with William Trimble. Dr. Lewis served as president and director of an international conference on the history of civil and commercial aviation in Switzerland in 1992. He received the Leonardo da Vinci Medal for lifetime achievement from the Society for the History of Technology in 1993. He served on the NASA History Advisory Committee between 1986 and 1990. Dr. Lewis has recently published a definitive biography of aviation pioneer Eddie Rickenbacker.

EDWARD JAY PERSHEY is Director of Education and Research at the Western Reserve Historical Society (WRHS) and Task Force Director for the new Crawford Museum of Transportation and Industry in Cleveland, Ohio. Pershey joined WRHS in 1995 as the Director of Education and Curator of Urban and Industrial History. From 1987 to 1995, he was the founding Director of the Tsongas Industrial History Center, an innovative hands-on museum education program in Lowell, Massachusetts. From 1981 to 1987, he served as Supervisory Museum Curator at the Edison National Historic Site, the home and laboratory of the American inventor Thomas Alva Edison in West Orange, New Jersey. In 1982, he received his doctorate in the history of technology from Case Western Reserve University in Cleveland, where he also served briefly as Associate Curator at the Dittrick Museum of Medicine.

SUSAN WARE is affiliated with the Radcliffe Institute for Advanced Study, Harvard University, where she is editing the next volume of the biographical dictionary *Notable American Women* (Boston: Harvard University Press, 2004). She is also a visiting lecturer in the History Department at Harvard. A specialist in 20th-century American history and women's history, her books include *Still Missing: Amelia Earhart and the Search for Modern Feminism* (New York: W. W. Norton, 1993) and *Letter to the World: Seven Women Who Shaped the American Century* (New York: W. W. Norton, 1998). She has just completed a biography of radio talk show pioneer Mary Margaret McBride.

List of Acronyms

AAA	American Automobile Association
ABMA	Army Ballistic Missile Agency
AFRS	Auxiliary Flight Research Station
ARPA	Advanced Research Projects Agency
ATA	Air Transport Association
ATC	Air Transport Command
CAA	Civil Aeronautics Authority
EOR	Earth Orbital Rendezvous
EVA	Extravehicular Activity
FDR	Franklin Delano Roosevelt
FEAF	Far Eastern Air Force
GM	General Motors
IRBM	Intermediate Range Ballistic Missile
JPL	Jet Propulsion Laboratory
KAC	Kellett Autogiro Company
LOR	Lunar Orbital Rendezvous
MIT	Massachusetts Institute of Technology
MSFC	Marshall Space Flight Center
NAAC	North American Aviation Corporation
NACA	National Advisory Committee for Aeronautics
NASA	National Aeronautics and Space Administration
NASM	National Air and Space Museum
NATO	North Atlantic Treaty Organization
NBS	National Bureau of Standards
PARD	Pilotless Aircraft Research Division

RFC	Reconstruction Finance Corporation
SAC	Strategic Air Command
SIOP	Single Integrated Operational Plan
UNIA	United Negro Improvement Association
USAAF	United States Army Air Forces
USAF	United States Air Force
VfR	Verein fur Raumshiffahrt (Society for Space Travel)
WRHS	Western Reserve Historical Society

Index

A

B

Index

C

D

E

F

G

H

I

J

K

L

M

N

O

P

Q

R

S

T

U

V

W

X

Y

Z

The NASA History Series

REFERENCE WORKS, NASA SP-4000:

Grimwood, James M. *Project Mercury: A Chronology.* NASA SP-4001, 1963.

Grimwood, James M., and C. Barton Hacker, with Peter J. Vorzimmer. *Project Gemini Technology and Operations: A Chronology.* NASA SP-4002, 1969.

Link, Mae Mills. *Space Medicine in Project Mercury.* NASA SP-4003, 1965.

Astronautics and Aeronautics, 1963: *Chronology of Science, Technology, and Policy.* NASA SP-4004, 1964.

Astronautics and Aeronautics, 1964: *Chronology of Science, Technology, and Policy.* NASA SP-4005, 1965.

Astronautics and Aeronautics, 1965: *Chronology of Science, Technology, and Policy.* NASA SP-4006, 1966.

Astronautics and Aeronautics, 1966: *Chronology of Science, Technology, and Policy.* NASA SP-4007, 1967.

Astronautics and Aeronautics, 1967: *Chronology of Science, Technology, and Policy.* NASA SP-4008, 1968.

Ertel, Ivan D., and Mary Louise Morse. *The Apollo Spacecraft: A Chronology, Volume I, Through November 7, 1962.* NASA SP-4009, 1969.

Morse, Mary Louise, and Jean Kernahan Bays. *The Apollo Spacecraft: A Chronology, Volume II, November 8, 1962–September 30, 1964.* NASA SP-4009, 1973.

Brooks, Courtney G., and Ivan D. Ertel. *The Apollo Spacecraft: A Chronology, Volume III, October 1, 1964–January 20, 1966.* NASA SP-4009, 1973.

Ertel, Ivan D., and Roland W. Newkirk, with Courtney G. Brooks. *The Apollo Spacecraft: A Chronology, Volume IV, January 21, 1966–July 13, 1974.* NASA SP-4009, 1978.

Astronautics and Aeronautics, 1968: Chronology of Science, Technology, and Policy. NASA SP-4010, 1969.

Newkirk, Roland W., and Ivan D. Ertel, with Courtney G. Brooks. *Skylab: A Chronology.* NASA SP-4011, 1977.

Van Nimmen, Jane, and Leonard C. Bruno, with Robert L. Rosholt. *NASA Historical Data Book, Volume I: NASA Resources, 1958–1968.* NASA SP-4012, 1976, rep. ed. 1988.

Ezell, Linda Neuman. *NASA Historical Data Book, Volume II: Programs and Projects, 1958–1968.* NASA SP-4012, 1988.

Ezell, Linda Neuman. *NASA Historical Data Book, Volume III: Programs and Projects, 1969–1978.* NASA SP-4012, 1988.

Gawdiak, Ihor Y., with Helen Fedor, compilers. *NASA Historical Data Book, Volume IV: NASA Resources, 1969–1978.* NASA SP-4012, 1994.

Rumerman, Judy A., compiler. *NASA Historical Data Book, 1979–1988: Volume V, NASA Launch Systems, Space Transportation, Human Spaceflight, and Space Science.* NASA SP-4012, 1999.

Rumerman, Judy A., compiler. *NASA Historical Data Book, Volume VI: NASA Space Applications, Aeronautics and Space Research and Technology, Tracking and Data Acquisition/Space Operations, Commercial Programs, and Resources, 1979–1988.* NASA SP-2000-4012, 2000.

Astronautics and Aeronautics, 1969: Chronology of Science, Technology, and Policy. NASA SP-4014, 1970.

Astronautics and Aeronautics, 1970: Chronology of Science, Technology, and Policy. NASA SP-4015, 1972.

Astronautics and Aeronautics, 1971: Chronology of Science, Technology, and Policy. NASA SP-4016, 1972.

Astronautics and Aeronautics, 1972: Chronology of Science, Technology, and Policy. NASA SP-4017, 1974.

Astronautics and Aeronautics, 1973: Chronology of Science, Technology, and Policy. NASA SP-4018, 1975.

Astronautics and Aeronautics, 1974: Chronology of Science, Technology, and Policy. NASA SP-4019, 1977.

Astronautics and Aeronautics, 1975: Chronology of Science, Technology, and Policy. NASA SP-4020, 1979.

Astronautics and Aeronautics, 1976: Chronology of Science, Technology, and Policy. NASA SP-4021, 1984.

Astronautics and Aeronautics, 1977: Chronology of Science, Technology, and Policy. NASA SP-4022, 1986.

Astronautics and Aeronautics, 1978: Chronology of Science, Technology, and Policy. NASA SP-4023, 1986.

Astronautics and Aeronautics, 1979–1984: Chronology of Science, Technology, and Policy. NASA SP-4024, 1988.

Astronautics and Aeronautics, 1985: Chronology of Science, Technology, and Policy. NASA SP-4025, 1990.

Noordung, Hermann. *The Problem of Space Travel: The Rocket Motor.* Edited by Ernst Stuhlinger and J. D. Hunley, with Jennifer Garland. NASA SP-4026, 1995.

Astronautics and Aeronautics, 1986–1990: A Chronology. NASA SP-4027, 1997.

Astronautics and Aeronautics, 1990–1995: A Chronology. NASA SP-2000-4028, 2000.

MANAGEMENT HISTORIES, NASA SP-4100:

Rosholt, Robert L. *An Administrative History of NASA, 1958–1963.* NASA SP-4101, 1966.

Levine, Arnold S. *Managing NASA in the Apollo Era.* NASA SP-4102, 1982.

Roland, Alex. *Model Research: The National Advisory Committee for Aeronautics, 1915–1958.* NASA SP-4103, 1985.

Fries, Sylvia D. NASA *Engineers and the Age of Apollo.* NASA SP-4104, 1992.

Glennan, T. Keith. *The Birth of NASA: The Diary of T. Keith Glennan.* J. D. Hunley, editor. NASA SP-4105, 1993.

Seamans, Robert C., Jr. *Aiming at Targets: The Autobiography of Robert C. Seamans, Jr.* NASA SP-4106, 1996.

Garber, Stephen J., editor. *Looking Backward, Looking Forward: Forty Years of U.S. Human Spaceflight Symposium.* NASA SP-2002-4107, 2002.

Chertok, Boris. *Rockets and People, Volume I.* NASA-SP-2005-4110, 2005.

Laufer, Dr. Alexander, Todd Post, and Dr. Edward J. Hoffman. *Shared Voyage: Learning and Unlearning from Remarkable Projects.* NASA-SP-2005-4111, 2005.

PROJECT HISTORIES, NASA SP-4200:

Swenson, Loyd S., Jr., James M. Grimwood, and Charles C. Alexander. *This New Ocean: A History of Project Mercury.* NASA SP-4201, 1966; rep. ed. 1998.

Green, Constance McLaughlin, and Milton Lomask. *Vanguard: A History.* NASA SP-4202, 1970; rep. ed. Smithsonian Institution Press, 1971.

Hacker, Barton C., and James M. Grimwood. *On the Shoulders of Titans: A History of Project Gemini.* NASA SP-4203, 1977.

Benson, Charles D., and William Barnaby Faherty. *Moonport: A History of Apollo Launch Facilities and Operations.* NASA SP-4204, 1978.

Brooks, Courtney G., James M. Grimwood, and Loyd S. Swenson, Jr. *Chariots for Apollo: A History of Manned Lunar Spacecraft.* NASA SP-4205, 1979.

Bilstein, Roger E. *Stages to Saturn: A Technological History of the Apollo/Saturn Launch Vehicles.* NASA SP-4206, 1980, rep. ed. 1997.

SP-4207 not published.

Compton, W. David, and Charles D. Benson. *Living and Working in Space: A History of Skylab.* NASA SP-4208, 1983.

Ezell, Edward Clinton, and Linda Neuman Ezell. *The Partnership: A History of the Apollo-Soyuz Test Project.* NASA SP-4209, 1978.

Hall, R. Cargill. *Lunar Impact: A History of Project Ranger.* NASA SP-4210, 1977.

Newell, Homer E. *Beyond the Atmosphere: Early Years of Space Science.* NASA SP-4211, 1980.

Ezell, Edward Clinton, and Linda Neuman Ezell. *On Mars: Exploration of the Red Planet, 1958–1978.* NASA SP-4212, 1984.

Pitts, John A. *The Human Factor: Biomedicine in the Manned Space Program to 1980.* NASA SP-4213, 1985.

Compton, W. David. *Where No Man Has Gone Before: A History of Apollo Lunar Exploration Missions.* NASA SP-4214, 1989.

Naugle, John E. *First Among Equals: The Selection of NASA Space Science Experiments.* NASA SP-4215, 1991.

Wallace, Lane E. *Airborne Trailblazer: Two Decades with NASA Langley's Boeing 737 Flying Laboratory.* NASA SP-4216, 1994.

Butrica, Andrew J., editor. *Beyond the Ionosphere: Fifty Years of Satellite Communication.* NASA SP-4217, 1997.

Butrica, Andrew J. *To See the Unseen: A History of Planetary Radar Astronomy.* NASA SP-4218, 1996.

Mack, Pamela E., editor. *From Engineering Science to Big Science: The NACA and NASA Collier Trophy Research Project Winners.* NASA SP-4219, 1998.

Reed, R. Dale, with Darlene Lister. *Wingless Flight: The Lifting Body Story.* NASA SP-4220, 1997.

Heppenheimer, T. A. *The Space Shuttle Decision: NASA's Search for a Reusable Space Vehicle.* NASA SP-4221, 1999.

Hunley, J. D., editor. *Toward Mach 2: The Douglas D-558 Program.* NASA SP-4222, 1999.

Swanson, Glen E., editor. *"Before this Decade Is Out . . .": Personal Reflections on the Apollo Program.* NASA SP-4223, 1999.

Tomayko, James E. *Computers Take Flight: A History of NASA's Pioneering Digital Fly-by-Wire Project.* NASA SP-2000-4224, 2000.

Morgan, Clay. *Shuttle-Mir: The U.S. and Russia Share History's Highest Stage.* NASA SP-2001-4225, 2001.

Leary, William M. *"We Freeze to Please": A History of NASA's Icing Research Tunnel and the Quest for Flight Safety.* NASA SP-2002-4226, 2002.

Mudgway, Douglas J. *Uplink-Downlink: A History of the Deep Space Network 1957–1997.* NASA SP-2001-4227, 2001.

Dawson, Virginia P., and Mark D. Bowles, *Taming Liquid Hydrogen: The Centaur Upper Stage Rocket, 1958–2002.* NASA SP-2004-4230, 2004.

CENTER HISTORIES, NASA SP-4300:

Rosenthal, Alfred. *Venture into Space: Early Years of Goddard Space Flight Center.* NASA SP-4301, 1985.

Hartman, Edwin P. *Adventures in Research: A History of Ames Research Center, 1940–1965.* NASA SP-4302, 1970.

Hallion, Richard P. *On the Frontier: Flight Research at Dryden, 1946–1981.* NASA SP-4303, 1984.

Muenger, Elizabeth A. *Searching the Horizon: A History of Ames Research Center, 1940–1976.* NASA SP-4304, 1985.

Hansen, James R. *Engineer in Charge: A History of the Langley Aeronautical Laboratory, 1917–1958.* NASA SP-4305, 1987.

Dawson, Virginia P. *Engines and Innovation: Lewis Laboratory and American Propulsion Technology.* NASA SP-4306, 1991.

Dethloff, Henry C. *"Suddenly Tomorrow Came . . .": A History of the Johnson Space Center.* NASA SP-4307, 1993.

Hansen, James R. *Spaceflight Revolution: NASA Langley Research Center from Sputnik to Apollo.* NASA SP-4308, 1995.

Wallace, Lane E. *Flights of Discovery: 50 Years at the NASA Dryden Flight Research Center.* NASA SP-4309, 1996.

Herring, Mack R. *Way Station to Space: A History of the John C. Stennis Space Center.* NASA SP-4310, 1997.

Wallace, Harold D., Jr. *Wallops Station and the Creation of the American Space Program.* NASA SP-4311, 1997.

Wallace, Lane E. *Dreams, Hopes, Realities: NASA's Goddard Space Flight Center, The First Forty Years.* NASA SP-4312, 1999.

Dunar, Andrew J., and Stephen P. Waring. *Power to Explore: A History of the Marshall Space Flight Center.* NASA SP-4313, 1999.

Bugos, Glenn E. *Atmosphere of Freedom: Sixty Years at the NASA Ames Research Center.* NASA SP-2000-4314, 2000.

GENERAL HISTORIES, NASA SP-4400:

Corliss, William R. *NASA Sounding Rockets, 1958–1968: A Historical Summary.* NASA SP-4401, 1971.

Wells, Helen T., Susan H. Whiteley, and Carrie Karegeannes. *Origins of NASA Names.* NASA SP-4402, 1976.

Anderson, Frank W., Jr. *Orders of Magnitude: A History of NACA and NASA, 1915–1980.* NASA SP-4403, 1981.

Sloop, John L. *Liquid Hydrogen as a Propulsion Fuel, 1945–1959.* NASA SP-4404, 1978.

Roland, Alex. *A Spacefaring People: Perspectives on Early Spaceflight.* NASA SP-4405, 1985.

Bilstein, Roger E. *Orders of Magnitude: A History of the NACA and NASA, 1915–1990.* NASA SP-4406, 1989.

Logsdon, John M., editor, with Linda J. Lear, Jannelle Warren-Findley, Ray A. Williamson, and Dwayne A. Day. *Exploring the Unknown: Selected Documents in the History of the U.S. Civil Space Program, Volume I, Organizing for Exploration.* NASA SP-4407, 1995.

Logsdon, John M., editor, with Dwayne A. Day and Roger D. Launius. *Exploring the Unknown: Selected Documents in the History of the U.S. Civil Space Program, Volume II, Relations with Other Organizations.* NASA SP-4407, 1996.

Logsdon, John M., editor, with Roger D. Launius, David H. Onkst, and Stephen J. Garber. *Exploring the Unknown: Selected Documents in the History of the U.S. Civil Space Program, Volume III, Using Space.* NASA SP-4407, 1998.

Logsdon, John M., general editor, with Ray A. Williamson, Roger D. Launius, Russell J. Acker, Stephen J. Garber, and Jonathan L. Friedman. *Exploring the Unknown: Selected Documents in the History of the U.S. Civil Space Program, Volume IV, Accessing Space.* NASA SP-4407, 1999.

Logsdon, John M., general editor, with Amy Paige Snyder, Roger D. Launius, Stephen J. Garber, and Regan Anne Newport. *Exploring the Unknown: Selected Documents in the History of the U.S. Civil Space Program, Volume V, Exploring the Cosmos.* NASA SP-2001-4407, 2001.

Siddiqi, Asif A. *Challenge to Apollo: The Soviet Union and the Space Race, 1945–1974.* NASA SP-2000-4408, 2000.

MONOGRAPHS IN AEROSPACE HISTORY, NASA SP-4500:

Launius, Roger D. and Aaron K. Gillette, compilers, *Toward a History of the Space Shuttle: An Annotated Bibliography.* Monographs in Aerospace History, No. 1, 1992.

Launius, Roger D., and J. D. Hunley, compilers, *An Annotated Bibliography of the Apollo Program.* Monographs in Aerospace History, No. 2, 1994.

Launius, Roger D. *Apollo: A Retrospective Analysis.* Monographs in Aerospace History, No. 3, 1994.

Hansen, James R. *Enchanted Rendezvous: John C. Houbolt and the Genesis of the Lunar-Orbit Rendezvous Concept.* Monographs in Aerospace History, No. 4, 1995.

Gorn, Michael H. *Hugh L. Dryden's Career in Aviation and Space.* Monographs in Aerospace History, No. 5, 1996.

Powers, Sheryll Goecke. *Women in Flight Research at NASA Dryden Flight Research Center, from 1946 to 1995.* Monographs in Aerospace History, No. 6, 1997.

Portree, David S. F. and Robert C. Trevino. *Walking to Olympus: An EVA Chronology.* Monographs in Aerospace History, No. 7, 1997.

Logsdon, John M., moderator. *Legislative Origins of the National Aeronautics and Space Act of 1958: Proceedings of an Oral History Workshop.* Monographs in Aerospace History, No. 8, 1998.

Rumerman, Judy A., compiler, *U.S. Human Spaceflight, A Record of Achievement, 1961–1998.* Monographs in Aerospace History, No. 9, 1998.

Portree, David S. F. *NASA's Origins and the Dawn of the Space Age.* Monographs in Aerospace History, No. 10, 1998.

Logsdon, John M. *Together in Orbit: The Origins of International Cooperation in the Space Station.* Monographs in Aerospace History, No. 11, 1998.

Phillips, W. Hewitt. *Journey in Aeronautical Research: A Career at NASA Langley Research Center.* Monographs in Aerospace History, No. 12, 1998.

Braslow, Albert L. *A History of Suction-Type Laminar-Flow Control with Emphasis on Flight Research.* Monographs in Aerospace History, No. 13, 1999.

Logsdon, John M., moderator. *Managing the Moon Program: Lessons Learned From Apollo.* Monographs in Aerospace History, No. 14, 1999.

Perminov, V. G. *The Difficult Road to Mars: A Brief History of Mars Exploration in the Soviet Union.* Monographs in Aerospace History, No. 15, 1999.

Tucker, Tom. *Touchdown: The Development of Propulsion Controlled Aircraft at NASA Dryden.* Monographs in Aerospace History, No. 16, 1999.

Maisel, Martin D., Demo J. Giulianetti, and Daniel C. Dugan. *The History of the XV-15 Tilt Rotor Research Aircraft: From Concept to Flight.* Monographs in Aerospace History, No. 17, 2000.

Jenkins, Dennis R. *Hypersonics Before the Shuttle: A Concise History of the X-15 Research Airplane.* Monographs in Aerospace History, No. 18, 2000.

Chambers, Joseph R. *Partners in Freedom: Contributions of the Langley Research Center to U.S. Military Aircraft in the 1990s.* Monographs in Aerospace History, No. 19, 2000.

Waltman, Gene L. *Black Magic and Gremlins: Analog Flight Simulations at NASA's Flight Research Center.* Monographs in Aerospace History, No. 20, 2000.

Portree, David S. F. *Humans to Mars: Fifty Years of Mission Planning, 1950–2000.* Monographs in Aerospace History, No. 21, 2001.

Thompson, Milton O., with J. D. Hunley. *Flight Research: Problems Encountered and What They Should Teach Us.* Monographs in Aerospace History, No. 22, 2000.

Tucker, Tom. *The Eclipse Project.* Monographs in Aerospace History, No. 23, 2000.

Siddiqi, Asif A. *Deep Space Chronicle: A Chronology of Deep Space and Planetary Probes, 1958–2000.* Monographs in Aerospace History, No. 24, 2002.

Merlin, Peter W. *Mach 3+: NASA/USAF YF-12 Flight Research, 1969–1979.* Monographs in Aerospace History, No. 25, 2001.

Anderson, Seth B. *Memoirs of an Aeronautical Engineer—Flight Tests at Ames Research Center: 1940–1970.* Monographs in Aerospace History, No. 26, 2002.

Renstrom, Arthur G. *Wilbur and Orville Wright: A Bibliography Commemorating the One-Hundredth Anniversary of the First Powered Flight on December 17, 1903.* Monographs in Aerospace History, No. 27, 2002.

No monograph 28.

Chambers, Joseph R. *Concept to Reality: Contributions of the NASA Langley Research Center to U.S. Civil Aircraft of the 1990s.* Monographs in Aerospace History, No. 29, 2003.

Peebles, Curtis, editor. *The Spoken Word: Recollections of Dryden History, The Early Years.* Monographs in Aerospace History, No. 30, 2003.

Jenkins, Dennis R., Tony Landis, and Jay Miller. *American X-Vehicles: An Inventory—X-1 to X-50.* Monographs in Aerospace History, No. 31, 2003.

Renstrom, Arthur G. *Wilbur and Orville Wright Chronology.* Monographs in Aerospace History, No. 32, 2003.

Bowles, Mark D., and Robert S. Arrighi. *NASA's Nuclear Frontier: The Plum Brook Reactor Facility, 1941–2002.* Monographs in Aerospace History, No. 33, 2004.

McCurdy, Howard E. *Low-Cost Innovation in Spaceflight.* Monographs in Aerospace History, No. 36, 2005.

Seamans, Robert C. *Project Apollo: The Tough Decisions.* Monographs in Aerospace History, No. 37, 2005.

Lambright, Henry W. *NASA and the Environment: The Case of Ozone Depletion.* Monographs in Aerospace History, No. 38, 2005.

ELECTRONIC MEDIA, NASA SP-4600:

Remembering Apollo 11: The 30th Anniversary Data Archive CD-ROM. SP-4601, 1999.

The Mission Transcript Collection: U.S. Human Spaceflight Missions from Mercury Redstone 3 to Apollo 17. SP-2000-4602, 2001.

Shuttle-Mir: The United States and Russia Share History's Highest Stage. SP-2001-4603, 2002.

U.S. Centennial of Flight Commission Presents Born of Dreams—Inspired by Freedom. SP-2004-4604, 2004.

Of Ashes and Atoms: A Documentary on the NASA Plum Brook Reactor Facility. NASA SP-2005-4605, 2005.

Taming Liquid Hydrogen: The Centaur Upper Stage Rocket Interactive CD-ROM. SP-2004-4606, 2004.

Fueling Space Exploration: The History of NASA's Rocket Engine Test Facility DVD. NASA SP-2005-4607, 2005.

CONFERENCE PROCEEDINGS, NASA SP-4700:

Dick, Steven J. and Cowing, Keith L., editors. *Risk and Exploration: Earth, Sea and the Stars.* NASA SP-2005-4701, 2005.